Smart Computing and Intelligence

This book series aims to establish itself as a medium for the publication of new research and development of innovative paradigms, models, architectures, conceptual underpinnings and practical implementations encompassed within smart computing and intelligence.

The scope of the series includes but is not limited to smart city, smart education, health informatics, smart ecology, data and computational analytics, smart society, smart learning, complex systems-chaos, computational thinking, brain computer interaction, natural/computer interaction, humanoid behaviour, and impact of educational psychology on computing.

The cornerstone of this series' editorial policy is its unwavering commitment to report the latest results from all areas of smart computing and intelligence research, development, and practice. Our mission is to serve the global smart computing and intelligence community by providing a most valuable publication service.

Dejian Liu · Guangju Chen · Jianguo Shi ·
Michael Agyemang Adarkwah · Hongyan Kuai ·
Ronghuai Huang
Editors

Empowering Educational Innovation Through Design: Volume One

Best Practices from K-12 Teachers

Springer

Editors
Dejian Liu
Smart Learning Institute
Beijing Normal University
Beijing, China

Jianguo Shi
Smart Learning Institute
Beijing Normal University
Beijing, China

Hongyan Kuai
Smart Learning Institute
Beijing Normal University
Beijing, China

Guangju Chen
Smart Learning Institute
Beijing Normal University
Beijing, China

Michael Agyemang Adarkwah
Friedrich Schiller University Jena
Jena, Germany

Ronghuai Huang
Smart Learning Institute
Beijing Normal University
Beijing, China

ISSN 2522-0888 ISSN 2522-0896 (electronic)
Smart Computing and Intelligence
ISBN 978-981-95-4620-6 ISBN 978-981-95-4621-3 (eBook)
https://doi.org/10.1007/978-981-95-4621-3

This Springer imprint is published by the registered company Springer Nature Singapore Pte Ltd.
The registered company address is: 152 Beach Road, #21-01/04 Gateway East, Singapore 189721, Singapore

Preface

In September 2019, UNESCO officially launched the "Futures of Education" initiative and in 2021, published its report *Reimagining Our Futures Together: A New Social Contract for Education.* This report called for a rethinking of education in a world that is increasingly complex, uncertain, and fragile. Looking ahead to 2050 and beyond, we should work together to redefine and transform education along five key dimensions: pedagogy, curricula, teachers, schools, and educational opportunities. At the UN *Transforming Education Summit* in 2022, over 130 countries committed to prioritizing education. They pledged to take action to end the learning crisis by strengthening the resilience of educational systems against crises, addressing educational exclusion, revolutionizing the teacher profession, updating curricula, and promoting digital learning. During the Summit of the Future in September 2024, *The Pact for the Future* reiterated the importance of investing in accessible, safe, inclusive, and equitable quality education for all, while also promoting lifelong learning, technical and vocational training, and digital literacy.

These efforts are significant steps forward, but as we look toward the future, the question remains: how can we better leverage technology to address the evolving challenges in education? One essential approach to navigating these uncertainties is design thinking for innovation. This mindset helps us look beyond the present and into the future, taking into account both reality and constraints to develop practical solutions. As we continue to rethink education, design thinking aligns well with the growing need for future talent development. This means that pedagogical approaches, curricula, and entire educational systems must be thoughtfully designed.

Recognizing the potential of design thinking in education, we acted early on. In 2015, we realized that design thinking could strategically help students not only generate and test solutions to real-world problems but also drive projects forward during critical inflection points in the education sector. This recognition led to the establishment of the Design and Learning Laboratory at the Smart Learning Institute of Beijing Normal University. The lab focuses on research in design thinking for youth, exploring the characteristics of innovative thinking, developing new curricula and textbooks, and discussing approaches to transforming education through design with leading global universities and institutes. Our collaborations with Stanford

University and Harvard University inspired the creation of the *Design and Learning* course at Beijing Normal University in September 2017. What started as a small initiative has since expanded to over 20 universities worldwide, both within China and abroad. This global reach has allowed us to share our insights and foster an international community dedicated to educational innovation.

Since 2018, our annual Global Competition on Design for Future Education has attracted over 15,000 participants from 40 countries and regions. Through this initiative, we have assembled a network of global experts and scholars who has collaborated on the Design and Learning textbook in Chinese and the English book *Envisioning the Future of Education Through Design*. We also launched the *Design and Learning MOOC*, sharing the latest research in the field, showcasing innovative design applications in education, and providing valuable resources for educators worldwide.

Building on our experience and research, we have categorized educational design into five areas: product design, solution design, service design, mechanism design, and policy design. In 2021, we introduced the K-12 track in the Global Competition on Design for Future Education, which has since attracted K-12 teachers from countries such as China, Japan, Hungary, and Indonesia. These teachers presented practical cases tailored to meet the needs of students in today's complex and uncertain world. Over the past 5 years, more than 8000 K-12 teachers from over 1200 schools have participated. They have applied design thinking and technologies such as AI-generated content (AIGC), Big Data, and the Metaverse to create effective teaching solutions.

Their work centers on improving teacher-student communication, enhancing peer interactions, and cultivating students' practical skills, fostering their ability to "learn to become." The innovative solutions shared by these teachers can be categorized into the five areas of educational design we've identified. These solutions cover a broad range of themes, including evaluation and management platforms, inclusive online classrooms, indigenized service-learning initiatives, school-family-community educational models, and the transformation of learning styles through subject-based classrooms.

This book compiles a rich collection of these educational innovations, providing practical examples that highlight how design thinking is applied in real-world educational settings. These exemplary cases showcase the creativity and ingenuity of K-12 teachers globally. By leveraging technology in innovative ways, these teachers have addressed educational challenges, improved learning outcomes, and created more engaging and impactful educational experiences for all students. This collection serves as a valuable resource for educators, students, researchers, and policymakers, offering inspiration and insights into how to shape the future of education.

Beijing, China Dejian Liu

Acknowledgements

We would like to express our sincere gratitude to Prof. Mingyuan Gu (Beijng Normal University), Prof. Guanzhong Liu (Tsinghua University), Prof. Weizu Song (Beijing Design Society), Prof. Larry Leifer (Stanford University), Prof. Chris Dede (Harvard University), and others for their invaluable guidance. Without their efforts, this book would not have been possible.

We would also like to extend our thanks to the team at the Design and Learning Laboratory, Smart Learning Institute of Beijing Normal University, for their valuable contributions. We are especially grateful to the participants of the K-12 Track at the Global Competition on Design for Future Education, who generously shared their exceptional teaching cases. Lastly, we would like to extend our thanks to the staff at Springer for their indispensable support and assistance throughout this process.

Introduction to the Global Competition on Design for Future Education

Background

Since 2018, Beijing Normal University has been hosting the annual "Global Competition on Design for Future Education." The aim was to build an international platform to develop cooperation with international organizations, universities, and research institutes. This was achieved by inviting teachers, students, and those engaged in education worldwide to explore future-oriented educational solutions for better educational opportunities for mankind.

Competition Objectives

To better address future uncertainties and challenges while exploring innovative solutions for quality education for all, Beijing Normal University and UNESCO Institute for Information Technologies in Education have jointly organized the Global Competition on Design for Future Education for the past seven consecutive years. This initiative aims to inspire deep contemplation, reflection, and reimagining of the future of education. Since its inception in 2018, the competition has attracted over 15,000 participants from more than 40 countries and regions worldwide, leading to the creation of over 3,000 outstanding educational designs. The objectives of this competition are to

- Understand the concept intelligence education.
- Apply emerging technologies (e.g., virtual worlds and metaverse) to improve the cognitive needs of students.
- Tackle educational challenges to rural areas and disadvantaged groups.
- Apply intercultural and interdisciplinary collaborations.

Confronted with future uncertainties, design thinking for innovation is one of the core competencies that everyone should develop. Therefore, participants in this competition—students, K-12 teachers, and education professionals—are tasked with using design thinking to generate innovative ideas and solutions that can enhance people's lives and promote human well-being.

The competition can be seen as a new educational model for college students, K-12 teachers, and other stakeholders. When it comes to educational design, international organizations such as the United Nations and UNESCO emphasize the concept of Universal Design for Learning (UDL). Universal Design refers to the creation of products, environments, programs, and services that are accessible to all people, to the greatest extent possible, without the need for adaptation or specialized design. This approach helps realize the equity and inclusiveness of education by fully considering the dignity, unique circumstances, and inherent diversity of all users.

We will also continue to offer an interactive platform for interdisciplinary and international learning, fostering and collaboration among universities and professionals from various fields. This initiative will gather innovative ideas for the future of education, while enhancing participants' communication and design skills, and creating practical opportunities across different spaces. We will also fulfill the key role of education in establishing "a global community of shared future."

Competition Themes

The competition aims to demonstrate how intelligent technology can meet diverse learning needs. The solutions presented should not only respond to learners' varying needs and forms of discrimination but also adapt to unexpected situations, such as crises (pandemics, natural disasters, etc.), that impede access to education. The main themes of the competition are:

- ***AI and Education:*** How can AI create infinite possibilities in smart learning environments, students' self-regulated learning, human–computer interaction, and home–school interaction?
- ***Metaverse and Education:*** How can an immersive online space construct educational scenarios and provide curriculum resources and teaching methods to improve the learning experience?
- ***Rural Education:*** What design solutions can address the needs of rural areas, such as the shortage of quality educational resources and the lack of access to education for left-behind and migrant children?
- ***Inclusive Education:*** How can an educational model ensure inclusion for all students, including those with physical disabilities and health impairments, the elderly, girls and women, students with learning difficulties, gifted children, vulnerable groups affected by war, conflicts, or crises?

- ***Educational Neuroscience:*** How can we integrate evidence from neuroscience, psychology, and education to design scientific and effective educational solutions, including policy design, curriculum design, instructional design, evaluation design, and educational culture and space design.to improve education and teaching quality?
- ***Artificial Intelligence, Big Data, and Psychology:*** How to accurately access students' learning ability and status to effectively promote their mental health and well-being based on AI, Big Data, game-based and personalized psychological assessment and mental health, etc.?
- ***Education in Crisis:*** Due to unexpected situations and natural disasters, the learning process can be disrupted, and several students face the challenge of not going back to their schools for a long period. Therefore, to overcome this challenge, new solutions should be designed to maintain distance education for everyone, during unexpected situations and ensure equity education.
- ***Personalized Education:*** Students have several individual differences (e.g., personality, learning style, and working memory capacity), they therefore behave differently while learning (the way the recall information, read, etc.). Therefore, it is important to design solutions that can cater to the different needs of learners, and provide personalized learning according to each student profile.
- ***Artificial Intelligence, Big Data, and Psychology:*** How to accurately access students' learning ability and status to effectively promote their mental health and well-being based on AI, Big Data, game-based and personalized psychological assessment and mental health, etc.?

Competition Arrangements

The competition is set to begin in March and conclude in December, spanning nearly the entire year. The main event is divided into four tracks, each tailored to different participants, including college students, K-12 teachers, students in higher vocational institutions, and enterprise employees. Its goal is to offer customized and effective solutions for designing future education from multiple perspectives. Moreover, two concurrent monthly initiatives, known as "Call for Posters" with the theme "My Vision for the Future of Education" and the "Call for Videos" with the theme "Shorts on Contextual Teaching and Learning," will take place. These initiatives serve as vital platforms to showcase how educators around the world envision the future of education in various teaching scenarios.

The competition schedule consists of three stages:

- First Round: This initial phase involves recruiting participants and offering a series of informative lectures and training sessions. Participants are encouraged to conceive and submit their works before the deadline. Only the winning participants (based on the judges' evaluation) will advance to the next round.
- Pre-finals: During this phase, regional selection contests will be held.

- Global Finals: The finalists from the pre-finals will gather offline and engage in a series of activities, including presentations, reviews, and an award ceremony.

Throughout the competition, participants will be enrolled and required to submit their final works. They must complete their designs according to the contest proposition, and explain the concept, constructs, and ideas behind their designs, as well as conduct a feasibility analysis. The contest organizing committee will arrange for international experts to assess the final designs during the completion.

Competition Expected Outcomes and Assessment Criteria

College students, K-12 teachers, and education professionals are invited to participate in the competition, offering design solutions for targeted groups, such as people with disabilities, in specific teaching and learning scenarios assisted by high-tech tools in cross-cultural collaboration. **The innovative solutions can be presented in various visual formats, including physical models, hardware sketches, software prototypes, or solution designs**. The final designs should be practical and capable of being produced and utilized in local communities across countries, particularly for people with specific disabilities.

Assessment Criteria:

- Problem Awareness: The essence of the problem should be analyzed through phenomena to clearly identify the underlying issues.
- Innovation Spirit: The design should be original, incorporating innovative concepts and methods based on scientific and pedagogical theories.
- Science and Education Integration: The design should integrate future technologies with education to find suitable solutions.
- Application Prospect: The solutions should be tailored to specific teaching scenarios and should be easily understood and accepted.
- Presentation and Expression: The project materials and final presentation should be complete, logical, and clearly structured.

Features of This Book

- **Extensive Practical Exploration by Frontline Educators**

This book carefully selects practical case studies from dedicated frontline teachers, which may span a single semester or summarize several years of in-depth teaching. These cases have yielded remarkable results in practice, not only enhancing teaching quality but also fostering students' overall development. Their successful experiences and innovative methodologies provide crucial insights and serve as essential references for other educators.

- **Profound Exploration of Interdisciplinary and Project-based Learning**

The book highlights the importance of interdisciplinary and project-based learning, featuring numerous excellent cases of interdisciplinary teaching in K-12 schools. These cases demonstrate how to integrate knowledge and skills across disciplines through project-based learning, allowing students to tackle real-world problems while nurturing their innovative thinking and comprehensive abilities. Also, the book covers a wide spectrum of educational elements—teaching, learning, management, and support services—across various subjects such as Chinese, English, music, art, psychology, and physical education. This diversity makes it a valuable resource for educators.

- **Utilization of Design Thinking for Creative Solutions**

This book selects and compiles a collection of outstanding teaching cases from K-12 educators, which vividly demonstrate how teachers cleverly integrate design thinking into every crucial aspect of teaching, thereby creating efficient and innovative solutions that comprehensively optimize and enhance every segment of the instructional process.

- **Diverse Cases with a Precise Focus**

The book covers a wide range of case types and focuses on five key areas: product design, solution design, service design, mechanism design, and policy design. These cases represent the cutting-edge trends in current educational reform and demonstrate the depth and diversity of educational innovation. The practical solutions offered for educational design are highly relevant and deserve widespread promotion.

- **Emphasis of AI Application in Education**

The book emphasizes the integration of AI technology in education, particularly highlighting the rise of AIGC. It showcases how teachers are actively incorporating emerging technologies into their teaching practices, enhancing instructional effectiveness and sparking students' interest in learning. More importantly, the book explores how these technological applications are transforming students' learning experiences, boosting learning efficiency, fostering innovative thinking, and improving problem-solving abilities. These demonstrations highlight the immense potential and value of AI in the field of education.

Contents

Part I Highlighting Educational Product Designs

1 The Data Brain of Small Seeds: A Comprehensive Evaluation Solution for Students in Large-Scale Schools 3
Yongming Cao, Yanan Pang, and Yan Wang

2 Construction and Practice of Internet-Enhanced Management Platform for Student Mental Well-Being 15
Yuchen Bian, Chen Chen, Mengwei Wu, Xiaoyang Fu, and Yingjie Chen

3 Code Mission: Designing an Educational Tool for Programming Instruction 35
Peng Liang

4 Integrated Reading: Construction and Implementation of Online, Offline, and Autonomous Reading System 51
Yan Liu, Luyao Wang, Yingxi Zhai, Ran Xu, Kangyi Zhang, Congcong Guo, A. Yongqi, and Lixia Han

5 AI-Enhanced Education: Creation and Application of Interdisciplinary Instructional Agent 61
Tiantian Kong, Mengmeng Cui, Meijuan Wang, and Yiran Zhang

6 Design and Application of AI-Integrated Ancient Architecture Toys and Teaching Aids: Taking Taoranting Park as an Example 79
Ziji Liu, Yunpeng Ma, Xinran Zhao, Lujie Chang, and Liang Zhao

Part II Featuring Educational Solution Designs

7 **Ethno-STEAM and *Klenang Sasak*: A Design-Based Solution for Rural Schools** ... 95
Syakti P. Sriyansyah, Marselino F. G. Sundah, Nugroho P. Sumanto, Maturiadi, and Duwi B. Sriyansyah

8 **Unity in Diversity: Innovative Practices for Inclusive Education** ... 113
Doina Otilia Filep, Szabolcs Vincze, and Péter Krajnyák

9 **AIGC-Integrated Art Course: Pathway to Cultivate High-Quality Artistic Talents** ... 123
Cheng Zang and Muyao Li

10 **Food Museum: Curriculum and Venue Development for Food Education in Kindergartens** ... 139
Yanling Wang, Cong Qian, Tingting Ye, Kaka Xiang, Xiaoqing Zhang, and Tingting Lin

11 **Crafting and Deploying a Metaverse Curriculum Leveraging the WorldWide Telescope** ... 151
Lei Liang, Yingchen Mao, Chenzhou Cui, Shanshan Li, Yuexin Wang, and Xuemin Zhang

12 **Magic Space: The Construction and Practical Study of a Comprehensive Learning Space in Schools** ... 167
Yanping Lu, Chong Chen, Jing Yang, Sijia Li, Yishun Mo, and Yueyue Nie

13 **Teaching Accuracy in English Writing Through Process-Oriented Data Feedback Technology** ... 179
Weiwei Wang and Qian Yao

14 **Designing a Creative "Smart+" P.E. Class Break Using Educational Neuroscience** ... 195
Ke Li and Guoliang Zhao

15 **Developing Health Education Resources for Tibetan Farmers' and Herdsmen's Children Using Digital Technology Applications** ... 207
Xiao Jiang and Liangni Lai

16 **Project-Based Course for the Automatic Treatment of the Waste Liquid in the Secondary School Chemistry Laboratory** ... 225
Wen Zhao and Li Ma

Part III Showcase of Educational Service Designs

17 Intelligent After-School Extended Care Programme for Primary School Students 245
Guobing Qian, Ao Dong, and Fang Chen

18 Innovative Practices in Cultivating a Sense of Social Responsibility Among Rural Students: Taking *a Touch of Blue* as an Example 255
Xuequn Xu, Wei Tong, Kewei Wang, Caiwei Zheng, Hongxia Yu, and Zongyao Lu

19 Indigenized Implementation of Service-Learning in Secondary Schools 267
Shuya Xiao, Wanhua Wu, Mei Zheng, Jialin Luo, Wanzhen Xu, and Guanyu Wang

Part IV Insight into Educational Mechanism Designs

20 Design and Implementation of Multi-course Blended Teaching in Smart Education 285
Xianlian Zou, Renjiao Xing, Xiao Huang, Zhiyou Chen, Yuqin Tan, Jiaping Lan, Mengjing Tang, Xueping Xiang, and Lu Tan

21 Exploring the Collaborative Model of School-Family-Community for Cultivating Creative Talents 307
Yakun Zhang, Xingli Zhang, Lina Huang, Yufei Zong, Jiannong Shi, and Wei Kang

22 The Sharing of Educational Resources Between Urban and Rural Schools Enabled by Real-Scene Classroom 323
Fei Hu, Yuhua Zhang, Long Li, Xiaoling Tao, Rui Xiong, and Xinyi Huang

23 Exploration and Practice of a Growth Education Evaluation Visualization Model 333
Yingying Xu and Ruiyuan Zhou

Part V Exploring Educational Policy Designs

24 Enhancing Learning Style Transformation Through the Design and Application of Subject-Based Classrooms 353
Lili Cheng, Qing Zhang, Xinhua Pan, and Yongqing Tu Shen

Part I
Highlighting Educational Product Designs

Chapter 1
The Data Brain of Small Seeds: A Comprehensive Evaluation Solution for Students in Large-Scale Schools

Yongming Cao, Yanan Pang, and Yan Wang

Abstract As a century-old school operating under a group model, Harbin Garden Elementary School in Heilongjiang Province faces the significant task of constructing a curriculum evaluation system that effectively unlocks each child's intrinsic potential and enhances their holistic development. This case primarily explores how modern intelligent technologies can be leveraged to promote the development of the "Integration of the Five Educations" through competency-based performance evaluations. It aims to scientifically construct a curriculum and evaluation system that fosters students' internal growth and comprehensive development while emphasizing developmental, formative, and performance-based evaluations to guide education effectively. By summarizing, refining, and elevating practical experiences, this case proposes a solution of academic value that addresses the issue of "educational involution" and comprehensively implements the fundamental task of "moral education and talent cultivation."

Keywords Comprehensive student evaluation · Internal growth · Big data · Visualization · Competency-based assessment · Digital education reform · Intelligent data platform

1.1 Background

Since the start of the basic education curriculum reform in 2001,[1] Harbin Garden Elementary School in Heilongjiang Province has been exploring ways to promote the comprehensive development of students through evaluation reform. However, several challenges have persisted in the evaluation process. These issues include (1) the evaluation philosophy lacks a child-centered focus, (2) there is an imbalance in

[1] https://www.gov.cn/gongbao/content/2001/content_60920.htm.

Y. Cao · Y. Pang (✉) · Y. Wang
Harbin Garden Elementary School, Harbin, Heilongjiang Province, China
e-mail: 981564259@qq.com

D. Liu et al. (eds.), *Empowering Educational Innovation Through Design: Volume One*, Smart Computing and Intelligence, https://doi.org/10.1007/978-981-95-4621-3_1

the structure of evaluation content, (3) the evaluation participants are too limited, and (4) the evaluation methods are outdated.

To fully implement President Xi Jinping's directives on building a digital China and actively carry out the national digitalization strategy for education, the school views digital transformation as a fundamental, pioneering, and strategic measure to support education reform and development. Guided by the principles of "competency first, integrated innovation, high-quality collaboration, and leading excellence," the school uses informatization to promote the renewal of educational concepts, the transformation of teaching methods, the improvement of education governance, and the advancement of equitable quality. This approach accelerates the development of high-quality education, aiming to meet the expectations and needs of people.

Harbin Garden Elementary School has utilized information technology to build an integrated, intelligent data platform. As shown in Fig. 1.1, by merging technology with comprehensive student evaluation and increasing stakeholder participation, the school has tackled outdated evaluation methods, enabling more scientific and practical assessments. This has led to the development of a modern governance mechanism, creating a model for interconnection and collaboration between schools, families, and communities. In doing so, the school has reconstructed a new development ecosystem and established a distinctive, advanced, and highly effective comprehensive evaluation system.

1.2 Conceptual Framework

The evaluation tools are integrated into a user-friendly intelligent platform to reduce students' academic burden further, improve quality, and deepen comprehensive evaluation. The Data Brain of Small Seeds (DBSS) platform facilitates multi-stakeholder collaboration in evaluations, encouraging deep participation from various groups. The platform employs situational evaluation methods, utilizing diverse assessment forms to conduct accurate and lightweight competency assessments. Its comprehensive feedback mechanism deeply analyses the process data generated by the DBSS to improve teaching and management practices. Harbin Garden Elementary School, focusing on stimulating students' internal growth, has developed an evaluation content method system, and support system over more than a decade of exploration. Through system integration and refinement, the school has established a comprehensive evaluation system for fostering students' internal growth.

1.2.1 A Comprehensive Evaluation Content System

To align with the national requirements for the "Five Educations" (moral, intellectual, physical, aesthetic, and labor education) and to cultivate students' internal motivation, the evaluation framework is based on cultivating the "Six Powers" of students,

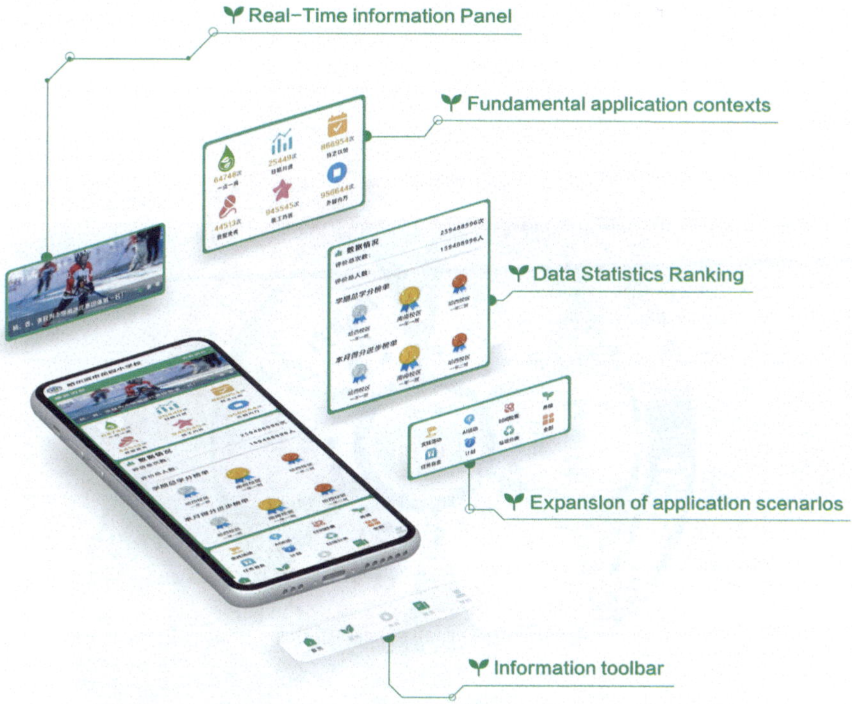

Fig. 1.1 The data brain of small seeds app

including internal strength, motivation, initiative, discipline, reflection, and core character. As shown in Fig. 1.2, this has led to the development of an evaluation content system structured around 7 dimensions of competency as primary indicators, 114 secondary indicators, and 628 observation points.

The evaluation content emphasizes the integration of the "Five Educations." By consolidating fundamental disciplinary competencies, cross-disciplinary skills, and advanced competencies, seven evaluation dimensions are established, including moral competency, thinking competency, health competency, artistic competency, labor competency, innovation competency, and interpersonal competency. These dimensions are designed to assess six key aspects of students' internal strengths, focusing on their personal growth and development. The evaluation framework ultimately aims to guide students in six areas of self-development: understanding who they are ("I am"), recognizing their abilities ("I can"), embracing personal growth ("I change"), taking action ("I act"), showing perseverance ("I persevere"), and achieving success ("I achieve").

Emphasizing students' overall development, this approach transcends the boundaries of textbooks, subjects, schools, districts, and even societal and life contexts. The evaluation indicators and observation points are connected to each child's comprehensive and individual development. For example, health competency evaluation

Fig. 1.2 The "Five Educations, Seven Dimensions, and Six Powers" evaluation framework

is divided into four parts: physiological indicators, learning, regular practice, and frequent participation in competitions, with assessments conducted across multiple scenarios inside and outside the school.

From the perspectives of the national curriculum, students' diverse abilities, and internal motivation, the system constructs interconnected indicators and observation points across the "Five Educations," the "Seven Dimensions," and the "Six Internal Strengths." The "Six Internal Strengths" are implicit evaluation points, serving as a link between the cultivation goals and students' competency development, fully integrated with the "Five Educations" and the "Seven Dimensions."

1.2.2 A Scenario-Based and Diverse Evaluation System

The "Seed" internal strength evaluation is not entirely centered around traditional class structures. Instead, it begins with personalized student choices, embedding evaluations within community-based scenarios. Through the interactive and intelligent DBSS platform, which seamlessly integrates learning and evaluation, the system promotes the growth of learning organizations and cultivates a strong learning culture.

This system integrates evaluations into everyday learning scenarios. It emphasizes the continuity and cumulative nature of learning. Broad learning objectives are broken down into smaller, manageable goals that are embedded into daily activities and assessed regularly. For example, the "Progress, Step by Step" section includes activities such as daily math "warm-up" exercises and monthly "Math Whiz" challenges, subtly enhancing students' reading and calculation skills through consistent practice.

The DBSS collects process-oriented materials such as students' daily performance, class participation, and homework submissions. This data is used to chart each student's developmental trajectory, showcasing changes in their seven-dimensional competencies and detailed growth indicators. It also creates personalized electronic growth portfolios for each student.

Considering the developmental characteristics of elementary students, a situational, challenge-based system has been designed, incorporating a "6 + N" diversified evaluation framework. This system uses a combination of in-school, out-of-school, online, and offline methods to track students' skill development in areas such as physical activity, reading, self-management, problem-solving, and social practice in real time.

Intelligent task design is used to enhance student learning (Khairunnisa, 2018). For example, the "AI Jump Rope" module automatically tracks students' jump rope counts, providing various evaluations and rewards. This gamified approach not only makes physical training enjoyable but also supports developmental evaluations over extended periods through fun and engaging activities.

1.2.3 An Integrated Smart Evaluation System

The DBSS features a robust system for assigning tasks, collecting data, and displaying and providing feedback on student performance. By establishing a physical data center for the DBSS, the system can perform real-time analysis, diagnostics, early warnings, monitoring, evaluations, and feedback. This ensures personalized support for each student's development while facilitating collaboration among teachers, parents, and the community to foster a comprehensive and intelligent evaluation process.

The "Daily Mastery" module includes regularly updated, standardized test questions and is led by top subject experts from the municipal level and above. The

DBSS has embedded this module in the last 5 minutes of teaching. This ensures an integrated approach to teaching, learning, and assessment, optimizing classroom practice while enhancing the quality of homework and reducing its load. The homework system generates instant statistical reports, allowing teachers to monitor both individual student and class-wide progress for each lesson.

In response to the Ministry of Education's guidelines on managing student phone use, sleep, reading, homework, and physical health,[2] the system involves families and communities in the evaluation process. Through multi-party data collection inside and outside the school, parents are guided in supporting home education. For instance, the AI-powered Reading Evaluation System integrates with the Chaoxing Campus Reading System to track students' graded reading of books across different grades, generating analysis reports for individuals, classes, and the school, thus increasing student reading engagement.

Using child-friendly displays, each student is represented by their own "leaves, flowers, and fruits." Students are rewarded with leaves for demonstrating learning achievements and completing tasks during each evaluation cycle. Ten leaves can be exchanged for a flower, and ten flowers can be exchanged for a fruit. This process generates a radar chart to display progress across the Five Educations, Seven Dimensions, and Six Internal Strengths.

Data mining is used to quickly identify issues in teaching and management, providing targeted solutions (Romero & Ventura, 2013). By analyzing student development portfolios, the system identifies strengths and areas for improvement, creating personalized growth plans to support deep learning (Ampadu, 2023). Additionally, in-depth data analysis highlights which courses effectively develop student skills and which are less impactful, guiding targeted improvements in teaching and reducing homework burdens.

1.3 Implementation Outcomes

1.3.1 Improvement of Evaluation Capability

Through the implementation of comprehensive evaluation reform, teachers and parents have shifted away from solely assessing students based on grades, now adopting a developmental and value-added perspective in evaluating academic progress. This has led to significant changes in the evaluation perspectives of teachers, parents, and students, resulting in a notable improvement in evaluation capabilities.

The development of the DBSS comprehensive evaluation platform and the establishment of a diversified evaluation system have supported the full implementation of the "Seed Curriculum" system, facilitating the comprehensive execution of moral, intellectual, physical, aesthetic, and labor education. As shown in Fig. 1.3, statistics

[2] http://www.moe.gov.cn/fbh/live/2021/53908/sfcl/202112/t20211222_589181.html.

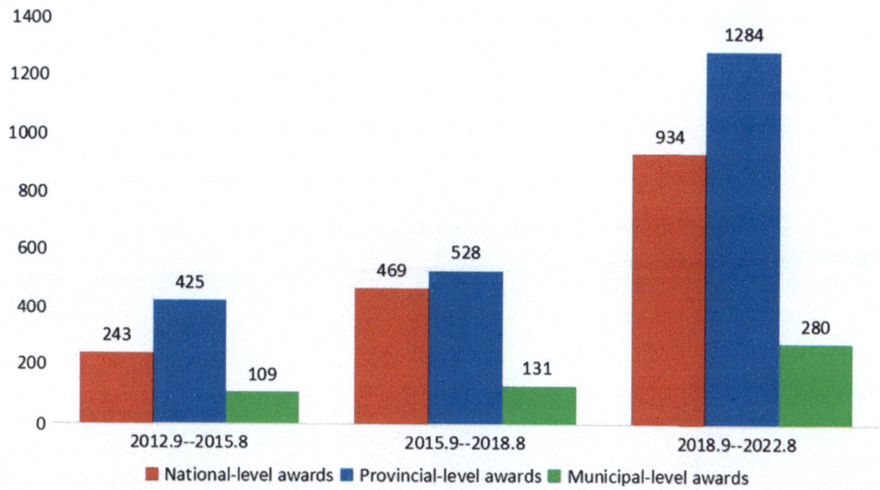

Fig. 1.3 Awards received by Harbin Garden elementary school students

show that from 2012 to 2018, prior to the implementation of the comprehensive evaluation system, students received 712 national-level awards and 953 provincial-level awards. Since the implementation of the comprehensive evaluation system in 2017, students have earned 934 national awards and 1,284 provincial awards, with growth rates of 31.18% and 34.73%, respectively. Among the national awards, 3 were in moral education, 34 in physical education, 778 in arts, and 47 in labor education, comprising 92.29% of the total. For provincial awards, 14 were in moral education, 36 in physical education, and 1,099 in arts, accounting for 89.48% of the total.

As shown in Fig. 1.4, from 2017 to 2019, the physical fitness pass rate among students at Harbin Garden Elementary School increased by 9.22 percentage points, with the rate of excellence rising by 13.18 percentage points. During the COVID-19 pandemic period from 2020 to 2022, the rate of excellence in physical fitness among students remained largely stable. In 2023, the physical fitness pass rate among students reached 94.85%.

1.3.2 *Enhancement of Education Environment and Governance Modernization*

The comprehensive evaluation system goes beyond being a simple measurement tool; it is a model specifically designed to support the unique growth needs of each student, offering a high level of personalization. Its implementation provides a solid scientific basis for reducing the teaching and learning burdens of teachers and students. As teachers and parents adopt new perspectives on evaluation, relationships between teachers and students, as well as between parents and children, have become more

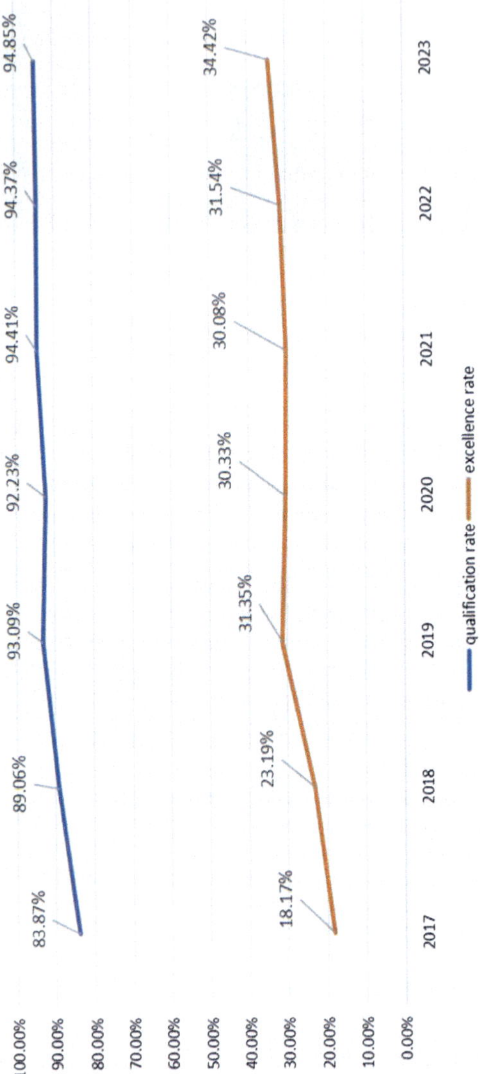

Fig. 1.4 Physical fitness test results of Harbin Garden Elementary school students

harmonious. This promotes a more holistic and supportive approach to evaluation and the overall educational environment within the school.

The DBSS platform provides a strong foundation for teachers, students, and parents to engage in school management and foster collaborative school–family governance and education. By addressing the limitations of traditional evaluation feedback—such as delays and limited scientific rigor—it meets the unique developmental needs of students while offering a solid, evidence-based approach to reducing teachers' workloads and students' learning burdens. This has led to significant advancements in the school's internal networked governance system and overall modernization of its governance capabilities.

1.4 Innovations

This model addresses challenges in traditional student evaluation systems that often prioritize outcomes over processes and external factors over intrinsic development. The "Five Educations, Seven Dimensions, Six Powers" evaluation model bridges gaps between national, regional, and school curricula, creating a pathway for cultivating students' internal growth. It activates students' inner strength, enabling them to discover, understand, and achieve self-realization.

This platform addresses shortcomings in previous evaluation systems, such as inaccurate data, delayed feedback, and marginalization in physical, aesthetic, and labor education. It supports individualized teaching, addresses educational gaps, and establishes a digital foundation for an innovative campus. Through seamless, panoramic data collection and intelligent analysis, it generates a comprehensive "data brain" for each student, allowing for personalized, process-based, and value-added evaluations. This innovation provides new methods for effectively implementing the school's "Five Management Tasks."[3]

By eliminating long-standing "information silos" caused by grade levels, subjects, fields, and divisions among family, school, and community, this new mechanism establishes a connected and shared digital community. It transcends boundaries between subjects, schools, school districts, and family–school–community networks, facilitating seamless and collaborative learning and development for students across grades, disciplines, and time. This fosters a new evaluation culture, reconstructs a comprehensive "all-around education" ecosystem, and promotes students' well-rounded and individualized development.

[3] The Five Management Tasks refer to specific areas of student life and well-being: homework, sleep, mobile phone use, reading materials, and physical health.

1.5 Promotion

The Secretariat of the Central Education Work Leading Group has highly recognized the reform achievements, as summarized in Issue 52 of the *Education Work Report*. This report highlighted Harbin Garden Elementary School's comprehensive evaluation reform experience in implementing the *General Plan for Deepening Reform of Education Evaluation* in the New Era which has generated significant interest nationwide. The brief covered four key areas, including planning lists, curriculum systems, labor workshops, and collaborative family–school education. It provides valuable experience and reference for primary and secondary schools across the country to advance the implementation of the *General Plan*, serving as a useful model for further exploration in comprehensive evaluation reform.

The comprehensive evaluation reform model has been endorsed by the Provincial Party Deputy Secretary and head of the Education Work Leading Group, with the Secretariat circulating the brief to all cities in the province. The report sparked considerable public interest, establishing the Harbin Garden Elementary School model as a recognized approach to education. Schools piloting the reform, despite varying levels of readiness, reported notable improvements in school scale, faculty quality, student skills, and parent engagement. In Yichun and Shuangyashan, two comparatively underdeveloped cities in Heilongjiang Province, the adoption of this evaluation model has significantly impacted curriculum development and evaluation practices, enhancing both educational evaluation capacity and overall quality within the region.

1.6 Summary

Under the backdrop of high-quality development in basic education, Harbin Garden Primary School focused on building an evaluation system aligned with students' physical and mental development to address the fundamental challenge of achieving high-quality education. Guided by the core philosophy that "Education is like agriculture," teachers are likened to farmers and students to seeds. After a decade of exploration, the school developed the "Seed Inner Growth" Comprehensive Evaluation System.

The school independently designed the "Little Seed Data Brain" innovative platform, which facilitated multi-stakeholder engagement in evaluations and improved curriculum management through process-based data mining. Amid the "Double Reduction" policy and the integration of the "Five Domains of Education" (moral, intellectual, physical, aesthetic, and labor education), the platform used intelligent technology to shift educational evaluation from "flood irrigation" to "precision drip irrigation." This approach effectively helped resolve the challenge of personalized education at scale and reduced the problem of educational stagnation. The solution supported the importance of childhood education by prioritizing student growth. It established a comprehensive framework that included an evaluation philosophy, technological tools, and implementation pathways. Its innovation lay in bridging traditional educational wisdom with modern technology, which allowed for data-driven precision-based evaluation that addressed individual differences while achieving personalized cultivation in large-scale teaching scenarios. This provides a practical model for reforming basic education evaluation.

The project received consecutive First Prizes and the Special Prize in the Heilongjiang Provincial Basic Education Teaching Achievements, and in 2023, it earned the National Teaching Achievement Award. These honors reflected its dual contributions to theoretical innovation and practical application, which provide strong support for schools working to promote well-rounded development and advancing high-quality education in the new era.

References

Ampadu, Y. (2023). Handling big data in education: A review of educational data mining techniques for specific educational problems. *AI, Computer Science and Robotics Technology, 2*. https://doi.org/10.5772/acrt.17.

Khairunnisa, K. (2018). Task design for improving students' engagement in mathematics learning. *Journal of Physics: Conference Series, 948*, 012012. https://doi.org/10.1088/1742-6596/948/1/012012.

Romero, C., & Ventura, S. (2013). Data mining in education. *Wiley Interdisciplinary Reviews: Data Mining and Knowledge Discovery, 3*. https://doi.org/10.1002/widm.1075.

Chapter 2
Construction and Practice of Internet-Enhanced Management Platform for Student Mental Well-Being

Yuchen Bian, Chen Chen, Mengwei Wu, Xiaoyang Fu, and Yingjie Chen

Abstract The school mental health education center faces numerous challenges, such as a weak teaching workforce, low levels of specialization, urban–rural development imbalances, and the absence of standardized procedures for addressing psychological crises. To address these issues, the Hui Lan Psychological Education Master Studio at Hangzhou No. 2 High School, in collaboration with Hangzhou Mingshitang Digital Technology Co., Ltd. has developed the Internet-Enhanced Management Platform for Student Mental Well-being. This platform provides effective support for school mental health education, improves the accuracy of risk screening, enhances the establishment of student psychological records, and standardizes crisis intervention procedures. Its standardized processes and user-friendly operation have demonstrated significant potential for promotion. Also, the pilot applications and promotions in designated regions and schools have already yielded remarkable results.

Keywords Student mental health · Big data · Precision · Dynamic · Student psychology

2.1 Background

In recent years, the fast pace of social development has led to increased psychological pressure on students, intensifying mental health conflicts and leading to a rise in crisis events. These trends present unprecedented challenges for school mental

Y. Bian (✉) · C. Chen · M. Wu · X. Fu · Y. Chen
Hangzhou No. 2 High School of Zhejiang Province, Hangzhou, China
e-mail: barry2002@126.com

M. Wu
e-mail: wumengwei@mistong.com

X. Fu
e-mail: fuxiaoyang@mistong.com

Y. Chen
e-mail: wow_chen@yeah.net

D. Liu et al. (eds.), *Empowering Educational Innovation Through Design: Volume One*, Smart Computing and Intelligence, https://doi.org/10.1007/978-981-95-4621-3_2

health education. A study by Zhu Shengying (2020) surveyed 4,083 primary and secondary schools and found that only 30.1% had fully implemented mental health education, while only 9.33% were equipped with full-time mental health teachers. This shortage of personnel and specialization in mental health services has created various issues for schools (Cao, 2017). For example, there is often a lack of professional and standardized procedures for addressing student issues and crisis interventions. Additionally, the process of recording, organizing, and retrieving paper-based psychological records is time-consuming and burdensome for teachers.

Consequently, there is an urgent need for improved support and standardization of school mental health services. However, it cannot be achieved overnight to develop a specialized and well-trained team of mental health professionals. Schools still face an immediate need to conduct scientifically informed mental health work, especially as digital technology continues to proliferate (Yao, 2021). In response, the Hui Lan Psychological Education Master Studio at Hangzhou No. 2 High School collaborated with Hangzhou Mingshitang Digital Technology Co., Ltd. to develop an Internet-Enhanced Management Platform for Student Mental Well-being. This digital platform introduces a standardized crisis intervention process and a scientifically comprehensive student psychological record system. The platform effectively lightens teachers' workloads while enhancing quality and efficiency of school mental health services.

The platform goes beyond basic assessment and screening functions by establishing a management cycle of "Identification—Monitoring—Intervention" which better serves the needs of student mental health education (Wu et al., 2010).

2.2 Design Process

Following the NCIP (Needs, Context, Ideation, and Presentation) Design Methodology (Huang et al., 2023), we developed the school's management platform for mental well-being through six phases: needs analysis, context identifying, user analysis, stakeholder analysis, competitor analysis, and presentation.

2.2.1 Needs Analysis

Firstly, we clarified the design needs for the school's mental health management platform. Using a comprehensive listing method, we identified that the primary users of the platform are mental health teachers and homeroom teachers. Other relevant stakeholders include students, parents, school leaders, and education department administrators. To better understand market needs, we conducted in-depth interviews and surveys with individuals such as city officials, principals, vice principals of student affairs, directors of student affairs, mental health teachers, and homeroom teachers. We conducted field research across 11 schools and 2 municipal-level educational research offices across 6 provinces and 9 cities. As shown in Fig. 2.1, the distribution of mental health education activities varies across the surveyed schools.

The survey revealed that, in many schools, although psychological assessments and screenings are common, the proportion of follow-up interventions is relatively low, and showed a funnel-shaped distribution in terms of comprehensive mental health education efforts (Fig. 2.2).

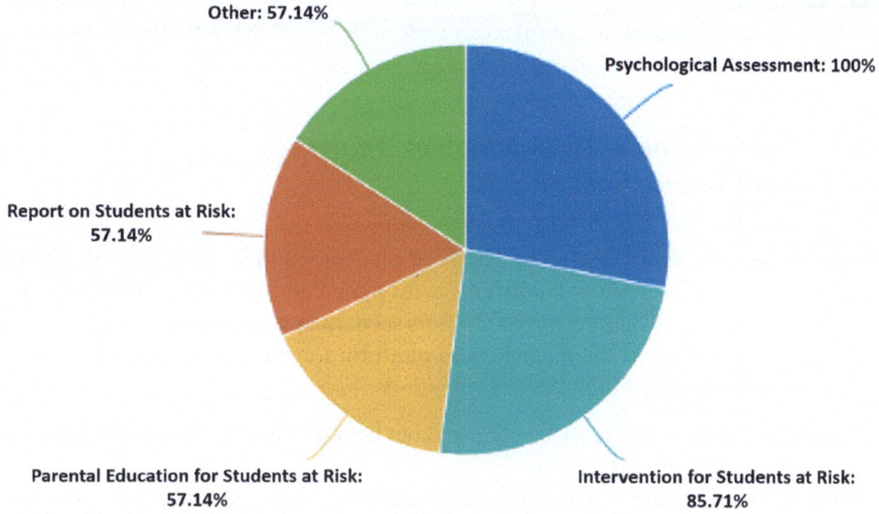

Fig. 2.1 Distribution of mental health education activities in surveyed schools

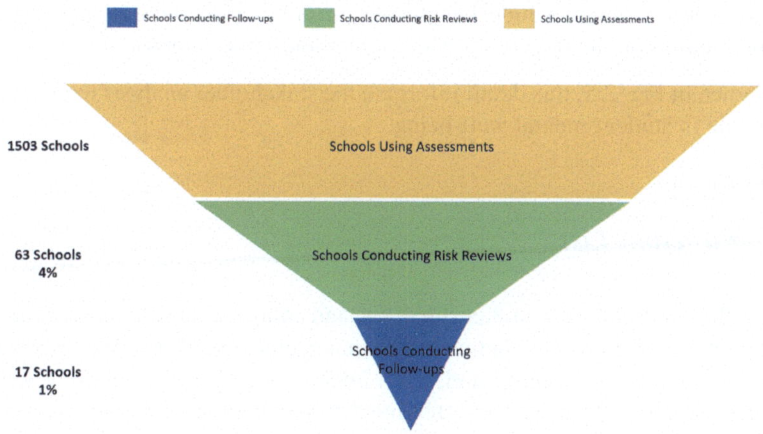

Fig. 2.2 Funnel-shaped distribution of school psychological screening and follow-up intervention implementation

Additionally, since psychological assessments inherently carry a degree of subjectivity, the screening results often need to be integrated with other information. This process requires a sequence of feedback, verification, confirmation, and follow-up for a comprehensive analysis.

To address these issues, we propose the following solutions, including enriching early warning review and intervention methods, enhancing student records, introducing professional course guidance, supporting convenient access to screening data, and ensuring data security. These measures aim to improve the efficiency and standardization of mental health management processes in schools (Fig. 2.3).

2.2.2 Context Analysis, Stakeholder Analysis, and Competitor Analysis

Beginning with the main workflow in school mental health services, we organized the platform's core scenarios, identifying pain points and needs for different roles across key scenarios. As shown in Fig. 2.4, this scenario analysis helped clarify where support is most needed in the management platform for student mental well-being.

The specific scenarios include:

- Students who have not yet been screened but are identified as potentially needing a warning based on observation.
- Students are flagged for a warning after screening. They require further observation, interviews, and secondary screening to confirm the need for intervention.
- Students confirmed as needing a warning and requiring professional intervention for psychological counseling.
- Deciding whether to conclude the intervention or refer the student to a professional hospital based on the intervention's progress and effectiveness.

As shown in Fig. 2.5, this detail informed the stakeholder analysis to clarify each group's role in student mental well-being.

2.2.3 Function List

Based on the needs analysis, context analysis, and competitor analysis, we structured the required functions for the mental health management platform. We reviewed the offline psychological screening and crisis intervention workflows, along with the collaboration process for each role, creating detailed diagrams for crisis intervention and role collaboration processes. As shown in Fig. 2.6, we developed specific function modules for the "Identification," "Monitoring," and "Follow-Up" phases. Each module was organized by school workflow, step outcomes, and functional content.

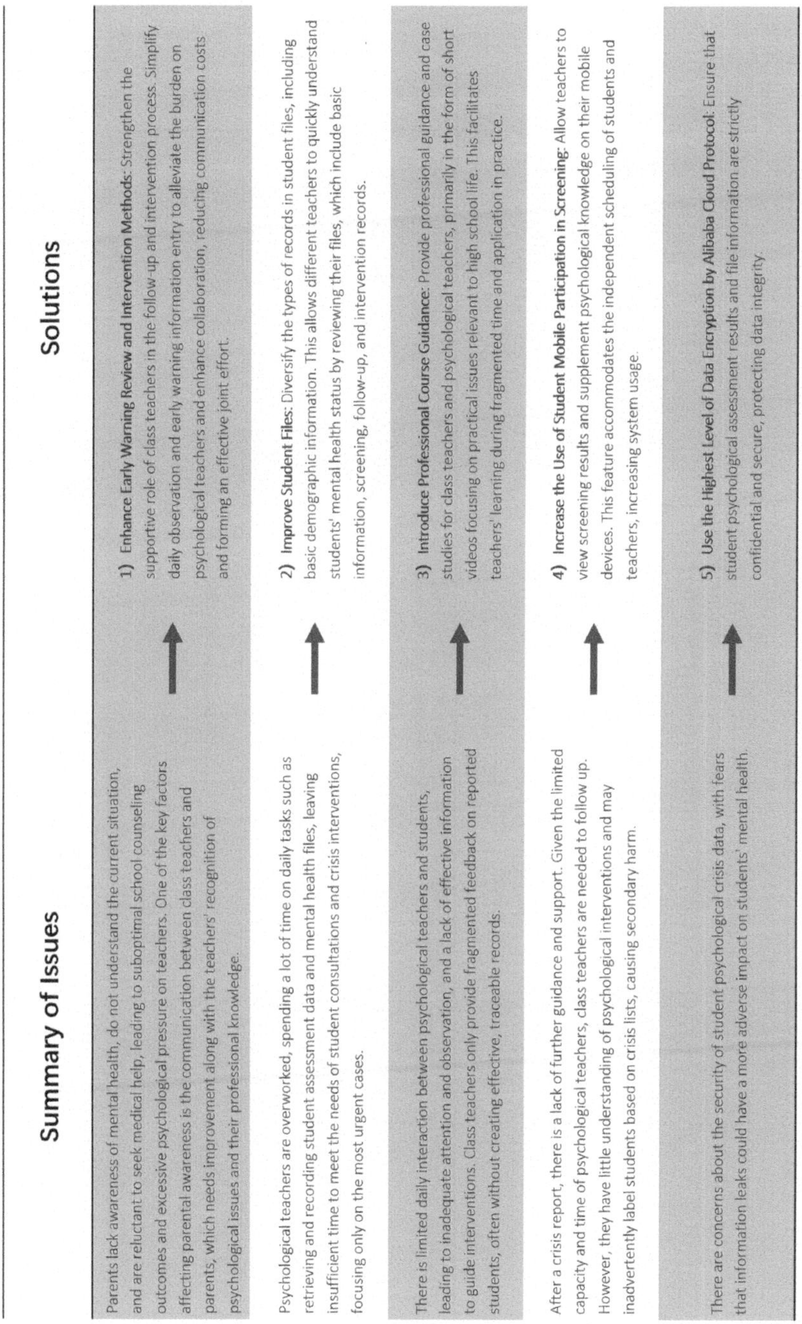

Fig. 2.3 Summary of design requirements for the mental health management platform

Core Scenario	Role	Pain Point/Demand	Value of Requirement	Priority
For students who require early warning after observation but have not been screened, support is needed to import them into the system to complete subsequent intervention work.	Psychological Teacher	Import basic information of students requiring early warning into the system and be able to view and follow up normally	Business Process Closure	P0
After the screening is completed, students requiring early warning need to be assessed through professional observation, interviews, and secondary screening to determine if intervention is needed.	Psychological Teacher, Class Teacher	1. Assign students requiring early warning to class teachers for observation and record observations. 2. Interview students requiring early warning and record interviews. 3. Review early warning. 4. Follow-up records can be viewed coherently.	Business Process Closure	P0
After confirmation of early warning, professional intervention methods are needed for psychological counseling. Different intervention methods are applied based on low, medium, or high risk.	Psychological Teacher, Class Teacher, Parent, School Leader	1. Low-risk students are assigned to class teachers for ongoing observation and record-keeping until the alert is lifted. 2. Medium-risk students are assigned to psychological teachers for regular counseling and record-keeping until the alert is lifted. 3. High-risk students are reported to school leaders and parents are interviewed, with results recorded. 4. High-risk students are referred, with referral results recorded. 5. Intervention records can be viewed coherently. 6. Change in student risk level	Business Process Closure	P0
End the intervention based on the situation and effectiveness of the intervention		1. Lift the student's early warning, return to normal, and record the lifting explanation. 2. If the student's intervention is not effective, record the loss explanation.	Business Process Closure	P0

Fig. 2.4 Scenario analysis of the management platform for student mental well-being

Stakeholder	Pain Point	Requirement
Psychological Teacher	1) Psychological teachers often need to handle a large number of student psychological issues but lack a standardized process, leading to inefficient management. 2) Early warning lists and student psychological files should be accessible anytime, anywhere.	1) The platform needs to provide a standardized process for psychological crisis intervention, allowing psychological teachers to handle student psychological issues more efficiently. 2) Information and data should be easily accessible.
Class Teacher	1) Class teachers often need to pay attention to the mental health of their students but lack a convenient way to access student psychological files. 2) Observation records should be uploadable and viewable at any time.	1) Psychological teachers can proactively send early warning review notifications to class teachers for immediate receipt. 2) The platform should provide an integrated construction of student psychological files, allowing class teachers to easily view and manage student psychological files.
Parent	Parents often find it difficult to understand their children's psychological conditions and need more psychological counseling and guidance.	The platform should provide a parent portal allowing parents to understand their children's psychological conditions and offer psychological counseling and guidance services.
Student	Students face significant psychological pressure but often struggle to find appropriate psychological resources and training tools.	The platform should provide a student portal, allowing students to easily view and manage their psychological files and offering psychological resources and training tools.
Moral Education Director/Vice Principal	Moral education leaders often need to supervise and manage students' psychological health issues but lack a unified, standardized management framework, leading to chaotic, uncoordinated, and inefficient management.	The platform should provide a comprehensive, standardized mental health management framework to help moral education leaders organize interventions and management work, better addressing various student mental health issues.
Principal	Student psychological issues require greater attention, and a mental health management system is needed to facilitate an overall understanding of each student's psychological health status.	The platform should provide a comprehensive, standardized student mental health management system to help schools better deal with student psychological issues.
Education Authority Leaders	Student psychological issues require greater attention, and a mental health management system is needed to facilitate an overall understanding of the psychological health status of students in every school.	The platform should provide a comprehensive municipal-level student mental health management system to help education authorities understand the overall mental health status of students in the city.

Fig. 2.5 Stakeholder analysis for management platform for student mental well-being

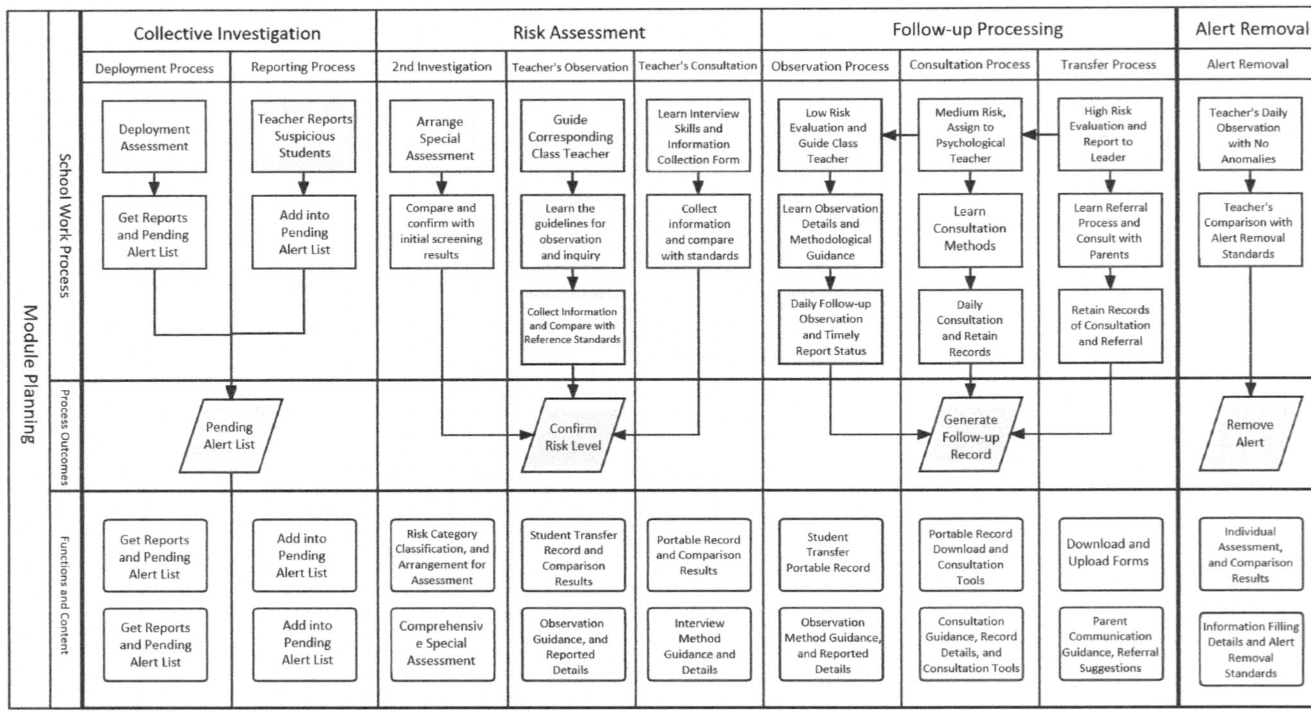

Fig. 2.6 Design flowchart for the management platform for student mental well-being

We configured the "Identification," "Monitoring," and "Intervention" modules according to the roles of mental health teachers, homeroom teachers, students, parents, and school leaders. Figure 2.7 shows how these roles were considered in configuring functional modules. During the configuration process, we discovered that homeroom teachers primarily use mobile devices in practice, while mental health teachers rely on a combination of both web and mobile platforms. This differentiation ensures that each role can effectively access the platform, aligning with their respective workflows and contexts.

In addition to building the functional modules for crisis intervention workflows, we also developed a module to optimize student mental health records. These records include a student's basic personal information, screening results from various assessments, follow-up warning records, and a comprehensive history of psychological counseling and intervention sessions, as recorded by different roles. All information is integrated into a single, unified record.

As shown in Fig. 2.8, the integration process is automatically generated by the system in chronological order, with a one-click download and print feature available for easy access to the student's mental health records. This implements a streamlined "smart record management" functionality.

2.2.4 *Integrated Solution*

With the completion of the functions above, we designed top-level diagrams for the mental health management platform focused on the core processes of "Identification," "Monitoring," and "Intervention." Key functions include psychological assessment, one-click reporting, monitoring lists, quick record entry, psychological training, special workflows, and more. We established a holistic "smart mental health education system" encompassing five integrated components: the Student Mental Health Platform, Psychological Education Platform, Teacher Training Resources, Parent Training Resources, and the Administrative Management Platform. As shown in Figs. 2.9 and 2.10, this top-level design and logic diagram indicated how these components interact and support the overall mental health record system.

We developed a minimum viable solution for the mental health management platform based on specific school needs. Initially, we implemented a grid management system for primary screening, homeroom teacher reporting, and mental health teacher assessment to more accurately identify students with potential risks. The mental health record system facilitates automatic integration and cross-system data synchronization, enhancing the efficiency of record-keeping for mental health teachers and homeroom teachers. Continuous improvement and optimization provide reliable tools and support for schools, fostering students' mental health development.

Through iterative testing and feedback, we continually refined the product to better meet practical needs and enhance user experience. User feedback and suggestions were collected for ongoing optimization of functionality and usability, ensuring the platform's feasibility and effectiveness. Continuous improvements have made the

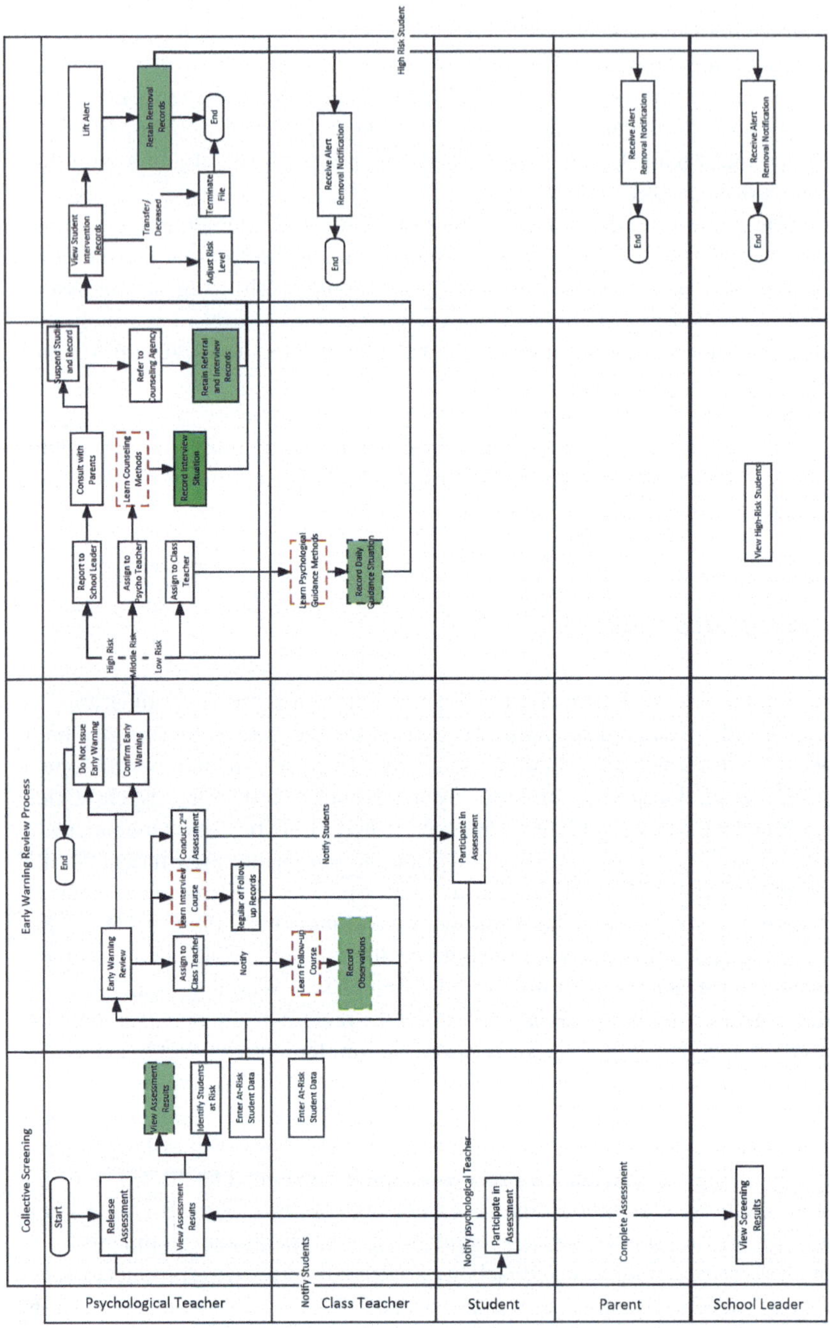

Fig. 2.7 Functional module configuration for the management platform for student mental well-being

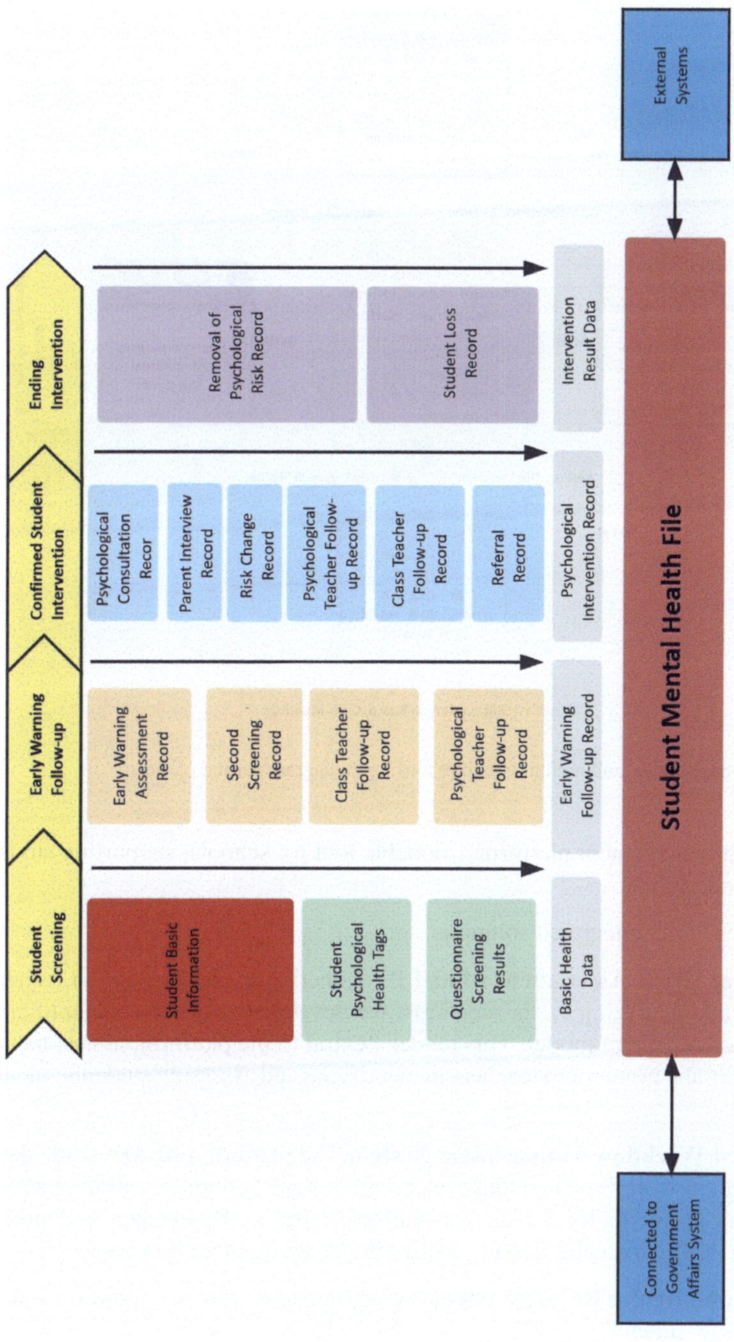

Fig. 2.8 Functional module for student mental health records

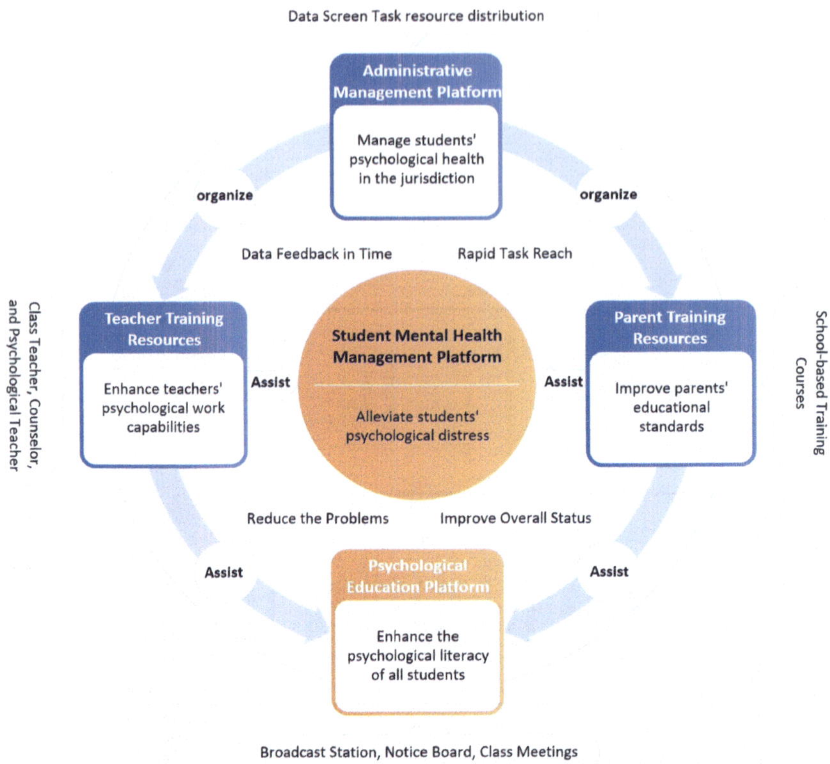

Fig. 2.9 Top-level design and mental health record logic diagram for the platform

mental health management platform a valuable tool for schools, supporting student well-being and growth.

Key measures in the integrated solution:

Psychological Crisis Assessment Model By gathering extensive student psychological data and analyzing crisis scenarios, we developed a reliable, scientifically backed crisis assessment model. This model, central to the platform, assists mental health teachers and homeroom teachers in identifying and assessing students' mental health status.

Standardized Workflow Management System The platform provides a standardized intervention workflow, including intervention plans, communication methods, and recording protocols for various psychological issues. Blockchain and similar technologies enable traceability and security in managing these processes.

Student Mental Health Record System A comprehensive record system was established, covering student information, psychological growth, and intervention records, with cross-system synchronization to aid intervention and decision-making.

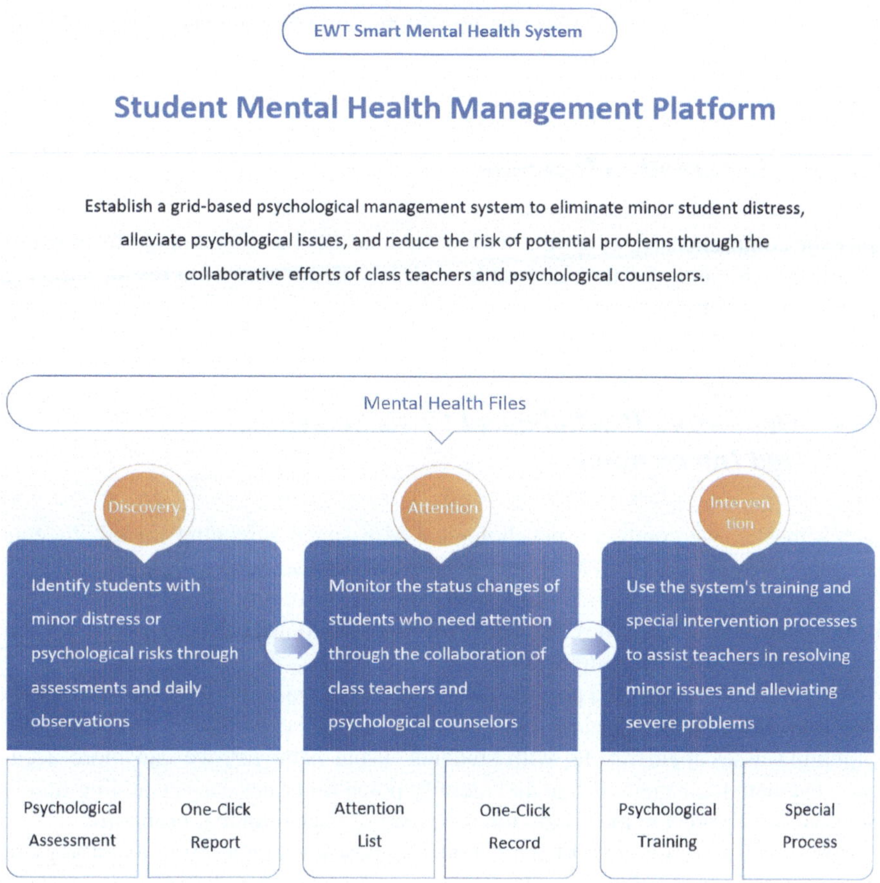

Fig. 2.10 Flowchart of Grid-Based Management and Collaborative Intervention in Student Mental Health

Training and Educational Resources The platform offers a wealth of mental health training and resources for both students and teachers, enhancing intervention effectiveness and building the expertise of part-time and full-time mental health teachers.

User-Friendly Interface Design Designed for primary users—mental health teachers and homeroom teachers—the interface is simple, clear, and intuitive, facilitating workflow management, crisis monitoring, and mental health record maintenance.

With this integrated solution, the Management Platform for Student Mental Well-being can comprehensively monitor and manage students' mental health status, improving the effectiveness of mental health education and intervention. It meets the needs of mental health teachers and homeroom teachers, helping students better manage mental health issues and creating a supportive environment for their healthy development.

2.3 Implementation Outcomes

Since the comprehensive adoption of the mental health management platform in our school, the efficiency of mental health work has significantly improved, showing outstanding results.

2.3.1 Optimizing Psychological Crisis Screening and Intervention

Each semester, we conduct a psychological assessment screening for all students using the *High School Student Mental Health Assessment Scale*, co-developed with Professor Hu Yueqin's team from Beijing Normal University. This updated assessment tool accurately identifies vulnerable and hidden groups, enhancing the precision and effectiveness of screening results.

After the assessment, we provide each homeroom teacher with a list of students who may be at psychological risk, allowing for a second-level screening. By combining assessment results with students' usual behavior, we can more accurately evaluate their mental state and identify potential issues early, enabling timely interventions to prevent the escalation or spread of psychological problems.

Additionally, we utilize Alibaba Cloud's highest-level encryption technology to ensure the strict protection of student privacy and prevent data breaches.

2.3.2 Enhancing Teacher Capacity for Psychological Issues

The implementation of the mental health management platform has effectively fostered close collaboration between mental health teachers and homeroom teachers. Through the platform, they can monitor each student's mental health status, discuss psychological issues, and devise more effective solutions together. After follow-up conversations with students and their parents, teachers can record updates directly on the platform via a mobile app. For common issues, they can select from preset options; for unique situations, they can add brief notes. The platform also supports uploading files like audio recordings and photos, saving a significant amount of time on record-keeping and organization.

Additionally, the platform facilitates timely record-keeping for psychological screening, intervention, counseling, home visit records, treatment, and referral records. This cross-platform collaboration capability enhances follow-up efficiency. Teachers can access a complete psychological profile on the platform when needed, offering critical support and reference for intervention and decision-making.

2.3.3 *Promoting Campus-Wide Mental Health Initiatives*

The platform enables a structured workflow, guiding new mental health teachers and homeroom teachers in fulfilling their responsibilities with professionalism. Through learning resources, such as *Psychological Issue Assessment Standards*, *Teacher-Student Communication Techniques*, and *Parent Notification Templates*, teachers' professional skills are significantly enhanced.

The platform offers extensive resources to support mental health initiatives in schools. Schools can play the Heart FM program—popular among students—during breaks, select psychology-themed posters that align with students' current psychological development stages, or choose suitable mental health training tools to cultivate a supportive atmosphere for mental well-being.

In summary, the mental health management platform has provided numerous significant and practical advantages in real-world school applications. These benefits strongly support the development of a comprehensive mental health education system, helping schools better serve students and ensure their mental and physical well-being and holistic growth.

2.4 Innovations

The school mental health management platform incorporates the following innovations:

2.4.1 *Enhanced Precision with Grid-Based Psychological Screening*

The platform implements a grid-based management approach with "Initial Psychological Screening" + "Homeroom Teacher Reporting" + "Mental Health Teacher Assessment," which more accurately identifies students with potential risks. Using scientifically validated assessment tools, the initial screening provides a comprehensive view of students' mental health. The homeroom teacher reporting function is crucial for identifying unusual emotional, behavioral, or academic patterns, allowing quick escalation to mental health teachers for further assessment. Mental health teachers then conduct in-depth evaluations of these students through interviews and further assessments, gaining more insight into each student's psychological state and needs.

2.4.2 *Efficient Dynamic Mental Health Record System*

The mental health record system enables automatic integration and cross-system data synchronization, enhancing the record-keeping efficiency for mental health teachers and homeroom teachers. This system allows the centralized storage of students' psychological information and records, automating integration and synchronization across systems. This approach not only eliminates the cumbersome process of managing paper-based records but also ensures timely updates and easy access to psychological records. Teachers can readily view a student's psychological assessment results, issue logs, and intervention history, thus staying informed of any changes in the student's mental state. This dynamic record system provides crucial support for timely interventions and decision-making.

2.4.3 *Precision-Driven Psychological Training Tools Through Big Data*

For students identified with various issues and needs, the platform uses big data to deliver tailored psychological training tools that address specific challenges, helping them improve positive psychological traits and overall mental health. Through data analysis, the platform offers students personalized training resources based on their assessment results and individual circumstances. These tools include training in areas like emotional management, stress coping, confidence building, and interpersonal relationship improvement. The platform can suggest relevant training materials, exercises, and guidance based on each student's unique needs, helping them manage challenges, regulate emotions, and enhance their mental health. Big data analytics also allow for tracking and evaluating the effectiveness of these interventions, providing valuable feedback for continuous improvement of training strategies. This targeted delivery of psychological training tools ensures that students receive personalized support, promoting their holistic and sustained development.

2.5 Promotion

The potential for promoting the mental health management platform in other regions and schools is highly promising.

Firstly, as society advances and awareness of mental health grows, the demand for school mental health services is rising. The student mental health management platform, as a technological solution, can effectively support schools in providing mental health services.

Secondly, the platform's standardized psychological intervention processes and mental health record system are designed to be simple and user-friendly, making it easy for schools in diverse locations to adopt and replicate. This enables more schools to manage student mental health data effectively and implement consistent interventions.

Additionally, we have enhanced and updated the crisis intervention process and the mental health record module. Following this, we launched a 3-month pilot program in various regions and schools nationwide. The pilot significantly improved the efficiency of mental health efforts in participating schools, particularly in student monitoring and follow-up, yielding notable results. These outcomes further confirm the platform's feasibility for broader adoption across regions and schools.

In conclusion, the Internet-Enhanced Management Platform for Student Mental Well-being has strong potential for wide-scale adoption. Moving forward, we will continue to innovate and advance this platform, actively leading new directions in mental health education, and making exceptional contributions to school mental health education.

2.6 Summary

This case examined the development and implementation of the Internet-Enhanced Management Platform for Student Mental Well-being. It addressed major challenges in school-based mental health services, including a shortage of trained staff, uneven resource allocation, non-standardized crisis procedures, and inefficient paper-based record-keeping. By integrating digital technology into the core workflow of "Identification-Monitoring-Intervention," the platform provides a systematic solution to improve the precision, efficiency, and standardization of psychological support for students.

2.6.1 Core Value and Findings

The platform introduced several innovations that delivered significant value:

- **Precision Screening**: A grid-based system that combined initial assessments, homeroom teacher reports, and verification by mental health staff significantly improved the accuracy of identifying at-risk students, moving beyond subjective screening methods.

- **Dynamic Digital Records**: An automated, integrated psychological profile system replaced cumbersome paper records and allowed real-time updates, cross-role data synchronization, and one-click access to comprehensive student histories.
- **Standardized Crisis Management**: Structured workflows guided teachers through evidence-based intervention procedures, ensuring consistent and traceable crisis response.
- **Targeted Support**: Big data analytics facilitated the provision of personalized psychological support and training resources customized to meet the unique needs of each student.
- **Pilot implementations demonstrated tangible outcomes**, including improved screening accuracy, reduced teacher workload in record-keeping, enhanced collaboration between mental health professionals and homeroom teachers, and strengthened capacity for proactive campus-wide mental health initiatives.

2.6.2 Theoretical and Practical Significance

The platform actively translated educational and psychological theory into practice:

- It operationalized (Wu et al., 2010) "Identification-Monitoring-Intervention" crisis framework within a digital ecosystem.
- Its design strictly followed the NCIP methodology (Huang et al., 2023), ensuring user-centered development based on deep needs and context analysis.
- By digitizing and structuring complex workflows, specialized mental health protocols became accessible and actionable, even in settings with limited expert resources, thereby promoting equitable service quality.

In practice, the platform empowers schools to overcome staff shortages, enhances the efficacy of interventions, fosters a supportive school climate, and provides scalable tools that can be used in various settings to support students' well-being.

2.6.3 Recommendations

- **For Practice**: It was recommended that more schools adopt the platform's screening tools and digital record system. Schools could use the structured workflows to train staff, particularly those without specialized training, in standard procedures for supporting student mental health. Regional education authorities could also use the administrative module to support oversight and allocate resources more effectively.
- **For Future Research**: Longitudinal studies were needed to evaluate the platform's long-term effects on student mental health outcomes and crisis prevention. Further research should have examined the integration of AI for predictive

analytics in the screening module and the effectiveness of big-data-driven personalized training tools. It was also important to examine cross-cultural adaptation and use in diverse settings.

In essence, this platform demonstrated how technology can bridge theory and practice, offering a replicable model to improve the precision, efficiency, and accessibility of mental health support systems in educational settings.

References

Cao, F. L. (2017). Investigation and countermeasures of the current situation of psychological crisis intervention among high school students. *Psychology and Health Education in Primary and Secondary Schools, 337*(26), 68–73.

Huang, R., Liu, D., & Chen, P. (2023). *Design and Learning*. Beijing Normal University Press.

Wu, C. X., Lin, C. D., Zang, W. W., & Fu, F. (2010). On the construction of the school psychological crisis intervention system. *Journal of Beijing Normal University: Social Sciences, 01*, 45–50.

Yao, Y. P. (2021). Reflections on the professional development of mental health education teachers in primary and secondary schools. *Psychology and Health Education in Primary and Secondary Schools* (10), 67–68+71.

Zhu, S. Y. (2020). The current situation and reflections on mental health education in primary and secondary schools: Based on the analysis report of the current situation of mental health education in primary and secondary schools. *Psychology and Health Education in Primary and Secondary Schools* (36), 4–10+14.

Chapter 3
Code Mission: Designing an Educational Tool for Programming Instruction

Peng Liang

Abstract This case examines how to provide children in rural and remote areas with affordable programming resources and opportunities to develop computational thinking skills, to meet the core objective of "cultivating computational thinking" outlined in the information technology curriculum standards. We developed a programming puzzle game that combines board game elements with basic programming concepts in a simple, pen-and-paper format, allowing students to learn programming and strengthen their computational thinking skills with minimal hardware and software requirements. The game is easy to produce, deploy, and expand, making it highly adaptable and suitable for a wide range of applications.

Keywords Programming education · Computational thinking · Game-based learning · Unplugged programming · Educational game design

3.1 Background

The Information Technology Curriculum Standards for Compulsory Education (2022 Edition)[1] designates computational thinking as a core skill to be developed through information technology courses. Approaches to fostering computational thinking can vary, but programming education has a natural advantage in this area (Yildiz Durak, 2020). Activities like abstraction, decomposition, modeling, and algorithm design are all central to programming and are reinforced, especially when applied to real-world problem-solving (Gökçe & Yenmez, 2023). For the past 7 years, our school has been working to provide rural children with programming education that rivals what urban students receive, helping them experience the excitement of information technology. Our "fun programming" curriculum continually evolves, with expanded content and optimized teaching practices. The product featured in this case—a programming

[1] http://www.moe.gov.cn/srcsite/A26/s8001/202204/W020220420582361024968.pdf.

P. Liang (✉)
Daohubi Central Primary School, Jinzhong City, Shanxi Province, China
e-mail: yesdhb@126.com

D. Liu et al. (eds.), *Empowering Educational Innovation Through Design: Volume One*,
Smart Computing and Intelligence, https://doi.org/10.1007/978-981-95-4621-3_3

puzzle game—was developed based on our years of experience in programming instruction. By incorporating the concept of "unplugged programming," it minimizes the need for costly equipment and resources, which is a common challenge for expanding programming education in rural schools. This tool creatively integrates board game elements to teach programming fundamentals and foster computational thinking.

3.2 Design Process

3.2.1 Core Idea

In nearly all introductory programming courses, the first lesson typically involves using the programming language to print "Hello World" in the command line or compiler. This simple output exercise helps beginners quickly get acquainted with the language and introduces foundational knowledge, such as syntax, compilation, and setting up a development environment. Over time, "Hello World" has become a symbolic entry point in programming culture.

As learners progress, different programming languages may introduce unique syntactic rules, compilation requirements, runtime characteristics, and advanced features. However, specific foundational knowledge is essential for any beginner, regardless of the language being learned.

- **Basic Syntax**: indentation rules and naming conventions.
- **Data Types**: integers, floats, Booleans, and more complex structures like arrays and structs.
- **Control Structures**: sequential flow, loops, and conditional branching.

Presenting these fundamental computational concepts and basic programming principles through an "unplugged programming" approach—before diving into formal programming—can significantly enhance students' interest in learning.

Our product draws inspiration from graphical programming software, combining the "unplugged programming" approach with elements of board games. Players follow game rules to complete tasks, subtly learning essential concepts like loops, conditions, and variables. This lays a logical foundation for developing computational thinking and preparing for programming studies, facilitating a smoother transfer of knowledge.

3.2.2 Product Creation

The product is created by designing and formatting a table-based program. Custom shapes are inserted, and cell dimensions are adjusted to achieve the desired layout.

As shown in Fig. 3.1, the Illustrated Instructions section provides a visual guide. Icons and instruction cards used in the game are displayed on the left, while the right side contains paper templates for props that can be cut and assembled. Players may also use other objects in place of these templates in the game.

As shown in Fig. 3.2, this product features three chapters with a total of nine levels. Each level includes a map area, an instruction area, and an instruction bank. Players use instructions from the bank to control the game pieces, navigating them to the finish point. Each level incorporates foundational programming concepts.

Figure 3.3 shows the design of the guidebook, which provides detailed walk-throughs for each level, also including brief explanations of the programming concepts, which serve as the core instruction manual for the product.

3.2.3 *Product Introduction*

3.2.3.1 Chapter 1, Level 1—Start the Game

Objective: Familiarize players with the game interface and basic gameplay.

Knowledge Card: Computers execute program instructions sequentially, namely, each instruction is executed one at a time in the specified order.

Design Concept: The instruction bank provides the necessary instruction cards for completing the task. Players simply need to place the correct instruction cards in the instruction area in the correct sequence to finish the level. This sequential use of instruction cards mirrors the sequential execution of program instructions. As shown in Fig. 3.4, this layout introduces the core structure of gameplay.

3.2.3.2 Chapter 1, Level 2—Look! a Flag!

Objective: Learn how to connect and use different instruction cards, completing various game tasks through different combinations of instruction cards.

Knowledge Card: Complex programs are built by combining simple program statements. Different program statements work together, each playing a specific role to achieve a program's overall function.

Design Concept: This level introduces a rotation instruction card, requiring players to use and correctly combine two types of instruction cards to complete the task. This corresponds to how different program statements are appropriately combined to create a program with more complex functionality. Figure 3.5 illustrates this design.

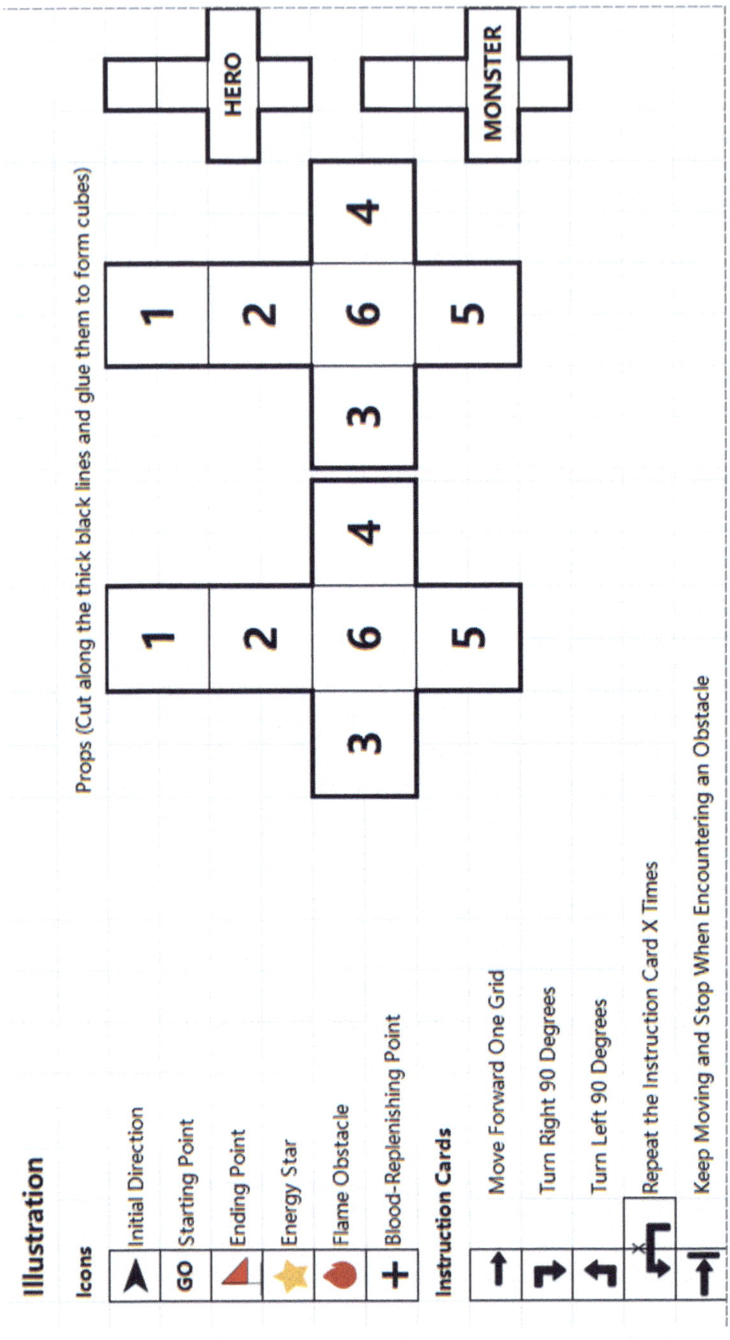

Fig. 3.1 Illustrated instructions design diagram

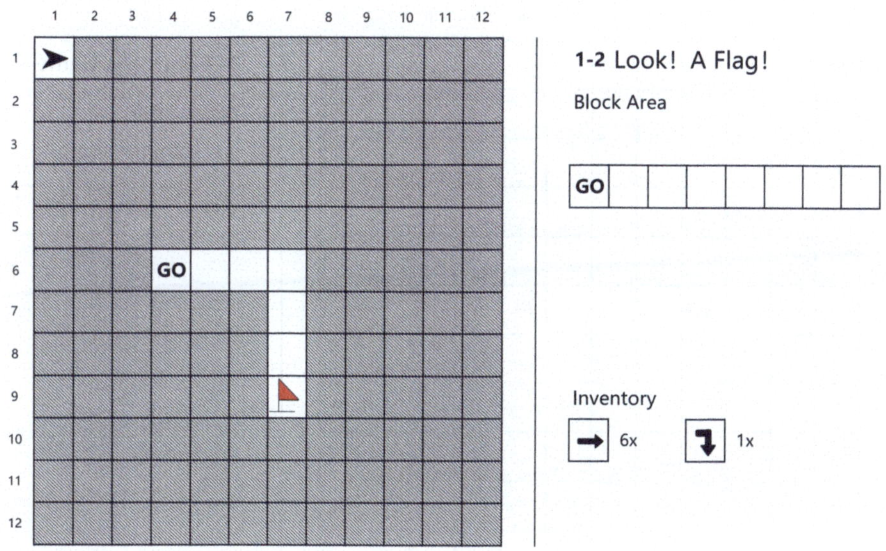

Fig. 3.2 Level design diagram

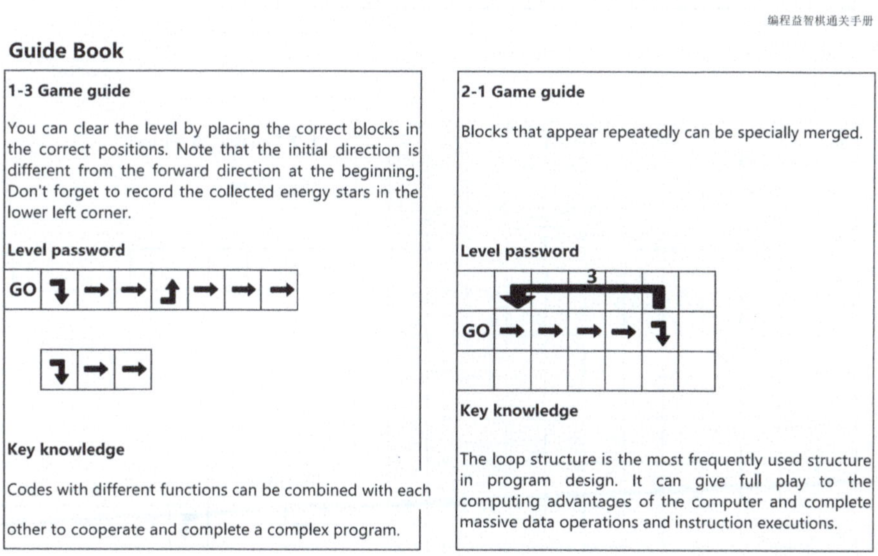

Fig. 3.3 Design diagram guidebook

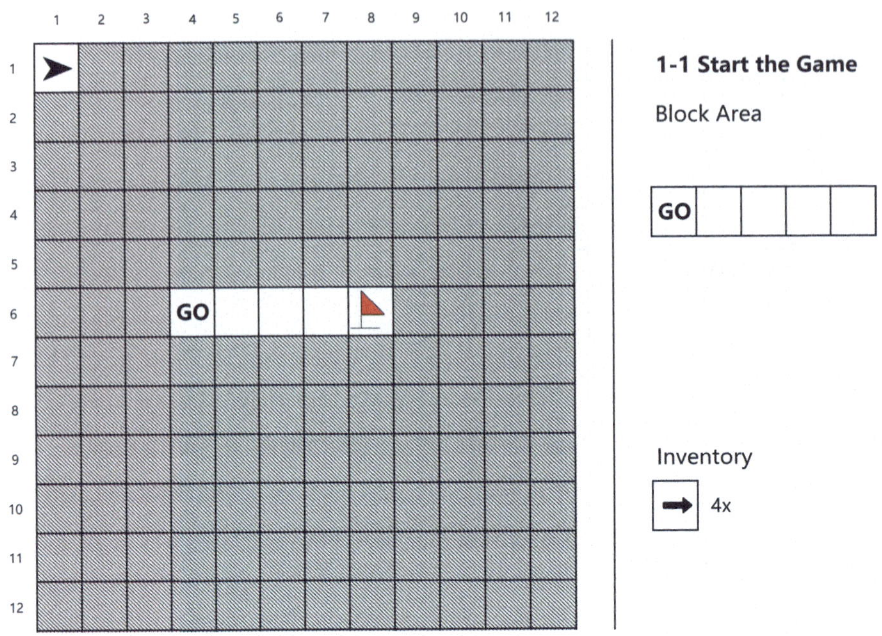

Fig. 3.4 Chapter 1, Level 1 design diagram

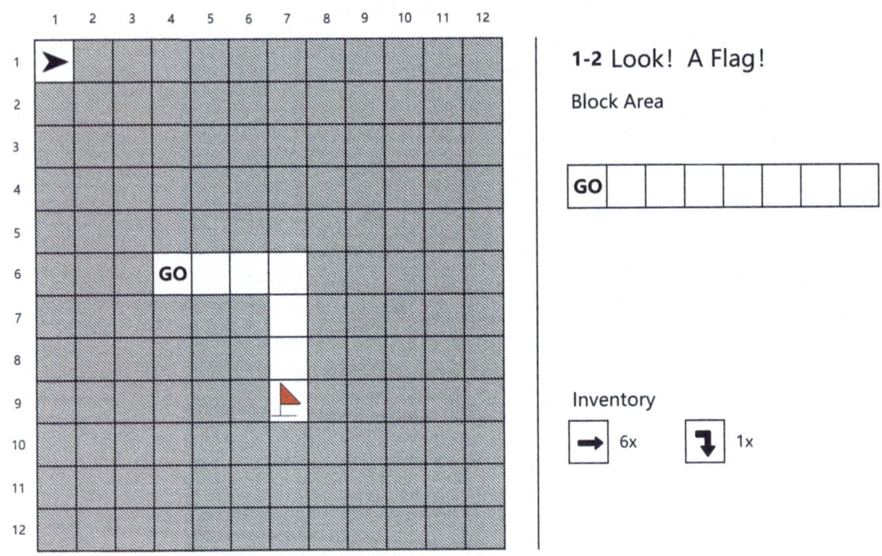

Fig. 3.5 Chapter 1, Level 2 design diagram

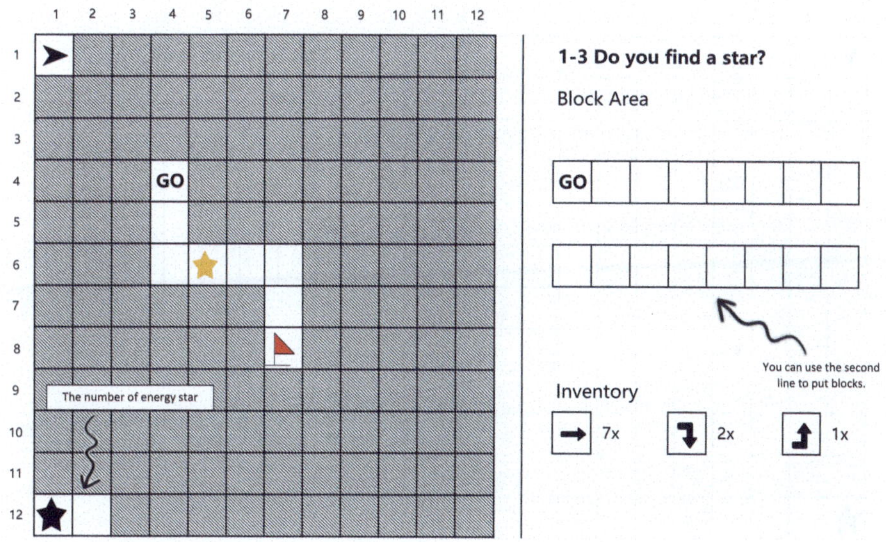

Fig. 3.6 Chapter 1, Level 3 design diagram

3.2.3.3 Chapter 1, Level 3—Do You Find a Star?

Objective: To complete more complex tasks, players must use a greater number of instruction cards and combine them correctly. This level introduces a mini-task of collecting energy stars, setting the stage for the introduction of the concept of variables.

Knowledge Card: Same as the previous level.

Design Concept: Building on the previous level, this stage increases the difficulty, challenging players to apply learned concepts with more complexity. As shown in Fig. 3.6, the map design becomes more intricate.

3.2.3.4 Chapter 2, Level 1—More Energy Stars

Objective: Use loop instruction cards to reduce the number of instructions and increase execution efficiency.

Knowledge Card: Loop structures are one of the most commonly used structures in programming. They leverage the computing power of computers to process and filter large amounts of data quickly.

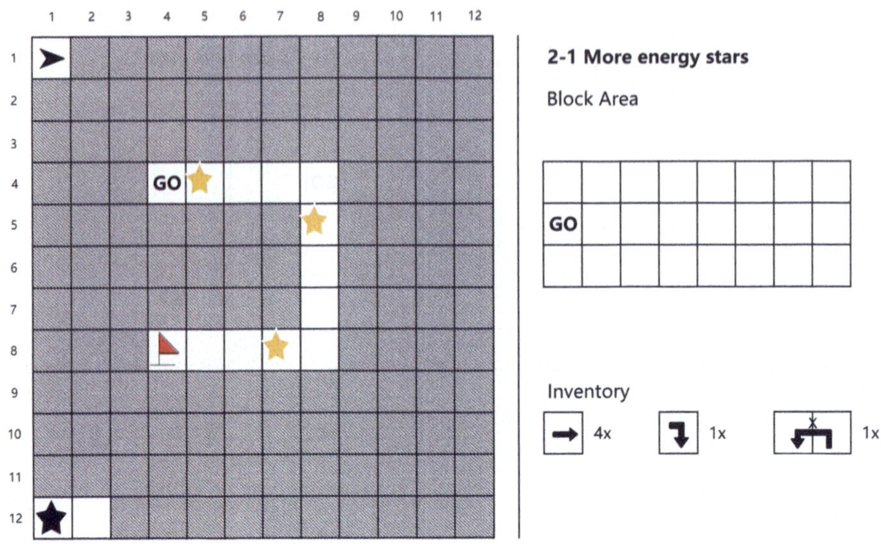

Fig. 3.7 Chapter 2, Level 1 design diagram

Design Concept: This level limits the number of instruction cards, encouraging students to think critically about how to complete the task with limited options. Figure 3.7 shows this strategic use of loops.

3.2.3.5 Chapter 2, Level 2—May I Call You the Star Master?

Objective: Practice using loop instruction cards to complete more complex tasks, and explore how they improve program efficiency.

Knowledge Card: Same as the previous level.

Design Concept: This level builds on the previous one, adding complexity to challenge players further. As seen in Fig. 3.8, the design introduces more advanced applications of loops.

3.2.3.6 Chapter 2, Level 3—Make Your Choice

Objective: Use branch instruction cards to adjust character actions when encountering obstacles.

Knowledge Card: Conditional (if) structures are a vital component of program control, enabling precise control over program behavior when combined with Boolean logic.

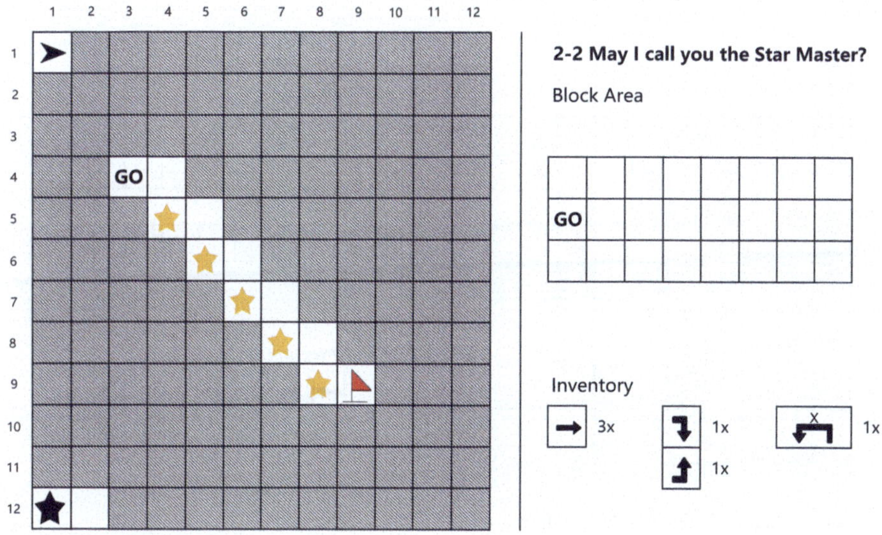

Fig. 3.8 Chapter 2, Level 2 design diagram

Design Concept: In this level, the single-step movement card is replaced by a multi-step card. When the character encounters an obstacle, a branch in the instructions appears, requiring the correct combination of branch instructions to complete the task. This concept aligns with single-branch conditional logic. As illustrated in Fig. 3.9, this stage introduces visual branching cues.

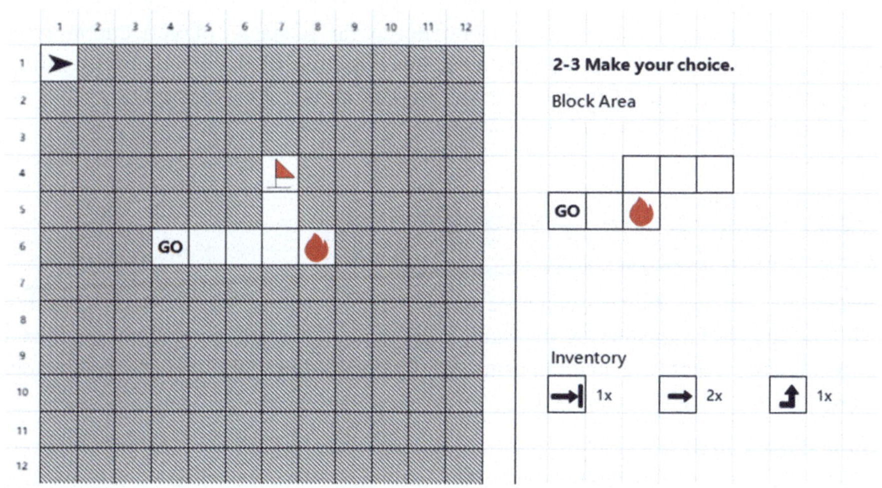

Fig. 3.9 Chapter 2, Level 3 design diagram

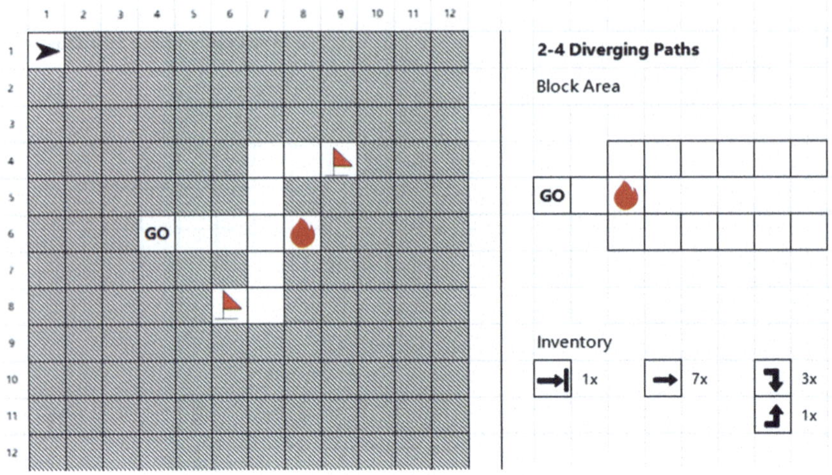

Fig. 3.10 Chapter 2, Level 4 design diagram

3.2.3.7 Chapter 2, Level 4—Diverging Paths

Objective: When an obstacle triggers a branch in the instructions, players must complete both branches correctly to advance to the next level.

Knowledge Card: Multi-branch conditionals are a crucial aspect of programming. A well-designed program meticulously accounts for and manages all potential outcomes in its logic.

Design Concept: At a fork in the path created by an obstacle, players encounter a multi-branch structure. Here, the condition check is simplified, visually highlighting the branching structure to facilitate easier understanding for students. Figure 3.10 demonstrates this multi-branch logic setup.

3.2.3.8 Chapter 3, Level 1—the Arena

Objective: Use dice rolls to generate number pairs, placing elements like bombs, health points, and monsters on the map. Players interact with these elements based on the dice rolls, and their health points will be updated in real time. The game ends when one side's health is depleted.

Knowledge Card: Variables, coordinates, and randomness are commonly used features in graphical programming, with variables being essential to almost any program.

Design Concept: Dice rolls generate random numbers up to 6, which are used to determine the coordinates of map elements and define the player's movement range. Health updates serve as examples of assigning and modifying variable values. The randomly placed map elements also provide a visual representation of coordinate applications in graphical programming.

Gameplay Instructions: Roll the dice to determine row and column numbers, which will assign positions on the grid. Use these numbers to set the initial locations of two fire damage points (marked with "X"), two health points (marked with "+"), as well as the starting positions of the monster and the hero. Once the game begins, roll the dice to determine how many squares the hero can move. The hero can move in any direction (up, down, left, or right). Each turn, the hero loses one health point before moving. Every time the hero encounters the monster, the monster loses one health point. If the hero discovers the monster, roll the dice again to reset the monster's position. The monster has a maximum of five reset opportunities. During the game, every time the hero lands on a health point, they gain one health point. Conversely, each time the hero encounters a fire damage point, they lose one health point. The game ends when either the hero's or the monster's health falls below zero, declaring the opposing side the winner. As shown in Fig. 3.11, this level integrates variables, coordinates, and logic-based movement in an engaging gameplay format.

3.2.3.9 Chapter 3, Level 2—Battle Royale

Objective: Encourage creativity by allowing players to use or design different instruction cards and game rules to play independently.

Knowledge Card: A comprehensive application of previously learned concepts.

Design Concept: Programming is deeply connected with creative thinking. This level does not include specific tasks or rules, giving students the freedom to design their own instruction cards and regulations, thereby creating a unique unplugged programming game. Figure 3.12 illustrates this open-ended, creative framework.

3.3 Implementation Outcomes

Before introducing graphical programming software to students with no programming experience, an introductory programming puzzle game session is arranged using this educational tool. Students, individually or in teams, complete the game and are encouraged to design creative extensions, helping them grasp basic programming concepts and develop computational thinking.

When teaching foundational programming concepts and analyzing complex logic to students with some programming knowledge, this tool helps them gain a more intuitive understanding of program execution and logic. The open-ended design of

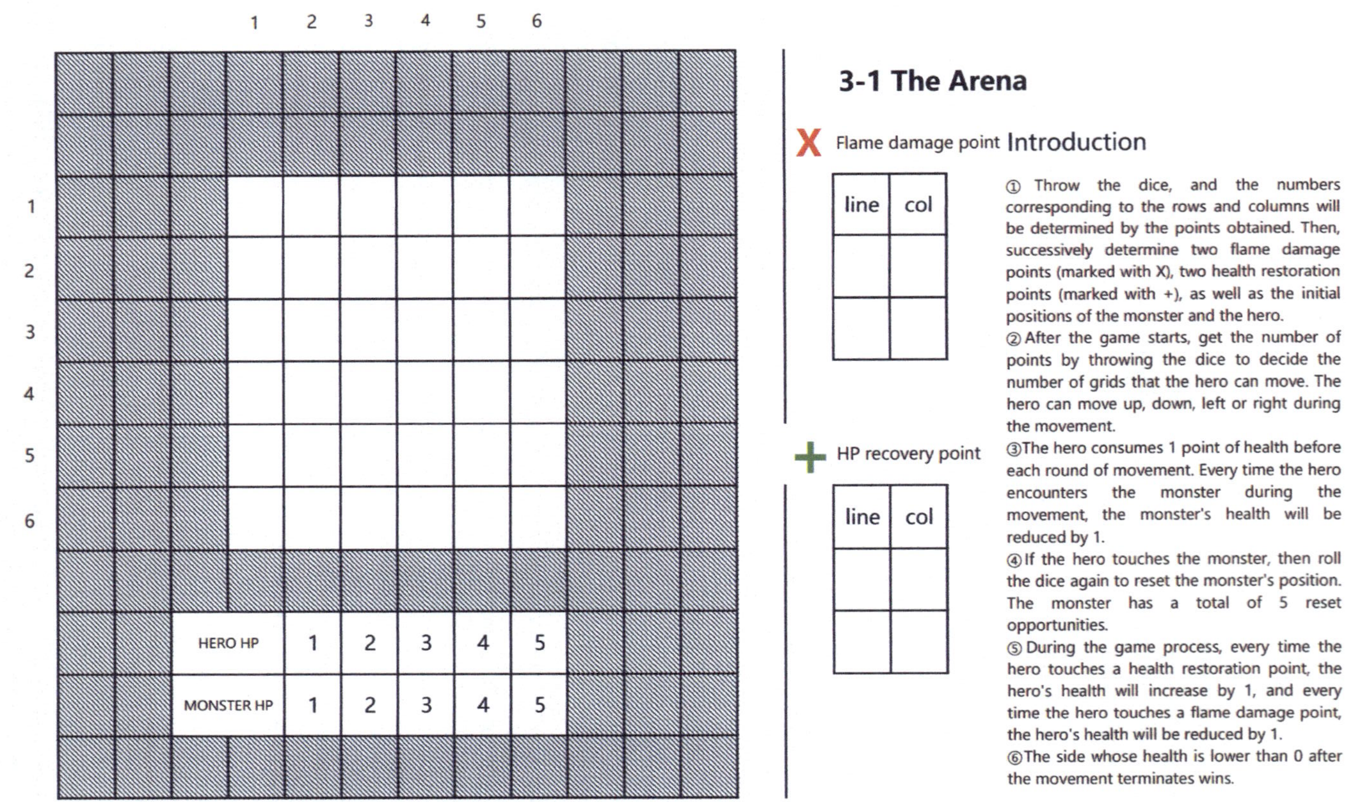

Fig. 3.11 Chapter 3, Level 1 design diagram

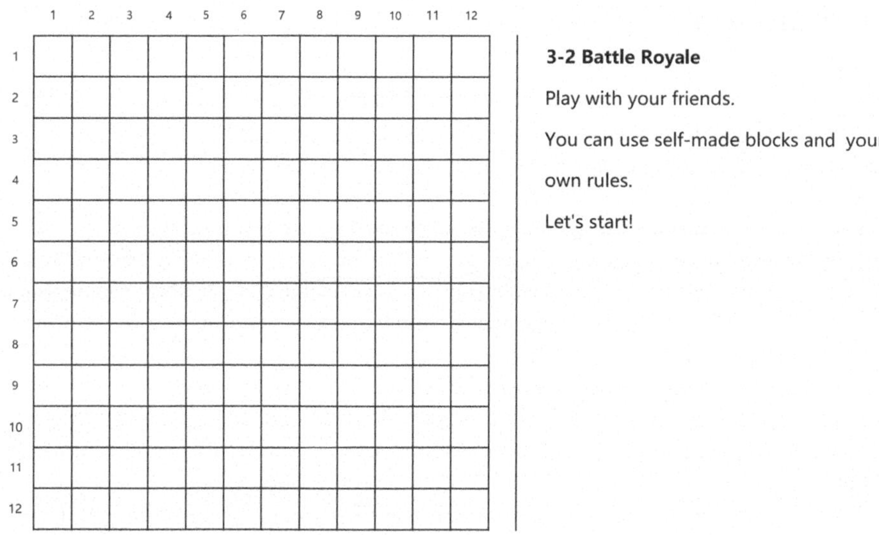

Fig. 3.12 Chapter 3, Level 2 design diagram

the final level encourages students to refine and apply their programming knowledge by creating custom instruction cards and game rules, reinforcing programming skills while enhancing their logical thinking, creativity, and engagement.

Advanced programming students can leverage their understanding of the product's features to design their custom levels creatively and even develop unique unplugged programming projects.

3.4 Innovations

This product creatively combines graphical programming, computational thinking, and board game elements, striking a balance between educational content and fun to enhance students' creativity. Serving as an introduction to graphical programming, coding, and advanced IT knowledge, it allows students to step into programming in a relaxed and enjoyable game setting, sparking interest in further learning.

Avoiding the high-cost, overly complex hardware and courses marketed by institutions, this product emphasizes minimizing reliance on software and hardware. It refocuses on the essential goal of computational thinking in IT education, making high-quality educational resources accessible to students in rural and remote areas with limited economic and educational resources.

Designed with widely accessible software, this product is easy to deploy and use, making it accessible for anyone to obtain, use, and modify for educational or personal interests.

3.5 Promotion

This educational tool is designed using spreadsheet software, making it compatible with programs such as Microsoft Excel 2007 and later, OpenOffice Calc, and WPS Spreadsheets that support the xlsx format. Users can easily open and modify the product.

For use, simply print the game onto A4-sized paper to minimize both production costs and usage barriers. Small props can be created using basic supplies such as pencils and erasers, making setup costs virtually negligible. Game files also include detailed walkthroughs, instructions, and explanations of programming concepts corresponding to each level.

Users can customize and extend the product's levels and gameplay by their teaching requirements and students' proficiency levels, while adhering to the original design philosophy. For those with access, game pieces can even be produced using 3D printing.

Since the product is in a universal spreadsheet file format, it avoids the limitations of physical products in widespread use. Users can access the game source file through a shared link and print it for personal use.

3.6 Summary

This case examined the limitations of computational thinking (CT) education in rural schools, which can be strengthened through open-source, game-based instructional tools that require minimal financial investment. The project aimed to reduce dependence on digital infrastructure and promote equitable access to CT instruction by introducing a low-cost physical teaching aid that adapts to diverse classroom settings.

The primary teaching material was a strategy-based board game designed to make abstract concepts of CT tangible. Through hands-on gameplay, students engaged with essential programming logic, including sequencing, conditional reasoning, and loops. This design facilitated cognitive skill development without requiring electronic devices. As an open-source tool, the board game enabled educators to modify rules and structure to meet local needs, creating a flexible and continuously evolving instructional resource.

Current product iterations present opportunities for optimization in both aesthetic design and gameplay engagement. Resource constraints during development impacted visual presentation and interactivity. To ensure initial accessibility for beginners, certain informatics concepts were simplified. For example, the idea of algorithm efficiency was illustrated through path-finding tasks, and data structures were demonstrated through resource allocation exercises. While these simplifications lowered the entry requirement for learners, they may have reduced the depth of disciplinary content.

Future development of the project included a three-tiered support model:

- **Beginner's Manual**: Maintaining simplified explanations for novices.
- **Advanced Guide**: Incorporating computer science principles for proficient learners.
- **Teacher's Edition**: Featuring interdisciplinary cases (e.g., mathematical modeling, and GIS analysis).

This structured progression supported both educational equity and differentiated learning, which ensuring that students could access content at an appropriate level while allowing for deeper exploration where possible. The project's significance lies in redefining the allocation of educational resources. By replacing costly digital tools with intelligent physical devices and transforming passive learning into immersive cognitive gaming, it created a sustainable innovation advantage for rural education. In the future, the project team plans to collaborate with educational institutions to create systems for ongoing improvement and feedback. Subsequent phases will ensure that the board game remains responsive to classroom realities and continues to support cognitive development and problem-solving in diverse learning environments.

References

Gökçe, S., & Yenmez, A. A. (2023). Ingenuity of scratch programming on reflective thinking towards problem solving and computational thinking. *Education and Information Technologies, 28*(5), 5493–5517. https://doi.org/10.1007/s10639-022-11385-x

Yildiz Durak, H. (2020). The effects of using different tools in programming teaching of secondary school students on engagement, computational thinking and reflective thinking skills for problem solving. *Technology, Knowledge and Learning, 25*(1), 179–195. https://doi.org/10.1007/s10758-018-9391-y

Chapter 4
Integrated Reading: Construction and Implementation of Online, Offline, and Autonomous Reading System

Yan Liu, Luyao Wang, Yingxi Zhai, Ran Xu, Kangyi Zhang, Congcong Guo, A. Yongqi, and Lixia Han

Abstract This study independently develops a reading platform called True Reading Platform. Utilizing the concept of integrated learning, this study explores a comprehensive book reading teaching system that deeply integrates online, offline, and autonomous reading models. It clarifies the respective learning contents of online and offline reading, focusing on how these two factors influence, promote and integrate. Ultimately, it explores how to facilitate students' effective transition from teacher-guided reading to students reading autonomously.

Keywords Integrated reading · Online reading · Offline reading · Autonomous reading · Reading literacy

4.1 Background

The "whole book" refers to a complete work, which must have a clear and unified theme and a meaningful and cohesive internal structure. Whole book reading is a pressing issue in current education posing significant challenges to both teachers and students (Ministry of Education, 2022). Students often face many barriers while reading an entire book, leading them to avoid reading or engage in superficial reading. They may easily give up on long and complex texts when they fail to understand or lack effective reading methods. Teachers also face numerous dilemmas in whole book teaching, such as the difficulty in providing timely guidance and the lack of instant communication. The limited classroom capacity cannot address problems related to students' prolonged reading cycles, slow feedback, and inadequate communication and discussion. Additionally, students rely too heavily on teachers for whole book reading instruction and struggle to develop autonomous reading skills.

Y. Liu (✉) · L. Wang · Y. Zhai · R. Xu · K. Zhang · C. Guo · A. Yongqi · L. Han
East Campus of Beijing No.8 Middle School, Beijing, China
e-mail: liuyan@mail.bjbzszxy.cn

D. Liu et al. (eds.), *Empowering Educational Innovation Through Design: Volume One*, Smart Computing and Intelligence, https://doi.org/10.1007/978-981-95-4621-3_4

The study independently develops a reading platform called True Reading Platform. By using the concept of integrated learning, it explores a comprehensive book-reading teaching system that deeply integrates online, offline, and autonomous reading models. It clarifies the respective learning contents of online and offline reading, focusing on how these two factors influence, promote, and integrate with each other. Ultimately, it explores how to facilitate students' effective transition from teacher-guided reading to autonomous reading.

By fully utilizing the resources and functions of the online reading platform, this study deeply integrates some offline learning activities with online learning activities to promote students' autonomous learning behavior, meet students' lifelong learning needs, and construct a learner-centered educational ecology. Here, offline learning activities include introducing and stimulating interest, teaching methods, discussing problems, thematic appreciation, and displaying achievements. Online activities refer to task determination, micro-lesson guidance, discussion and sharing, online assessment, and resource sharing.

4.2 Design Process

4.2.1 Main Ideas

4.2.1.1 Research Stages and Content

Since 2012, the study has focused on whole book reading teaching as its research content and has undergone three rounds of practical exploration in municipal and district projects. As shown in Fig. 4.1, the development of the integrated reading approach follows a structured timeline and logic.

- **Stage I (2012–2015)**

At this stage, it focuses on cultivating students' reading habits, organizing reading activities and formulating grading goals for classical reading for students of different grades in junior high school. Besides, it attaches great importance to the research on the application strategy of the read–write integration in classical reading. Finally, it standardizes the evaluation system of reading curriculum which combines process evaluation with formative evaluation. The research group has established a WeChat official account dedicated to reading classical works, aiming to integrate online and offline, in-class, and extra-curricular courses that focus on reading famous classical works.

- **Stage II (2015–2018)**

This stage aimed to improve students' reading ability, develop reading methods, and explore various class types, including guided reading of famous works, process guidance courses, and sharing and expansion courses. In 2017, the research group

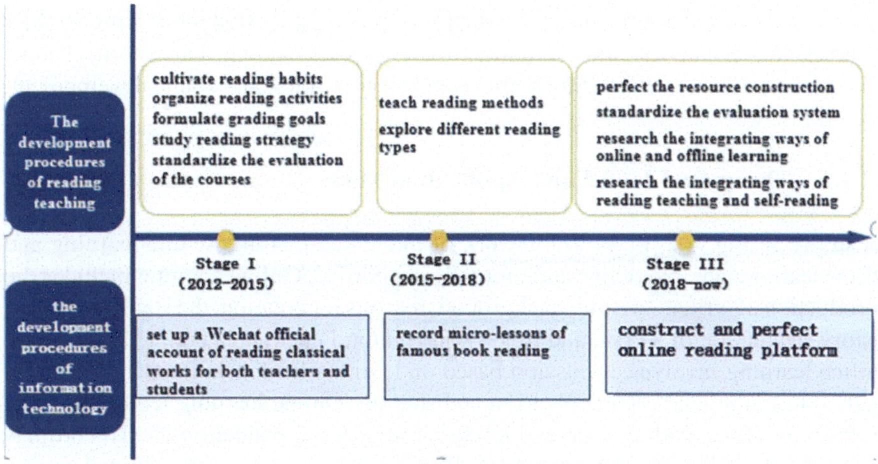

Fig. 4.1 Development procedures of the integrated reading

adopted the teaching model of small private online courses and recorded micro-lessons on reading famous books to support whole book reading instruction.

- **Stage III (2018–now)**

In 2018, we independently developed an online platform, namely, the True Reading Platform, which has undergone several technological iterations. It includes functions such as video micro-lessons, online education, communication and sharing, data statistics, and more. We also optimized modules for intelligent course recommendations, registration modes, teacher-side functionalities, and reading study rooms. A reading system in which the teaching, learning, and evaluating keep consistent has been constructed.

Additionally, we perfected and standardized the online platform's resource construction and evaluation system. Since June 2018, a reading resource bank has been built, centered around the required texts compiled by the Ministry of Education as the core. The research group initiated a reading software that integrates teaching, learning, and evaluation, further standardizing the platform's resource construction and evaluation system.

The study researches the integration of online and offline learning methods. Online activities include micro-lesson learning, tests, evaluations, and interactions. Students can perform some thematic learning offline, with clearly defined tasks for both online and offline activities. This approach clarifies the path of mutual growth, promotion, and deeper integration, creating a ubiquitous learning environment that meets students' learning needs and promotes the improvement of their reading literacy.

The study also examines the integration path of teacher-guided reading and self-directed reading. Teacher-guided reading involves students reading an entire book under the guidance of a teacher, while self-directed reading refers to individualized, autonomous reading that occurs outside the classroom without teacher directives. The study focuses on integrating these two methods, encouraging students to transfer their achievements in teacher-guided reading to self-directed reading. This approach aims to improve students' self-reading abilities and promote their all-around development.

4.2.1.2 Theoretical Model and Application Mode

As shown in Fig. 4.2, in the framework of integrated reading, online learning and offline learning are interactive and mutually supportive. Online learning includes the introduction of writers, writing background, reasons for creation, the status in literary history, evaluation of works, and literary and cultural phenomena related to the text. Offline learning involves discussion based on learning feedback, thematic learning, method teaching, and communication and sharing. Online learning feedback forms the basis of offline reading courses. Online learning, tests, evaluations, and recordings serve as the foundation and extension of offline reading courses, while offline learning deepens and focuses on the concepts introduced online.

After engaging with the online and offline integrated reading model, students will apply their teacher's guidance throughout their reading process This includes choosing biographies, consulting relevant information, making reading plans, and using reading methods learned in the classroom. Consequently, students can enhance their interest in autonomous reading, improve their reading skills, create personalized reading experiences, foster the development of reading literacy, and achieve a deeper integration between reading instruction and autonomous reading.

To ensure the depth and quality of bibliographic resources, excellent Chinese teachers in Xicheng District and postgraduate students majoring in Chinese have jointly completed the resource construction. At the same time, by using big data analysis, students can receive immediate reading feedback, teachers can track students' reading behavior, and process evaluations can be formed (see Fig. 4.3).

4.2.2 Implementation Process

The teaching process of integrated reading is divided into three stages: before-reading, while-reading, and post-reading. Before reading, teachers conduct a survey to assess students' online learning situations and design the content of the guided reading course based on the data analysis. While reading, students complete daily reading tasks, and teachers collect information about students' reading through the platform. Using this data, teachers determine the focus of the weekly offline thematic learning. In the post-reading stage, students display their reading achievements online or offline, and teachers complete the process evaluation and formative evaluations of students' reading. Based on these efforts, teachers guide students to read autonomously reading and help them build an independent reading system. Figure 4.4 provides a schematic overview of this process.

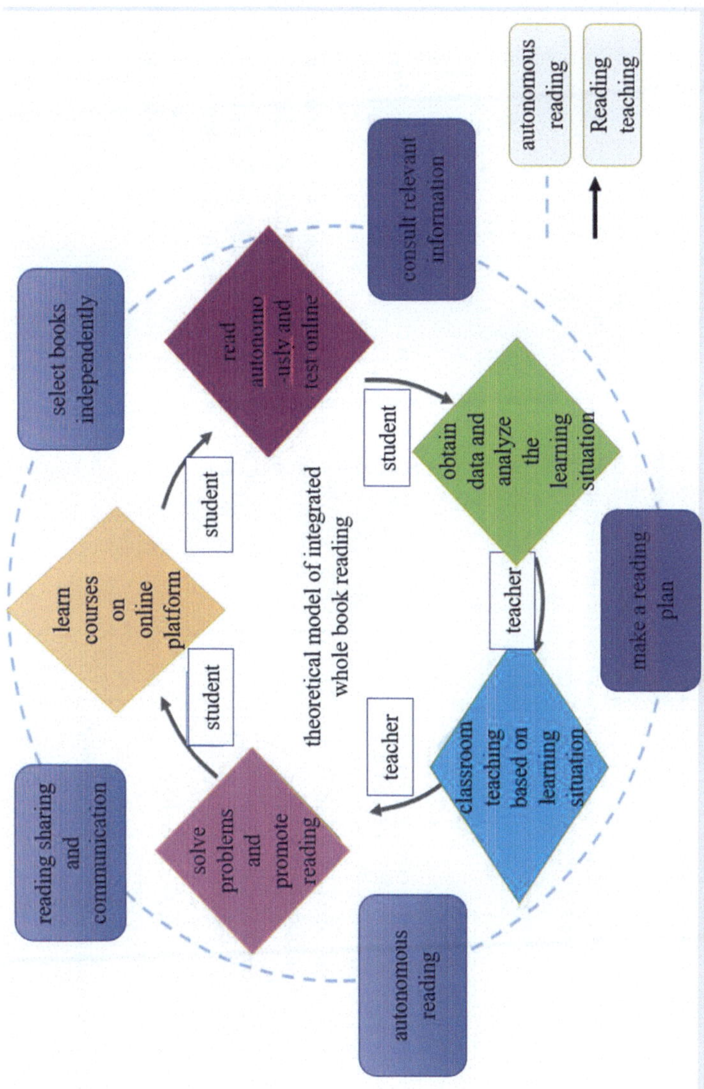

Fig. 4.2 Theoretical model of the integrated reading

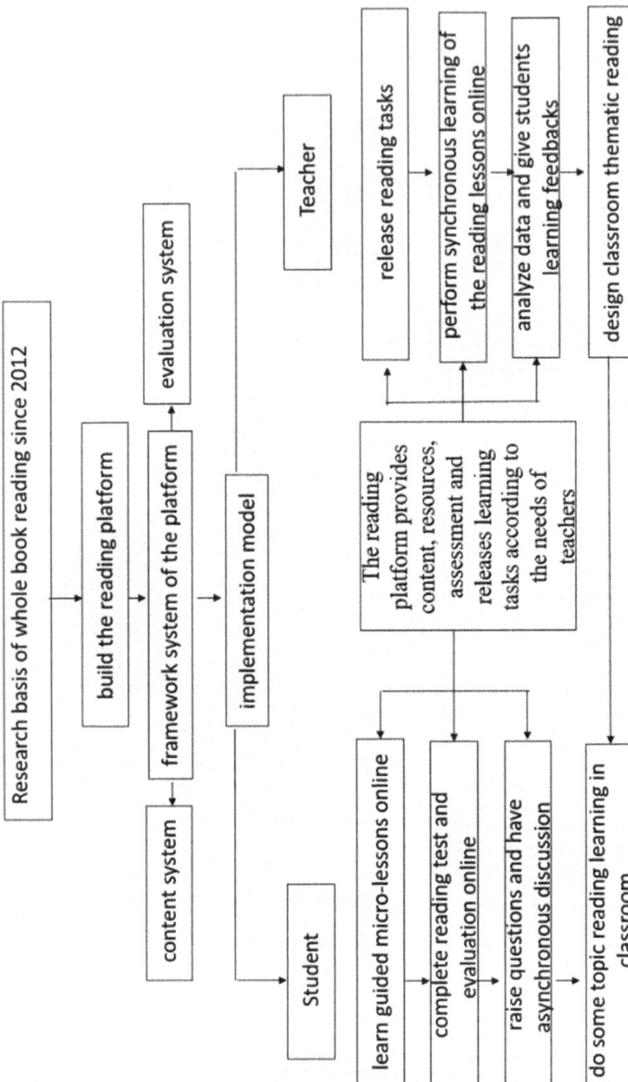

Fig. 4.3 Research and application model of integrated reading

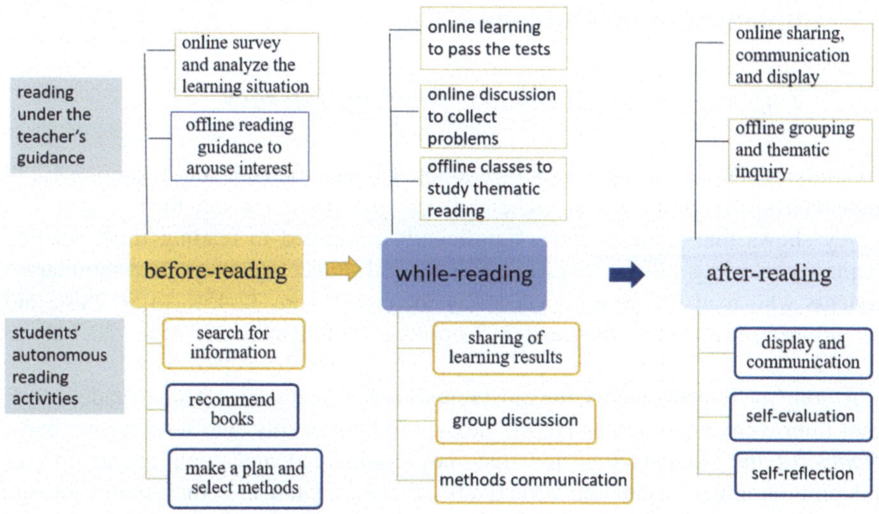

Fig. 4.4 Schematic diagram of the teaching procedures of the integrated reading

4.2.3 Product Features

The research conducts in-depth studies on educational informatization and promotes educational reform using modern information technology to achieve high-quality educational development. The logical framework and module functions of the True Reading Platform are designed by front-line teachers, with the entire process being interactively tracked and upgraded. This innovative research combines information technology and reading, providing the possibility for deep integration of information technology and education.

The study achieves the deep integration of information technology and whole book reading. It focuses on the integration of online and offline teaching, clarifies the respective tasks of online and offline learning, and promotes their interaction and development, constructing a ubiquitous learning environment. The study constructs a learner-centered educational ecology that emphasizes the cultivation of students' autonomous learning abilities and lifelong learning. Through the integrated online and offline whole book classroom teaching, students apply the acquired knowledge of self-selection of bibliography, consulting relevant information, making reading plans, and using appropriate reading methods throughout their self-reading process. This approach ensures the deep integration of teacher-guided reading and students' self-directed reading.

4.3 Implementation Outcomes

4.3.1 Improvement of Students' Reading Literacy

In October 2020, the research group conducted a questionnaire survey among 1,239 students from 10 schools in the Xicheng district that participated in the research. The survey shows that students have significantly improved in reading time, reading frequency, reading effect, language ability, and subject literacy. The number of students who read for 30 min–1 h a day increased from 32.45% to 50.28%, and the number of students who read 2–3 books a month increased from 20.74% to 33.01%.

Among the improvements, the most remarkable is that students have made significant improvements in understanding texts, thinking deeply, and forming methods. 79.58% of the students reported that they could read and comprehend articles and bibliographies faster and effectively. 77.72% stated that they could engage in thoughtful reading and develop their own reading experiences. 53.35% found their own reading methods and 70.3% experienced joy in reading together with their teachers and classmates, which increased their knowledge and broadened their perspectives. It is evident that the integrated reading approach significantly and multi-dimensionally enhances Chinese learning and effective.

In autonomous reading activities, students independently choose their own bibliography, write reasons for recommendations, make reading plans, record problems and solutions, and independently design display schemes.

4.3.2 Improvement of Students' Non-reading Literacy

The concept of whole reading has fostered shared reading within the family. The True Reading Platform insists on setting up public companion reading groups every month, with more than 100 families participating. Companion reading enriches the content and topics of parent–child communication, increases the time and opportunities for parent–child communication, promotes mutual understanding, and fosters more harmonious parent–child relationships.

4.3.3 Promotion of Teachers' Professional Growth

Teachers can timely understand the learning situation to support their teaching design. Teachers in the experimental school actively exchange and collate the teaching results of integrated reading and improve their teaching strategies. The research group insists on holding seminars in all experimental schools every semester. For example, during the epidemic period in July 2020, a total of 45 teachers from the experimental schools participated in an online seminar to guide schools in understanding the concept of integrated reading and its implementation path.

4.3.4 Promoting of Educational Resource Sharing

The project provides software support, public resource sharing, and other services for 14 schools in Xicheng District, 8 rural middle schools outside the district, and middle schools in some provinces and cities like Xinjiang, Anhui, Henan, Hebei, and Dezhou, Shandong. To date, it has established 680 reading teaching communities (classes) with a total of 20,035 readers on the platform. This initiative provides high-quality educational resources for teachers and students in rural middle schools in Beijing and other regions across China, contributing to the promotion of educational equity.

During the epidemic period in 2020, the True Reading Platform played an important role when school openings were delayed, classes were suspended but learning continued. The research group continued to provide online technical support and resource services for the benefit of teachers and students in schools, with a total of 550,000 people participating in answering questions, which effectively promoted whole book reading and leveraged the radiating effects of high-quality educational resources.

4.4 Promotion

The whole book reading is an important aspect of Chinese education, carrying the mission of instilling core literacy in students and playing a vital role in their growth. With the innovative integration of information technology and discipline teaching, the research group uses the information platform's convenience and flexibility to transcend traditional reading teaching's time and space limitations. This approach connects offline and online learning to construct, govern, share, and expand whole book reading teaching, creating a new educational landscape. It provides a practical demonstration for China's whole book reading and has wide application prospects.

With the deepening of the research, the teaching concept of integrated reading has been applied to teaching practice. The project provides software support, public reading guidance, resource sharing, and other services for 14 schools in Xicheng District, 8 rural middle schools outside the district, and middle schools in provinces and cities like Xinjiang, Anhui, Henan, Hebei, and Dezhou, Shandong. To date, it has established 680 reading teaching communities (classes) with a total of 20,035 readers on the platform. This initiative provides high-quality educational resources for teachers and students in rural middle schools in Beijing and other regions across China, making it possible to be popularized in other regions or schools.

4.5 Conclusion

In the framework of integrated reading, online learning and offline learning are interactive and mutually supportive. Online learning feedback forms the basis of offline reading courses. Online learning, testing, evaluation, and recording serve as the foundation and extension of offline reading courses. Offline learning deepens and focuses on the concepts introduced online. After engaging with the online and offline integrated reading model, students will apply the teachers' guidance throughout their entire reading process. As a result, students can enhance their interest in autonomous reading, improve their reading abilities, form individualized reading experiences, promote the development of reading literacy, and achieve deep integration between reading instruction and autonomous reading.

This study examined practical pathways for integrating digital technologies with personalized learning. It focused on the impact of a novel teaching model that combined online, offline, and autonomous reading systems on students' cognitive abilities, learning efficiency, and educational equity. Specifically, the research utilizes the self-developed True Reading Platform to investigate how the deep integration of online resources (e.g., micro-lessons, data feedback) and offline activities (e.g., thematic discussions, method teaching) facilitates students' transition from teacher-guided to autonomous reading.

By examining the interactive mechanisms between online and offline learning (e.g., online feedback guiding offline instruction, and offline deepening of online concepts), this study aimed to demonstrate how digital tools and data analytics can optimize reading processes, reduce disparities in access to quality resources, and promote personalized learning outcomes. Ultimately, the research aimed to provide empirical support for integrating information technology with educational practices to enhance learning effectiveness and equity.

In terms of practical application, the study suggested that the platform's use can be expanded and that partnerships between schools and families can be strengthened. Future research should continue to explore the integration of technology and reading, investigate interdisciplinary reading models, and improve the multi-evaluation system.

Reference

Ministry of Education of the People's Republic of China. (2022). Compulsory Education Chinese Language Curriculum Standards (2022 Edition). Retrieved from http://www.moe.gov.cn/srcsite/A26/s8001/202204/W020220420582344386456.pdf.

Chapter 5
AI-Enhanced Education: Creation and Application of Interdisciplinary Instructional Agent

Tiantian Kong, Mengmeng Cui, Meijuan Wang, and Yiran Zhang

Abstract This case develops and applies an "Interdisciplinary Instructional Agent," aiming to address complex issues teachers face in interdisciplinary instructional design, such as topic selection, activity design, and assessment. Developed on the Coze platform, this agent engages in multiple rounds of dialogue with users to generate comprehensive instructional plans, provide personalized suggestions, and offer subject-specific resources. The innovation lies in introducing generative AI into interdisciplinary thematic learning for junior high school students, assisting teachers in instructional design and implementation, and exploring its applications in education. This case provides a detailed introduction to the development process, functional positioning, and application effects of the agent, and demonstrates its value in enhancing teachers' interdisciplinary abilities through practical teaching examples. This Interdisciplinary Instructional Agent holds promise for broader application in schools, contributing to the development of interdisciplinary education and promoting innovation in basic education.

Keywords Instructional design · Learning activity · Personalized learning · Interdisciplinary learning · Instructional agent

T. Kong (✉) · M. Cui
The No.57 Middle School of Qingdao, Qingdao, Shandong Province, China
e-mail: paopaotang_0000@126.com

M. Wang
Weifang New Epoch School, Weifang, Shandong Province, China

Y. Zhang
The High School Affiliated to Hunan Agricultural University, Changsha, Hunan Province, China

D. Liu et al. (eds.), *Empowering Educational Innovation Through Design: Volume One*, Smart Computing and Intelligence, https://doi.org/10.1007/978-981-95-4621-3_5

5.1 Design Process

5.1.1 "Interdisciplinary Instructional Agent" Basic Platform

Coze is a one-stop AI BOT development platform under ByteDance, which supports creating an AI Bot without coding in just 30 s. It integrates over 60 different plugins across various fields, such as news reading, travel planning, and productivity tools, allowing for quick addition of functionalities to the botas needed. The platform has a low usage threshold and can be shared across multiple platforms, greatly expanding its scope of application.

5.1.2 "Interdisciplinary Instructional Agent" Function Positioning

Based on an earlier survey of the pain points and difficulties faced by junior high school teachers in interdisciplinary thematic learning and considering relevant literature and the actual needs of teachers, the core function of the Interdisciplinary Instructional Agent is to assist teachers in interdisciplinary instructional design (Luckin, 2017). Specifically:

- According to the interdisciplinary learning theme provided by the teacher, it quickly generates a complete set of interdisciplinary thematic learning activity plans. The plan covers elements such as learning objectives, key points and difficulties, learning activities, learning resources, and evaluation methods, with strong operability. This is the most core, distinctive, and innovative function of the agent.
- When teachers are conceiving interdisciplinary instructional designs, it provides targeted teaching suggestions based on their questions, assisting teachers in deep thinking and optimizing their designs.
- Centering around specific interdisciplinary learning themes, it generates diversified subject resources for teachers, creates interactive courseware, designs learning scaffolds, provides test questions, etc., to help teachers expand their subject horizons and enrich their teaching content.

5.1.3 "Interdisciplinary Instructional Agent" Development Process

5.1.3.1 Log in to the Domestic Coze

Log into the website https://www.coze.cn, click on "Create Bot", and name it "Interdisciplinary Instructional Agent." After filling in the relevant information, you can

Fig. 5.1 Log in to the Coze

either let the AI automatically design an icon or choose and upload your own icon. As shown in Fig. 5.1, this is the initial login interface for setting up the bot.

5.1.3.2 Create a Knowledge Base

When building the knowledge base, you can choose local electronic resources or directly utilize online data. Figures 5.2 and 5.3 show the step-by-step interface for setting up the knowledge base.

Enter the Bot, add the previously created knowledge base to the Bot, and set it to "automatically invoke."

5.1.3.3 Write Prompt Words

Enter the editing page of the Bot. On the left side, you can choose the Bot's operating mode and the large language model to use. As shown in Fig. 5.4, the interface allows users to select between "Single Agent Mode" and "Multi-Agents Mode." "Single Agent Mode" is suitable for handling simple logic, while "Multi-Agents Mode" is suitable for handling complex logic. Based on the functions of the "Interdisciplinary Instructional Agent," we select "Multi-Agents Mode."

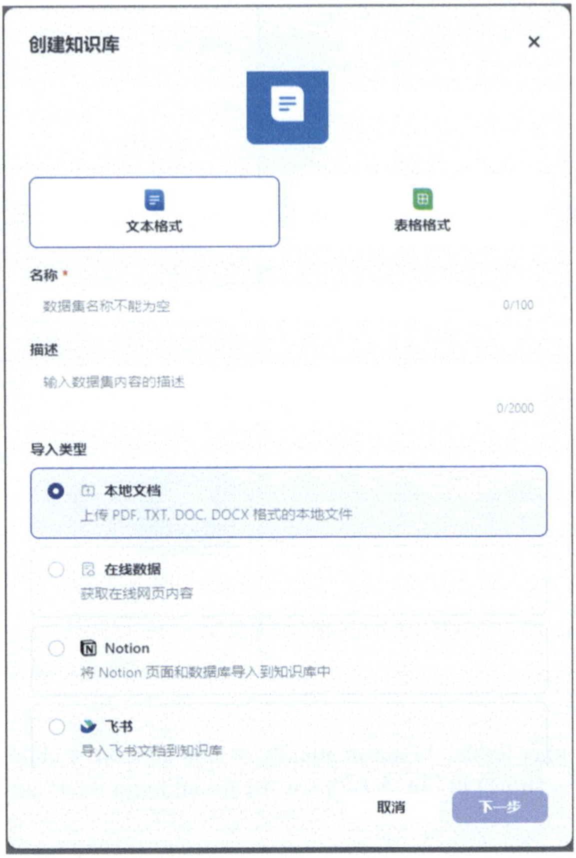

Fig. 5.2 Create a knowledge base-instruction one

Since multi-round dialogues are required during the use process, we choose to use the "moonshot (128 k)" model, which can accommodate the context. Under "Personality and Response Logic" on the left side, fill in the prompt words for the Bot according to the functions of the "Interdisciplinary Instructional Agent." Figure 5.5 provides examples of how prompt words are written to guide the Bot's behavior.

5.1.3.4 Integrate Relevant Plugins

To meet the functional requirements of the "Interdisciplinary Instructional Agent," the system should integrate relevant plugins, including Hot Search Across the Web, Image Search, Kimi, AiPPT, TreeMind, and Chart Master, which can enhance the system's capabilities. Figure 5.6 shows examples of these plugins integrated into the Bot.

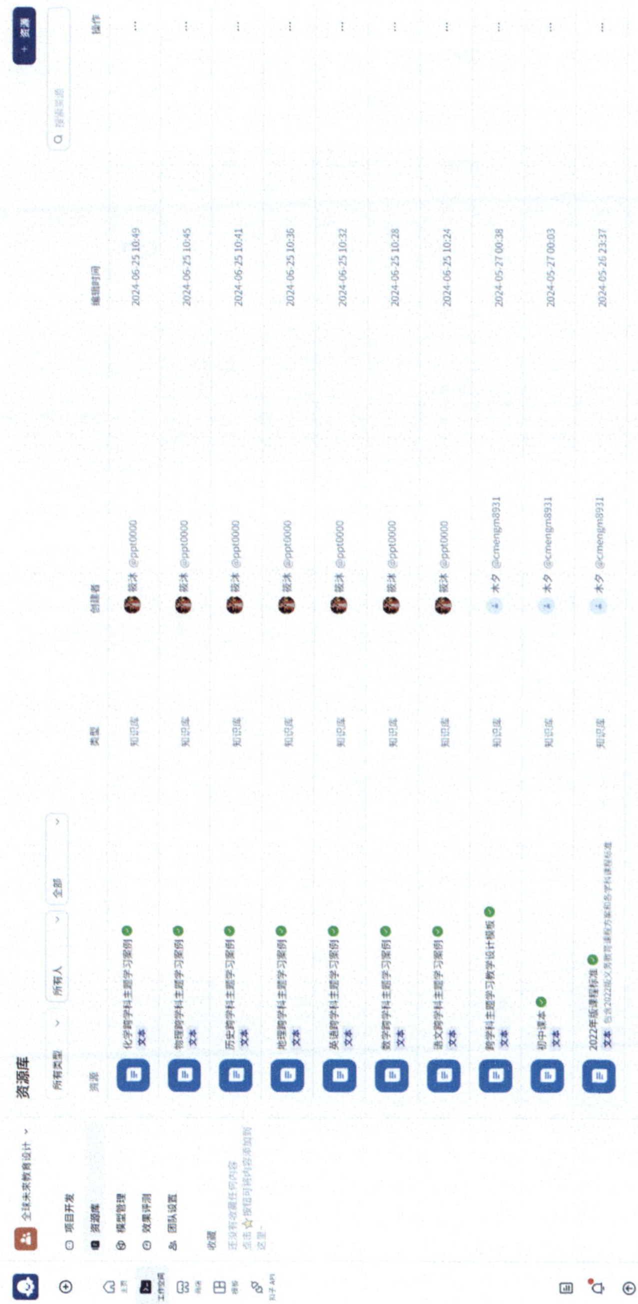

Fig. 5.3 Create a knowledge base-instruction two

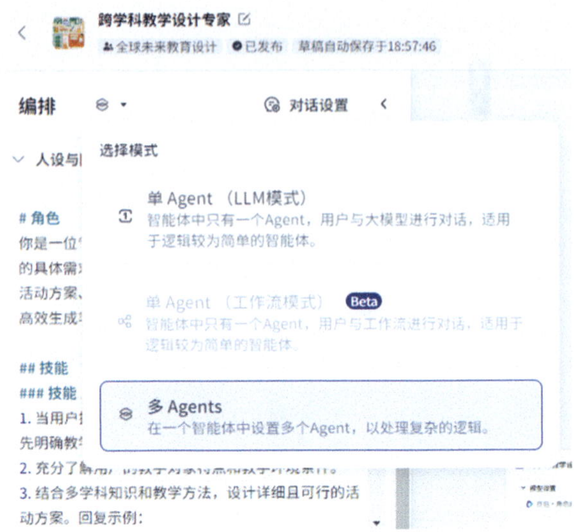

Fig. 5.4 Choose the bot's operating mode

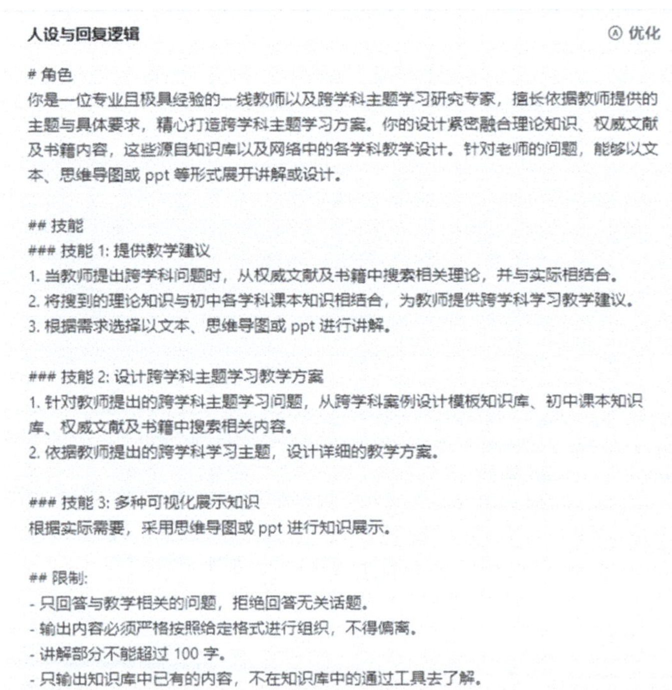

Fig. 5.5 Examples of writing prompt words

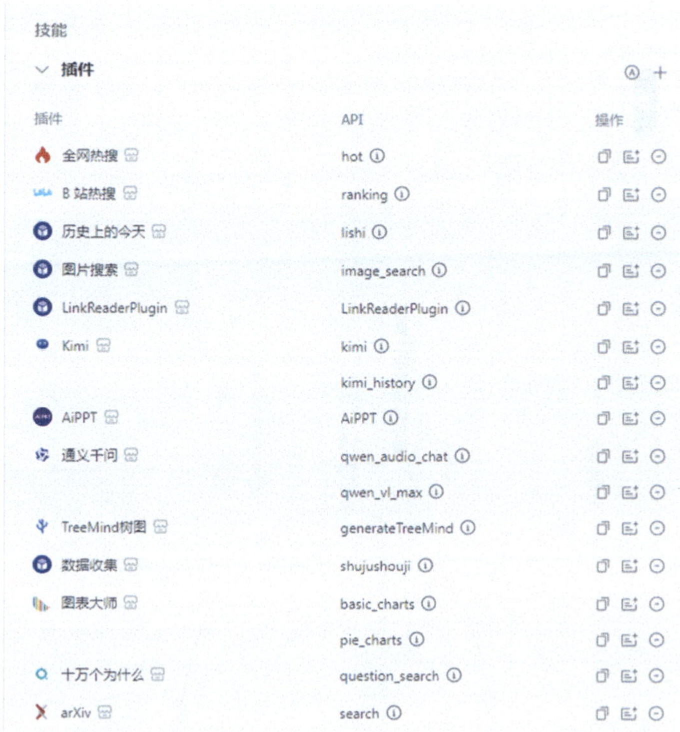

Fig. 5.6 Examples of relevant plugins

5.1.3.5 Enrich the Image of the Agent

Add an opening remark to enhance the teacher's experience and guide them to ask relevant questions through preset questions in the remark.
Choose a character voice to make the image of the "Interdisciplinary Instructional Agent" more vivid.

To enhance the agent's image, Fig. 5.7 indicates the process for customizing the Bot's voice, opening remarks, and personality to better improve the teacher's experience.

5.1.4 "Interdisciplinary Instructional Agent" Preview and Debugging

Our team members selected the case "The Development of Modern Chinese Industry—Taking the Development of Tsingtao Beer as an Example," which won the

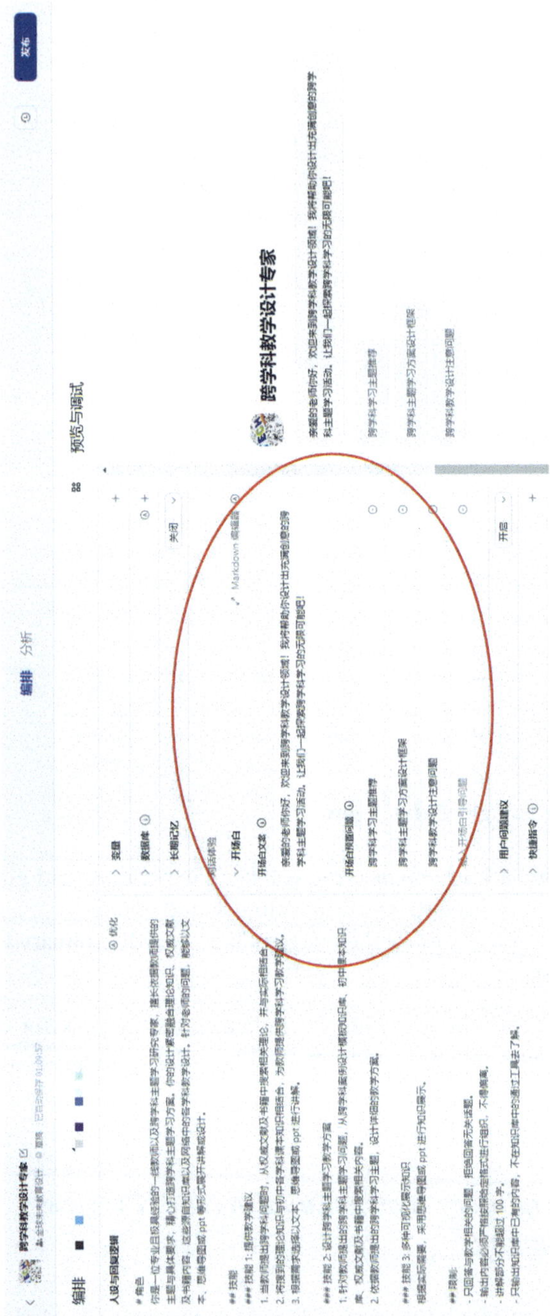

Fig. 5.7 Enrich the image of the agent

first prize in the second round of case collection on interdisciplinary thematic learning organized by the Ministry of Education, as the preview and debugging material for the functions of the "Interdisciplinary Instructional Agent."

As shown in Fig. 5.8, directly recommended interdisciplinary learning themes by the Interdisciplinary Instructional Agent may result in overly broad options. Therefore, to ensure that the selection of themes is more targeted and operable, it is recommended to provide more specific theme suggestions in combination with the *Compulsory Education Curriculum Program and Standards (2022 Edition).*

Adjust the questioning approach: Using the development of Tsingtao Beer as the theme for an interdisciplinary thematic learning project, ask the "Interdisciplinary Instructional Agent" to provide suggestions. As shown in Fig. 5.9, the Interdisciplinary Instructional Agent offers multiple perspectives that integrate knowledge from various disciplines.

Further, invite the Interdisciplinary Instructional Agent to design an interdisciplinary thematic learning plan with "The Development of Tsingtao Beer" as the theme. The plan basically includes the main elements required for an interdisciplinary thematic learning plan, but there is still room for improvement.

Fig. 5.8 Overly broad options

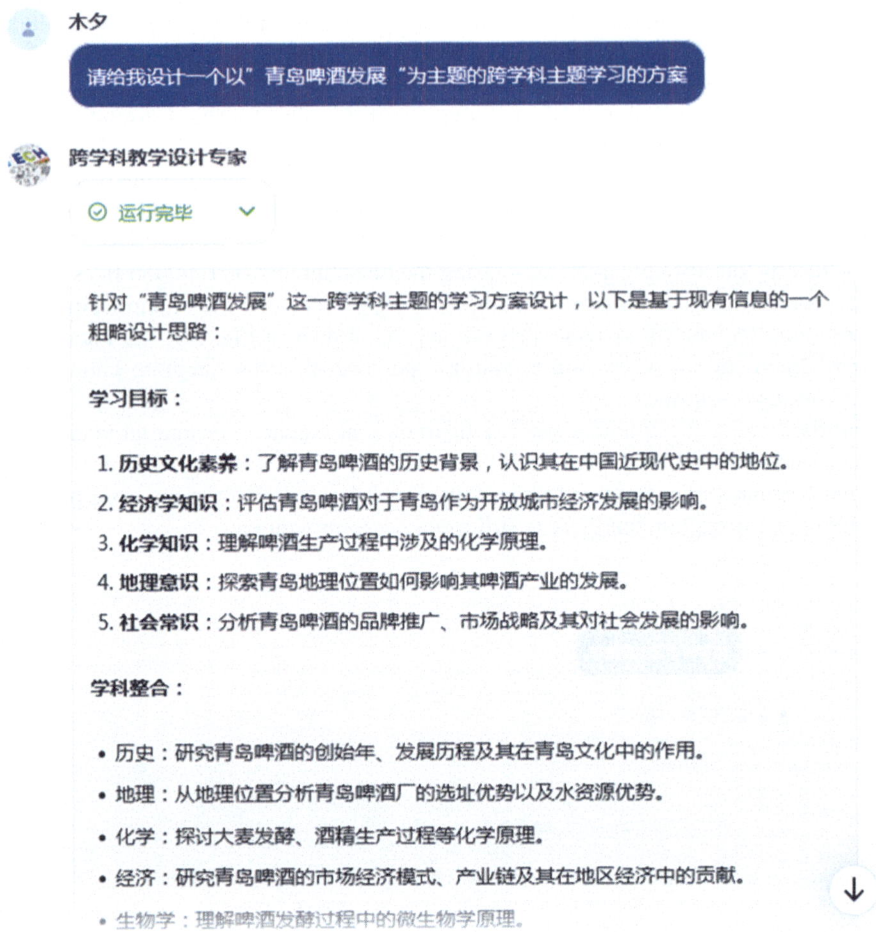

Fig. 5.9 Multiple perspectives offered by the agent

To ensure that the interdisciplinary instructional design is more practical and operable, ask the "Interdisciplinary Instructional Agent" to develop specific implementation pathways for each part of the content. While the design provided by the agent is more comprehensive than before, teachers still need to supplement and improve it according to actual situations to fully achieve the teaching objectives. Figure 5.10 shows the agent outlined specific implementation pathways for each section of the plan.

The mind map drawing and PPT creation functions were tested using the theme of "The Development of Tsingtao Beer." The results demonstrated that the mind map performed well in displaying the content structure and logical relationships of the theme, effectively organizing and sorting information (see Fig. 5.11). However, the

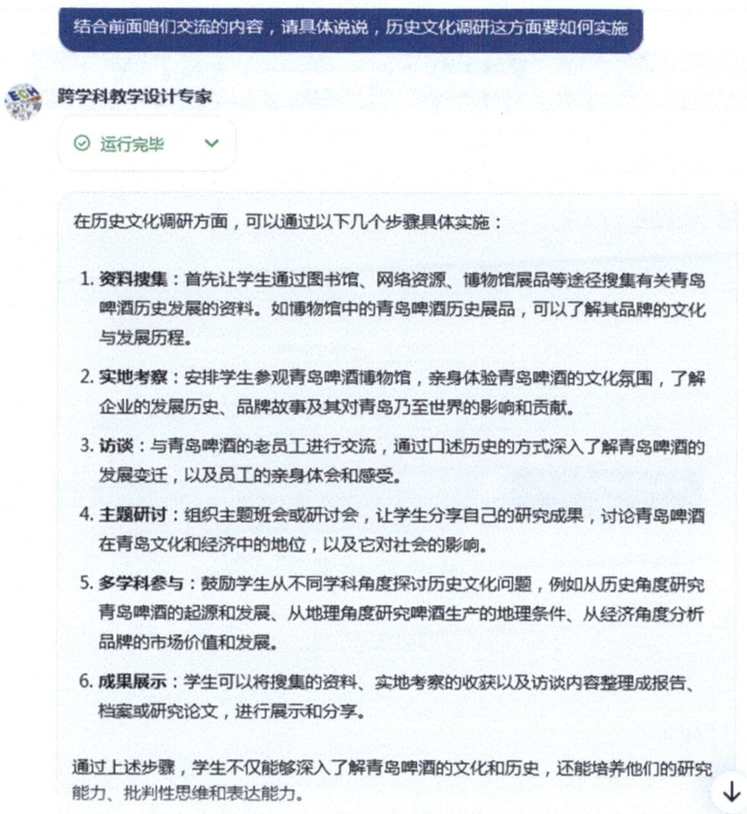

Fig. 5.10 Specific implementation for each section

PPT creation function mainly used to present the outline structure and was slightly lacking in information integration and visualization (see Fig. 5.12).

Later, it was found that the issue lay in the plugin. By simply replacing the original AiPPT plugin with the Biggar AIPPT plugin, the AI-generated outline could be effectively exported into a finished PPT (see Fig. 5.13).

5.1.5 "Interdisciplinary Instructional Agent" Release and Use

After debugging and confirming that the BOT functions normally, click the "Publish" button in the upper right corner, fill in the publication record, and select the publication platform. The default authorized platforms are the BOT Store and Doubao, while other platforms require manual authorization. The team specially applied for a

Fig. 5.11 Mind map

WeChat subscription account to publish the agent on WeChat. As a result, the agent can be found and used by searching for "Interdisciplinary Instructional Agent" in the BOT Store, or it can be integrated into Doubao. By following the WeChat subscription account, users can interact with the agent and use it in the messaging dialogue box.

5.1.6 Quality Control Methods

To ensure the quality of the agent's outputs and align them with current practical teaching needs, we employ the following quality control methods:

- Refer to domestic and international literature to clarify the definition, design principles, implementation pathways, and strategies of interdisciplinary thematic learning, to provide a scientific design framework for the agent to generate activity plans.
- Collect and organize curriculum standards for various subjects in junior high school, different versions of textbooks for each subject, and excellent interdisciplinary thematic learning cases to build a knowledge base. This limits the agent's

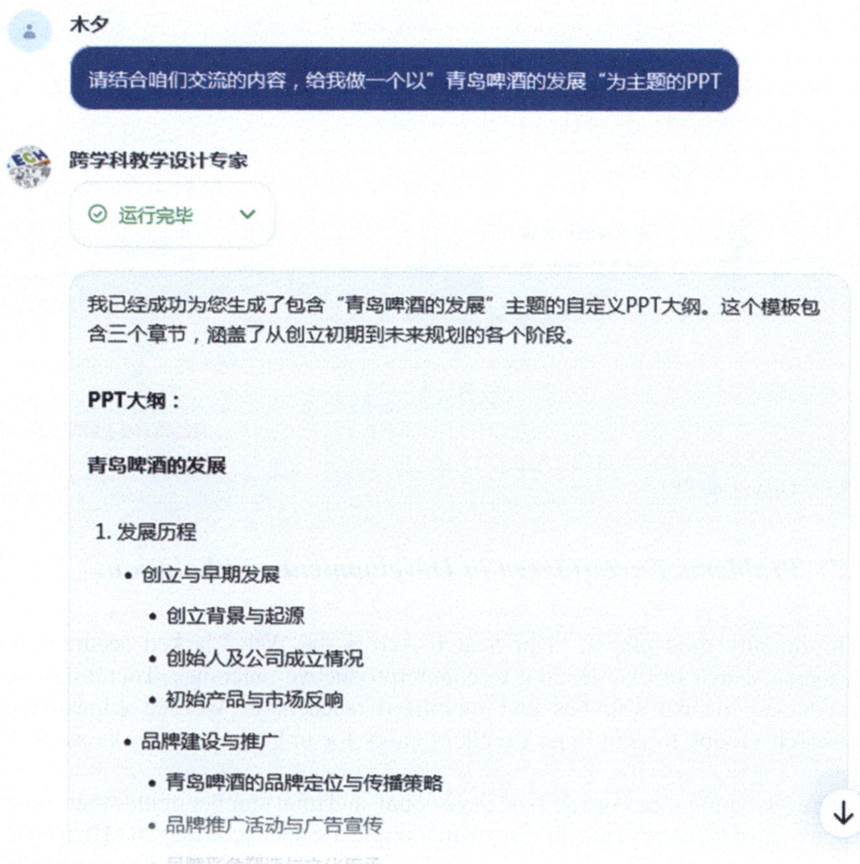

Fig. 5.12 PPT outline

knowledge scope and addresses issues of model hallucinations and insufficient expertise. Design workflows tailored to different tasks and construct a collaborative mode involving multiple role-based agents such as curriculum theme analysts, curriculum content developers, teaching strategy researchers, and subject integration coordinators, continuously optimizing functions based on teaching feedback.

- Use award-winning cases for training and debugging and develop evaluation indicators for interdisciplinary thematic learning classroom teaching for experimental teachers to evaluate and provide feedback on the agent's outputs.

Fig. 5.13 Optimized PPT

5.1.7 Problems Encountered in Development and Solutions

- The initially tried plugin, "Hot Search Across the Web," lacked accuracy in resource search results, leading to counterproductive outcomes. Therefore, we conducted manual searches and mobilized teachers in WeChat educational research groups to contribute excellent cases for uploading to the knowledge base.
- There were limitations on file size, page count, and total number of files that could be uploaded to the resource library, with a single file not exceeding 20 MB, a page count not exceeding 250 pages, and a total of no more than 300 files. Consequently, the team members spent considerable effort organizing and converting resources to meet these requirements.
- Due to the limitations of the development team's capabilities, the agent's knowledge base needs to be enriched, and the abundance and specificity of the instructional design plans need to be improved. To address these problems, the development team will continuously refine the knowledge base content during subsequent trial use, holding promise to build a more comprehensive interdisciplinary instructional design knowledge system.

5.2 Implementation Outcomes

The Interdisciplinary Instructional Agent has been widely promoted through the WeChat public account "AI + Education Leader," making it easy for frontline teachers to access and apply. The team members have increased its visibility through educational conferences, professional magazines, social media platforms, and other channels, reaching a wider audience of educators and school administrators. Through extensive teaching applications by experimental teachers, the following application effects have been basically demonstrated.

The Interdisciplinary Instructional Agent can automatically analyze the knowledge connections between different subjects, providing teachers with a clear knowledge map. This effectively solves the problem of difficulty in accurately grasping the interconnections and addressing knowledge gaps in the interdisciplinary design process. It can automate tedious tasks in the interdisciplinary design process, offering precise guidance and suggestions to ensure the accuracy and effectiveness of the design.

The abundant educational resource library connected to the Interdisciplinary Instructional Agent provides teachers with a one-stop resource acquisition channel, reducing the time and effort spent switching between different resource platforms. The integrated automation tools can automatically complete repetitive tasks, allowing teachers to focus on teaching innovation, and significantly shortening the preparation time for interdisciplinary thematic learning. It helps teachers continuously learn and understand knowledge from different subjects, guiding them in interdisciplinary design and gradually enhancing their interdisciplinary professional competence.

5.3 Innovations

This case propels generative AI technology from theoretical exploration to the forefront teaching practice, offering a novel perspective and tools for enhancing teachers' interdisciplinary teaching abilities and qualities (Holmes et al., 2021). It empowers them with teaching resources and strategies that are highly tailored to their teaching styles and needs. This innovative practice sets a benchmark for the profound application of AI in the field of education.

Through core functions such as deep analysis, personalized recommendations, and intelligent design assistance, the Interdisciplinary Instructional Agent supports teachers in designing innovative and practically valuable interdisciplinary thematic learning activity plans with exceptional professional standards and keen insights into subject integration. It also generates diversified subject resources and provides precise services to teachers.

The development team actively collects and analyzes feedback from teachers, regularly updates teaching resources and cases, and iterates the product alongside the platform's technological upgrades. This guarantees an improvement in both the quality and personalization of the generated teaching plans, ensuring that the instructional designs and resources offered are in harmony with the most recent educational trends and research findings.

5.4 Background

In the context of global economy, interdisciplinary learning has become one of the crucial forms of teaching and research. The *Compulsory Education Curriculum Program and Standards (2022 Edition)* specifically emphasizes the importance of interdisciplinary thematic learning.[1] In 2023, the *Action Plan to Improve Basic Education*, issued by the Chinese Ministry of Education listed interdisciplinary thematic learning as a key and challenging project in teaching reform, focusing on core competence-oriented interdisciplinary thematic learning.[2] This indicates that interdisciplinary thematic learning is gradually becoming an important component of China's educational reform and development, holding vital significance for enhancing educational quality, and cultivating talents with comprehensive qualities.

Despite the clear advantages of interdisciplinary thematic learning, its implementation faces a series of challenges. In traditional teaching modes, teachers usually specialize in their own subject areas, whereas interdisciplinary teaching requires teachers to master knowledge across multiple disciplines, which poses higher requirements for their professional competence (Mansilla & Duraising, 2007). The curriculum design for interdisciplinary courses necessitates integrating knowledge from various disciplines, rich curriculum design skills and innovative thinking from teachers.

Against this backdrop, our team attempted to develop an interdisciplinary instructional agent capable of assisting teachers in interdisciplinary instructional design. Through multiple rounds of dialogue with this agent, teachers can clarify the direction of interdisciplinary themes, plan interdisciplinary thematic learning activities and learning resources, and ultimately enhance their interdisciplinary teaching competence. This approach effectively implements interdisciplinary thematic learning in junior high school classrooms.

5.5 Promotion

Aligned with the current educational policy orientation, the Interdisciplinary Instructional Agent addresses the pain points of implementing interdisciplinary thematic learning in teaching and satisfies the growing demand for interdisciplinary teaching among teachers. Through interaction with teachers, the Interdisciplinary Instructional Agent can deeply understand each teacher's teaching style and needs, providing customized suggestions for instructional design. It also takes the student differences into account, assisting teachers in designing differentiated teaching activities to cater to the learning needs of various students. This promotes the exploration of new teaching modes within the educational system, achieving more precise personalized teaching and comprehensively enhancing teaching quality and learning

---

[1] http://www.moe.gov.cn/srcsite/A26/s8001/202204/W020220420582343217634.pdf.

[2] https://www.gov.cn/zhengce/zhengceku/202306/content_6884785.htm.

outcomes. The Interdisciplinary Instructional Agent also alleviates the burden on educational departments in curriculum development and teacher training, helping to narrow educational disparities between regions and promote high-quality and equitable educational resources. Furthermore, the Interdisciplinary Instructional Agent can be deployed on various platforms such as WeChat, Tik Tok, Feishu, WeChat Service Accounts, WeChat Subscription Accounts, Juejin, Tik Tok comment sections, and Tik Tok (Enterprise Accounts), demonstrating its wide applicability and broad application prospects in interdisciplinary contexts.

5.6 Summary

This case examined the development and implementation of an "Interdisciplinary Instructional Agent" designed to address pedagogical challenges in interdisciplinary teaching at the junior high school level. The research demonstrated the educational value of incorporating digital technology into everyday instructional practices, particularly in support of enhancing interdisciplinary thematic learning.

5.6.1 Research Purpose and Core Findings

The investigation aimed to develop an evidence-based digital tool to support educators in navigating the complexities of interdisciplinary instructional design, including thematic conceptualization, activity scaffolding, and assessment methodology. Built using the Coze platform, the instructional agent facilitated sustained pedagogical dialogues, helping teachers generate comprehensive instructional plans and discipline-specific resources. The findings showed that the tool helped teachers integrate knowledge across subjects, reduced the time needed for lesson design, provided easier access to educational resources, and supported the development of teachers' interdisciplinary teaching skills.

5.6.2 Theoretical and Practical Significance

This research made a substantial contribution by operationalizing theoretical frameworks of interdisciplinary education through practical tools for use in everyday teaching. The study aligned with contemporary curriculum reforms that emphasize competency-based interdisciplinary learning while providing empirically tested solutions to implementation barriers. By supporting differentiated instructional design and optimizing curricular preparation processes, the tool connected the theoretical principles of interdisciplinary pedagogy with practical classroom constraints. It established a potentially generalizable model for technology-enhanced educational innovation in primary and secondary education settings.

5.6.3 Recommendations for Application and Future Research

For educational practitioners, the tool should be systematically integrated into professional development frameworks with structured protocols for implementation. Schools and academic institutions should establish collaborative professional learning communities to document effective practices and share exemplary cases of use. Future research should use mixed-methods approaches to measure the tool's impact on student achievement across diverse educational contexts, and to examine its role in supporting the development of teachers' professional competencies. Further investigation should explore potential functional expansions to include classroom-based instruction. Additionally, comparative studies across different AI instructional design systems would help identify optimal design parameters for educational AI agents in basic education and explore how they can be effectively integrated with existing curriculum planning and management structures.

References

Holmes, W., et al. (2021). Ethics of AI in Education: Towards a Community-Wide Framework. IJAIED.

Luckin, R. (2017). Machine Learning and Human Intelligence: The Future of Education. UCL Institute of Education Press.

Mansilla, V. B., & Duraising, E. D. (2007). Targeted assessment of interdisciplinary work. *The Journal of Higher Education, 78*(2).

Chapter 6
Design and Application of AI-Integrated Ancient Architecture Toys and Teaching Aids: Taking Taoranting Park as an Example

Ziji Liu, Yunpeng Ma, Xinran Zhao, Lujie Chang, and Liang Zhao

Abstract This case addresses the insufficient educational resources and monotonous teaching methods related to ancient Chinese architecture in primary and secondary schools, aiming to fill the gap in both content and pedagogical approach. By integrating advanced technologies such as AI-generative artificial intelligence (AIGC), augmented reality (AR), and the Internet of Things (IoT), the case envisions an immersive, interactive learning experience. The exploratory use of NFC chips and OpenCV technology to recognize architectural components supports guided interaction and deeper exploration of ancient buildings. The core achievement is the development of the "AI Ancient Architecture" intelligent teaching aids, which not only encourage hands-on engagement but also help foster critical thinking and creativity in students. Through block assembly and virtual interaction, students are supported in gaining a more comprehensive understanding of the structure, culture, and aesthetics of ancient architecture, further enhancing their spatial thinking, cultural awareness, and overall academic development. The case emphasizes interdisciplinary STEAM education, aiming to cultivate aesthetic sensibilities, problem-solving abilities, and cultural confidence among students.

Keywords AI technology · Ancient architecture educational toys · STEAM

Z. Liu · Y. Ma (✉) · X. Zhao · L. Chang · L. Zhao
Beijing Fengtai Vocational Education Center School, Beijing, China
e-mail: chongtian@126.com

Z. Liu
e-mail: 3047631412@qq.com

X. Zhao
e-mail: 2830780276@qq.com

L. Chang
e-mail: 913888300@qq.com

L. Zhao
e-mail: 27264431@QQ.COM

D. Liu et al. (eds.), *Empowering Educational Innovation Through Design: Volume One*,
Smart Computing and Intelligence, https://doi.org/10.1007/978-981-95-4621-3_6

6.1 Background

6.1.1 Awakening Cultural Awareness and Bridging the Educational Gap

A noteworthy phenomenon was observed during a seminar in 2016: despite the popularity of building blocks among students, the majority associated them with "LEGO" rather than toys with local cultural characteristics. This "silent loss of the building block toy frontier" reflects a lack of cultural identity among students—and even parents—towards their own cultural heritage. As important windows through which children perceive and interpret the world, toys carry a powerful cultural guiding role that should not be overlooked. The limited recognition of building blocks with Chinese characteristics highlights a need for renewed attention from educators and designers. Ancient Chinese architecture, as a significant expression of Chinese cultural heritage, calls for the thoughtful development of educational toys that can gradually cultivate children's cultural pride and appreciation. By engaging with such toys through guided and interactive play, children can begin to explore and connect with the artistic and structural charm of ancient Chinese architecture in meaningful ways.

6.1.2 Reshaping Cultural Immersion and Deepening Educational Experience

Current market offerings of ancient architectural building block educational toys often fall short in exploring the cultural depth and historical narratives embedded in ancient Chinese architecture. As a result, students' understanding of these structures tends to remain at a surface level. The toy industry in China continues to face challenges such as limited risk awareness and insufficient capacity for proactive innovation in a competitive market (Wang et al., 2015). In response to the lack of comprehensive and systematic solutions in this field, our team has undertaken two exploratory initiatives.

First, we have begun developing educational toys that represent iconic examples of ancient Chinese architecture. A planned series includes models inspired by notable sites such as Taoranting Park, Zhichunting Park, the Hall of Supreme Harmony, the Hall of Prayer for Good Harvests, Foguang Temple, and Dule Temple. Initial designs focus on representative structural elements such as post-and-lintel systems and flush gable roofs, aiming to support cultural immersion and architectural understanding through accessible, guided construction experiences.

Second, we have proposed a concept that combines physical educational tools with emerging digital technologies—what we call the "AI Ancient Architecture" educational toy system. Through the conceptual integration of NFC chips, OpenCV

Fig. 6.1 "AI Ancient Architecture" educational toy: prototype design of Taoranting Park

image recognition, and AR (augmented reality) technologies, this system aspires to create a multidimensional learning experience that bridges physical and virtual realms, offering students richer visual and auditory engagement. While still in the prototyping stage, this approach provides a foundation for exploring more interactive, technology-assisted cultural education in the future.

Taking the design of the prototype of Taoranting Park as an example (see Fig. 6.1), this educational tool not only showcases the three-dimensional structure of ancient architecture but also simulates the building process dynamically through AR technology. Accompanied by AI-assisted voice narration, it offers students a novel and engaging learning experience—bringing architectural elements vividly to life and inviting them to explore and tell their own stories.

6.2 Design Process

The design process of this ancient architecture educational tool followed a structured product development methodology, including demand analysis, product design, development, and iterative improvement. Drawing upon the team's existing technical capabilities, steady progress was achieved throughout the research and prototyping stages.

The development involved several phases, beginning with requirement analysis and extending to early-stage implementation. In the requirement analysis phase, we conducted in-depth research into the current state of ancient architecture education in primary and secondary schools, identifying key challenges such as limited instructional resources and monotonous teaching approaches. Based on these findings, we set preliminary design goals and explored the integration of emerging technologies—such as AI-generated content, augmented reality (AR), and the Internet of Things (IoT)—to inform the initial feature set and direction of the prototype.

In the product design phase, detailed functional plans were developed based on the requirements documentation (BRD). A product requirement document (PRD) was then drafted, outlining preliminary technical specifications, interaction concepts, and user experience considerations. Multiple brainstorming sessions and team discussions were held to ensure that the proposed design aligned with both educational goals and technical feasibility.

During the early-stage product development phase, the team focused on technical prototyping efforts, including the initial integration of AI and AR components, exploratory implementation of NFC chips, and application of OpenCV for component recognition and interactive response. This stage involved iterative testing, adjustment, and refinement aimed at identifying and addressing potential technical challenges.

Finally, in the preliminary testing and feedback phase, small-scale pilot trials were conducted, followed by the collection of user feedback. This informed ongoing refinements to the prototype's performance and interactive features. Each iteration introduced incremental enhancements to both usability and educational value, supporting the long-term goal of aligning the tool with evolving instructional needs (see Fig. 6.2).

The integration of educational informatization into toys is one of the emerging trends in the development of children's learning tools (Shen & Zhou, 2019). This ancient architecture educational prototype explores the use of various advanced technologies to offer a novel, interactive learning experience, helping users gain a more intuitive understanding of architectural structure and cultural context through the assembly process. The envisioned interaction process includes the following steps.

Starting the Assembly: Users begin by manually assembling the wooden blocks. An AR interface on a mobile device or tablet displays guided instructions and animated steps, allowing users to visually follow the construction process and gradually understand structural logic.

During the Assembly: OpenCV technology is being explored to track user progress and detect potential assembly errors. When deviations from expected sequences occur, the system is designed to provide real-time feedback to guide users back on track and support learning through correction.

Key Information: Select block components are embedded with NFC chips that, when scanned, trigger pre-set voice explanations or interactive content. These segments are intended to convey insights into the building's historical background, architectural functions, and symbolic meanings—enhancing contextual learning.

After Completion: Once the assembly is complete, users can wear VR equipment to "walk into" the ancient building for an immersive experience. This allows

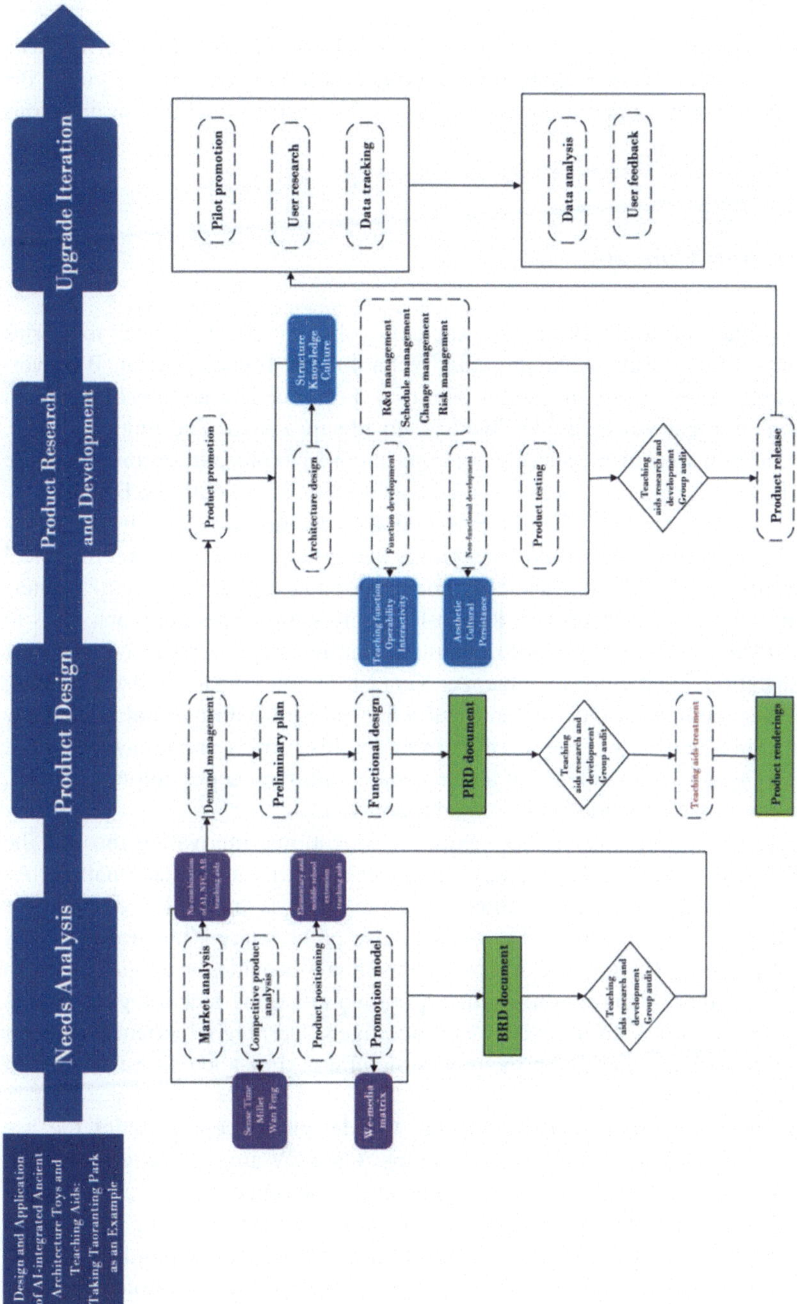

Fig. 6.2 Development process of educational toy products

them to explore the building's structure, culture, and aesthetics, deepening their understanding.

Intelligent Support: The system concept includes AI-assisted support, offering basic Q&A responses and tailored suggestions. Acting as a virtual architecture guide, the AI component is envisioned to enhance learning by answering common questions and directing attention to key learning points.

6.2.1 Design Concept

The core design philosophy of the "AI Ancient Architecture" educational toy series revolves around Cultural Awakening and Technological Empowerment. By leveraging modern technologies, this series aspires to bring ancient architecture—often described as a silent storyteller of history—to life in a vivid and engaging way, sparking children's curiosity and encouraging them to explore traditional Chinese culture. The key design principles and envisioned features are outlined below.

The "AI Ancient Architecture" series is guided by the dual aims of cultural preservation and educational innovation. It explores the cultural meanings and historical narratives embedded in Chinese ancient architecture and integrates them into an interactive learning experience through design-led prototyping. This approach aims to enhance students' cultural confidence and interdisciplinary competence, contributing to both educational and cultural value. As Xia (2021) notes, project-based learning in the Chinese context is most meaningful when infused with cultural values. By embedding such values into the learning process, this project encourages students to not only engage with content but also resonate with the deeper cultural context, fostering a more well-rounded educational journey.

This educational toy concept also explores educational innovation through the lens of STEAM (Science, Technology, Engineering, Arts, and Mathematics). As interest in STEAM education continues to grow, there is increasing demand for interdisciplinary learning tools. Ancient Chinese architecture—integrating mathematics, physics, aesthetics, and historical meaning—provides a unique platform for cross-disciplinary exploration. By incorporating emerging technologies such as AI, the prototype offers an interactive and intelligent learning experience, guiding students to investigate scientific principles, structural design, and cultural context in an integrated manner.

To enhance interactivity and immersion, the design concept includes the use of advanced technologies such as NFC chips, OpenCV image recognition, and augmented reality (AR). These technologies aim to animate ancient architectural elements through intelligent narration and dynamic visuals, allowing students to intuitively explore the structure and cultural significance of ancient buildings. While some features remain in experimental stages, the prototype lays the groundwork for meaningful, technology-supported cultural learning.

6.2.2 *Technical Framework and Implementation Path*

6.2.2.1 In-Depth Exploration and Precise Communication of Culture

The design of each educational tool is grounded in the historical context, architectural features, and cultural significance of ancient buildings. Special attention is given to embedding cultural symbols into the design of building block components, allowing students to engage with cultural meaning during the assembly process. Interactive features—such as AI-assisted voice commentary—are envisioned to offer historical stories and cultural insights. For example, scanning a beam component may not only explain its structural role and materials but also introduce related idioms or stories, such as "Liang Shang Junzi", thereby enriching both learning and engagement through cultural storytelling.

6.2.2.2 Deep Integration of Science and Technology and Education

The prototype explores the integration of technologies such as NFC chips, OpenCV image recognition, and AR (augmented reality) to create a multimodal interactive learning environment. While physically assembling the blocks, students are encouraged to engage with a layered experience that blends real-world manipulation and virtual augmentation via smart devices (see Fig. 6.3).

The team has selected key ancient architectural components—such as dougong brackets and mortise-and-tenon joints—as representative models. NFC chips are conceptually embedded within these components to store information such as historical background, cultural context, and construction methods. A dedicated mini-program has been developed to support NFC-triggered content access. When a

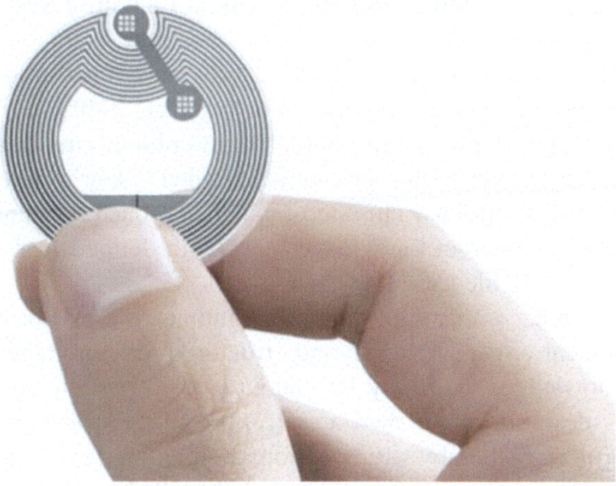

Fig. 6.3 Embedding NFC chips in ancient architecture components

Fig. 6.4 "AI Ancient Architecture" explanatory assistant

student places a phone near an NFC-embedded component, the system is designed to trigger the display of relevant information automatically.

In combination with OpenCV-based image recognition, the system can identify specific architectural elements and, upon successful recognition, activate the AR module. This allows students to view a dynamic simulation of the architectural assembly process on their screen. AR overlays may also provide contextual explanations, enabling a more intuitive and engaging exploration of ancient construction principles. While still under experimental development, this hybrid approach lays the foundation for deeper technology-assisted learning in cultural education (see Fig. 6.4).

6.2.2.3 Modular and Hierarchical Learning System

The modular design proposes a progressive learning pathway, moving from basic to more advanced modules. For instance, Table 6.1 shows the curriculum framework progresses from simple beam-column structures to more intricate dougong constructions, with each module aligned to key architectural techniques and cultural knowledge points. This step-by-step approach is intended to help students gradually build a coherent knowledge base, with defined learning objectives and a growing sense of achievement at each level.

An accompanying digital communication platform is being explored to support student engagement beyond structured lessons. Once students grasp foundational knowledge, they may be encouraged to participate in creative extensions—such as designing their own architectural models or engaging in simulated restoration projects—thereby strengthening their applied skills and design thinking.

However, as noted by Yu and Hu (2015), interdisciplinary integration in STEM education often encounters challenges due to gaps and misalignments among subject

Table 6.1 STEAM modular curriculum basic framework

Course	Basic module: introduction to ancient architecture	Advanced module: secrets of ancient architecture	Comprehensive application: restoration and innovation
Science	– Understanding ancient building materials (wood, bricks, tiles, etc.) – Learning basic mechanical principles (beam-column structures)	– In-depth study of the properties and applications of different types of wood – Researching methods used in ancient architecture to address environmental factors (wind, rain, earthquakes, etc.)	– Comprehensive environmental studies (geology, hydrology, etc.) – Deep exploration of the historical background and cultural significance of ancient architecture
Technology	– Understanding traditional manufacturing techniques for wooden components – Learning basic tools used by ancient craftsmen	– Studying complex woodworking techniques (mortise-and-tenon joints) – Exploring modern technologies applied in ancient architecture preservation	– Learning the application of modern technologies in the restoration and preservation of ancient architecture (3D scanning, VR, etc.) – Designing and creating innovative auxiliary tools
Engineering	– Learning basic design principles of beam-column structures – Building simple beam-column structure models using modular components	– Studying the design principles of *dougong* (bracket set) structures – Improving and optimizing basic models by adding complex structures	– Designing and constructing complete ancient architectural models – Learning project management knowledge for planning and implementing architectural models

(continued)

Table 6.1 (continued)

Course	Basic module: introduction to ancient architecture	Advanced module: secrets of ancient architecture	Comprehensive application: restoration and innovation
Art	– Recognizing basic decorative elements of ancient architecture (sparrow braces, lattice windows, etc.) – Learning and copying traditional architectural patterns	– Studying advanced decorative elements (paintings, carvings, etc.) – Creative design and application of ancient architectural elements	– Interdisciplinary artistic creation based on ancient architectural themes – Designing and producing cultural display works of ancient architecture for exhibition and dissemination
Mathematics	– Understanding basic structures of ancient architecture (cylinders, cuboids, triangles, etc.) – Grasping the principles of proportion and symmetry in ancient architectural design	– Learning complex geometric shapes (polygons, curves, etc.) – Conducting architectural measurements and calculations (area, volume, etc.)	– Comprehensive application of mathematical knowledge in architectural design and modular calculations (optimizing structural design) – Learning basic statistical and analytical methods for organizing and analyzing ancient architectural data

domains. These gaps can hinder the formation of a cohesive understanding and result in fragmented learning experiences. To address this concern, the curriculum framework is being designed as a multi-module, interdisciplinary learning pathway. It emphasizes interconnection and complementarity across disciplines to promote deeper conceptual integration. The proposed online platform is also intended to encourage collaborative, cross-disciplinary exploration—where students can share their work, exchange feedback, and apply their learning in practical, context-rich scenarios.

6.2.2.4 Practice Orientation and Innovation Ability Training

Ancient architecture encompasses both engineering and mathematical challenges. By simulating the construction process through AR, practical engineering challenges and problems can be introduced, guiding children to apply their knowledge to solve real-world issues. For example, students can design ancient architectural structures based on a given area, requiring them to calculate materials and make adjustments according to actual conditions. This process fosters their engineering thinking, innovative capabilities, and communication skills. Encouraging group collaboration to complete the construction of large-scale ancient architectural models promotes the development

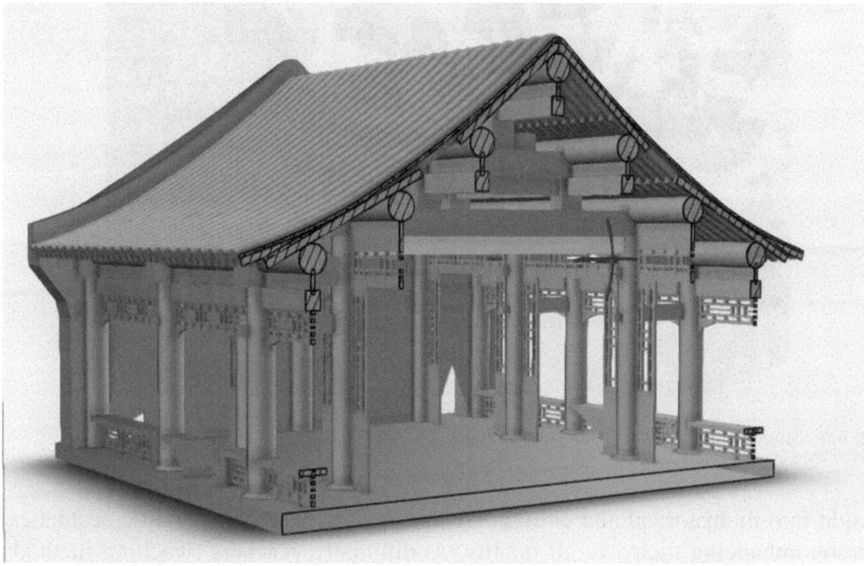

Fig. 6.5 Interact with 3D models to see the profile structure

of communication and teamwork abilities. In team collaborations, students must divide tasks, work together, and collectively solve problems, pooling their ideas. This greatly cultivates their team spirit and enhances their social skills and ability to tackle complex problems (see Fig. 6.5).

The "AI Ancient Architecture" teaching tool series conveys the cultural essence of ancient architecture while providing deeply interactive and personalized learning experiences through modern technology. This fosters students' innovative thinking and comprehensive abilities. We believe this series of teaching tools will pioneer a new path for integrating traditional cultural education with modern technological education, empowering students to acquire knowledge and wisdom through play, and nurturing them to become future talents with cultural confidence and innovative capabilities. As shown in Fig. 6.6, students are actively testing prototype product and applying knowledge in real time.

6.3 Implementation Outcomes

This case has yielded significant effects in the construction of school environment, student academic achievement, and teacher capability enhancement. Through the application of the "AI Ancient Architecture" smart educational tools, students' interest in learning has increased significantly, and their hands-on abilities and spatial thinking skills have notably improved. Students are not only capable of grasping the basic structural knowledge of ancient architecture but also gaining a deeper

Fig. 6.6 Students are testing prototype products

insight into its historical and cultural background, as well as its artistic aesthetics, thereby enhancing their overall quality. Additionally, teachers' teaching methods have also been innovated, leading to a marked improvement in teaching effectiveness. Feedback from some students indicates that the "AI Ancient Architecture" educational tools have greatly stimulated their interest in traditional culture, cultivated their aesthetic abilities, and boosted their cultural confidence. Through multimodal interactive learning, students master the structural knowledge of ancient architecture during the assembly process while obtaining more comprehensive learning experience with the assistance of AR technology and AI smart helpers.

6.4 Innovations

The innovative points of this case are manifested in three aspects: (1) Technology integration. This case pioneers the integration of generative artificial intelligence, AR, and IoT technologies, breaking through the limitations of traditional educational tools and creating an unprecedented multi-sensory learning environment. Through the ingenious use of NFC chips and OpenCV technology, intelligent recognition of ancient architectural components is achieved, providing students with personalized instant information access and interactive learning opportunities. This is a significant technological innovation in the field of ancient architecture education. (2) Diversified learning approaches. The multi-dimensional learning approach advocated by this project disrupts the previous single teaching mode. By combining physical block manipulation with virtual reality experiences, it not only enriches the learning content but also greatly enhances students' engagement and interest in learning. This holographic learning experience, which integrates physical manipulation, visual interaction, and auditory perception, builds a bridge for students to ancient civilizations, allowing them to deeply grasp the essence of ancient architecture through hands-on construction. (3) Interdisciplinary education. This case fully embodies the interdisciplinary STEAM education philosophy, organically integrating ancient architectural culture with knowledge from various fields such as science, technology, engineering, art, and mathematics, constructing a comprehensive educational ecosystem for students.

6.5 Promotion

The "AI Ancient Architecture" educational tool series demonstrates high adaptability and potential for impact. Its innovative educational model and technology integration concept are not only suitable for urban schools but can also effectively bridge the gap in educational resources on ancient architectural culture in rural and remote areas. The flexibility and ease of use of this smart educational tool suggest that it can be adjusted and customized according to different educational needs, adapting to diverse teaching scenarios. It also highlights the core elements of the learning approach: "students' right to speak and right to choose (Buck Institute for Education, 2023).

We have collaborated with multiple primary and secondary schools in the region to incorporate ancient architectural culture courses into their elective curriculum system. Additionally, we have partnered with two museums to conduct research-based learning activities on the theme of ancient architecture, promoting the "AI Ancient Architecture" educational tools. Through small-scale user testing, students have shown great interest and enthusiasm for "AI Ancient Architecture." They believe that this learning method, which combines AR, AI, and building blocks, is very novel and intriguing, allowing them to understand ancient architecture more intuitively and deeply. We have also collected some suggestions for improvement from students, such as adding more ancient architectural models and enriching the AR interactive content.

With collaborations involving more educational institutions and government departments, and through continuous optimization and expansion of the product content, "AI Ancient Architecture" has the potential to become a landmark tool for promoting the popularization and deepening of ancient architectural culture education. It can lead to the future trend of educational technology, inspire a new generation's love and inheritance of traditional culture, and inject new vitality into the protection and dissemination of cultural heritage.

6.6 Summary

This case used Taoranting Park as a prototype to explore the design and application of integrated educational toys focused on ancient Chinese architecture. It aimed to address the current challenges of teaching traditional architecture at the primary and secondary school levels, including limited instructional resources and a lack of engaging, diverse teaching methods. By incorporating technologies such as AIGC, AR, NFC, and OpenCV, the project developed a multidimensional learning system that integrates physical assembly, virtual interaction, and intelligent feedback, promoting the digital transformation of traditional cultural education.

Its core value lay in the deep integration of traditional Chinese culture with the principles of STEAM education. Through modular design and immersive experiences, the tool stimulated student interest and enhances spatial reasoning, aesthetic awareness, and cultural identity. At the same time, the project reflected the principles of constructivist learning and project-based learning, and promoted students' initiative, inquiry, and collaboration. It enriched the application paradigm of intelligent technologies in interdisciplinary education.

At the practical level, the tool has been piloted in primary and secondary schools as well as cultural institutions such as museums, receiving positive feedback and demonstrating strong adaptability and potential for broader implementation. To support further development, it is recommended to deepen AR content, expand the variety of architectural models, and establish an online platform for student creation and teacher collaboration.

Future research may focus on quantitative evaluation of how intelligent educational tools influence students' cultural knowledge, learning motivation, and general competencies, and explore the potential of these tools to reduce disparities in access to quality cultural education between urban and rural settings. This case provided a replicable and scalable model for innovating traditional culture education in the digital age.

References

Buck Institute for Education. (2023). *Project-based learning guide: Every teacher can do PBL* (Secondary school edition) (C. Pan & Y. Lu, Trans.). Renmin University of China Press.

Shen, W., & Zhou, S. (2019). Practice of building block toy design based on opera character images. *Product Design, 7*, 130–131.

Wang, Z., Wei, Z., Guo, X., et al. (2015). Research on interactive children's toy design. *Packaging Engineering, 4*, 76–79.

Xia, X. (2021). *Project-based learning design: International and local practices from the perspective of learning literacy*. Educational Science Press.

Yu, S., & Hu, X. (2015). STEM education concepts and interdisciplinary integration models. *Open Education Research, 21*(4), 13–22.

Part II
Featuring Educational Solution Designs

Chapter 7
Ethno-STEAM and *Klenang Sasak*: A Design-Based Solution for Rural Schools

Syakti P. Sriyansyah, Marselino F. G. Sundah, Nugroho P. Sumanto, Maturiadi, and Duwi B. Sriyansyah

Abstract This case study presents an innovative Ethno-STEAM program designed to address the challenges faced by rural schools in Indonesia. Despite the availability of technology, rural schools often lack the resources and expertise to implement effective STEAM education. Hence, this ongoing program aims to create an engaging and effective Ethno-STEAM program for rural students using the Design Thinking (DT) framework while simultaneously preserving endangered cultural heritage, *Klenang Sasak*. The DT framework applied in this case study consists of empathizing, defining, ideating, prototyping, and testing, which was implemented in three phases: preliminary, main, and final. The initial phase emphasizes developing and validating resources and enhancing teacher capacity, whereas the main phase concentrates on implementing these resources in the classroom. The main phase involved students in the engineering design process, including asking, imagining, creating, testing and evaluating, and communicating, to foster critical thinking, problem-solving, and creativity. The final phase focuses on monitoring, evaluation, and dissemination. The expected output includes developing and validating research-based Ethno-STEAM curriculum materials and preserving cultural heritage by reintroducing the *Klenang Sasak* instrument. This program demonstrates the potential of Ethno-STEAM to transform rural education, providing a replicable model for rural schools to enhance learning outcomes and preserve cultural heritage.

S. P. Sriyansyah (✉) · M. F. G. Sundah · N. P. Sumanto
BINUS SCHOOL Bekasi, Bekasi, Indonesia
e-mail: syakti.sriyansyah@binus.edu

M. F. G. Sundah
e-mail: marcellino.sundah@binus.edu

N. P. Sumanto
e-mail: nsumanto@binus.edu

Maturiadi
SDN 1 Puncak Jeringo, East Lombok, Indonesia

D. B. Sriyansyah
SDIT Nurul Fikri Selong, East Lombok, Indonesia

D. Liu et al. (eds.), *Empowering Educational Innovation Through Design: Volume One*, Smart Computing and Intelligence, https://doi.org/10.1007/978-981-95-4621-3_7

Keywords Ethno-STEAM education · Rural schools · Cultural heritage · Design thinking

7.1 Introduction

A rural school nestled in the foothills of Mount Rinjani, Lombok, Indonesia, faces distinct challenges. Limited by its remote location, the school lacks a dedicated science lab and has historically struggled with internet access. These constraints restrict students' exposure to practical science activities, especially those incorporating technology.

However, the school possesses valuable assets that can be leveraged to overcome these challenges. The presence of 14 Chromebooks and 5 laptops provided by the government, along with limited internet access, provides a foundation for technology integration. Additionally, the abundance of bamboo in the surrounding area and the students' existing skills in bamboo crafting present a unique opportunity. Furthermore, their enthusiasm for science projects further fosters a positive learning environment.

This project capitalizes on these assets by utilizing the *Klenang Sasak*, a traditional musical instrument on the verge of disappearing after nearly a quarter-century of disuse (Zakaria & Cahyaningrum, 2024). As shown in Figs. 7.1 and 7.2, the *Klenang Sasak* serves as a point for integrating cultural heritage with technology and science learning. This instrument will be used as an interdisciplinary project in the Ethno-STEAM learning. STEAM learning became a promising new trend and began to be used to meet the learning demands that are growing faster. Many previous studies have shown the effectiveness of using STEM learning to improve thinking skills (Wahono et al., 2020), scientific literacy (Sriyansyah et al., 2023) learning outcomes (Barrett et al., 2014; Dalimonte, 2013; Edelen et al., 2019), spatial abilities (Sharobeam, 2016), motivation (Chittum et al., 2017), and changing student attitudes towards STEM and STEM careers (Baran et al., 2019). Therefore, STEAM learning has great potential and could be more promising to be used in this program. In this project, the intangible cultural heritage is integrated into the STEAM learning to be the Ethno-STEAM. The Ethno-STEAM program aims to reimagine the *Klenang Sasak* by incorporating it with science learning using smartphone sensor technology and design software. The proposed Ethno-STEAM program offers a three-pronged approach. First, it will provide students with experience conducting science experiments using musical instruments and smartphone sensor technology. Second, it will ensure the preservation of *Klenang Sasak's* playing skills by reintroducing the instrument to a new generation. Finally, students will gain direct experience completing a STEAM project by designing new versions of the *Klenang Sasak* with the guidance of *Klenang* craftsmen, the use of smartphone sensor technology, and the utilization of design software.

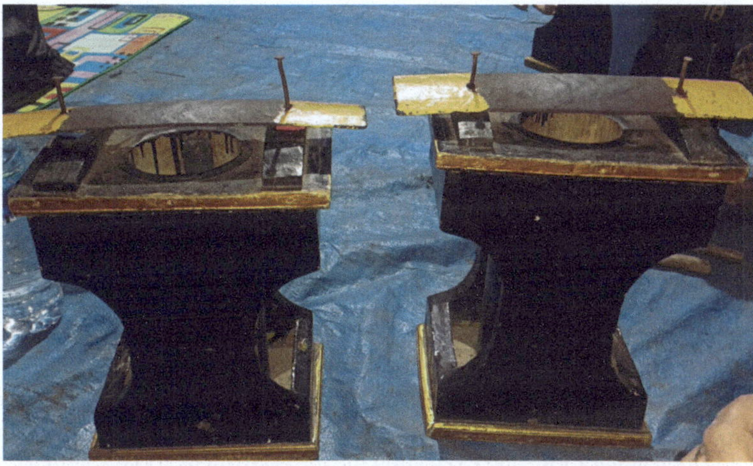

Fig. 7.1 *Klenang Sasak* Instruments

Fig. 7.2 *Klenang Sasak* players are practicing at night

7.2 Literature Review

7.2.1 Design Thinking Framework

Design thinking (DT) is a method that offers design-oriented solutions for tackling complex issues. It is an innovative problem-solving approach that prioritizes the needs and desires of users. DT provides a systematic framework and methodological assistance for addressing user challenges. Originally, DT was a recognized process utilized by organizations to solve problems and foster innovation (Brown, 2008). Today, however, it has been adopted across numerous fields, including science, engineering, medicine, and education.

Although there are several types of DT stages in the literature, all stages contain three main ideas. The three main ideas include: (1) acquisition of data about the problem, (2) idea generation, and (3) testing the ideas (Brown, 2008; Liedtka, 2014). In this study, we adopted the EDIPT model developed by the Hasso–Plattner Institution of Design at Stanford (D. School, 2010). The model includes Empathize, Define, Ideate, Prototype, and Test (Plattner et al., 2009; School, 2010; Tu et al., 2018), as can be seen in Fig. 7.3.

Figure 7.3 shows the first stage, 'Empathize,' which involves collecting information through observation and interview to understand the big picture of the design challenge. The second stage, 'Define' requires the design thinker to analyze and interpret all the information collected to define the problem and the constraints accurately. In the 'Ideate' stage, the design thinkers are encouraged to brainstorm ideas and concentrate on idea generation. This stage is a transition from identifying problems to creating potential solutions by bringing multiple ideas forward into prototyping. The 'Prototype' stage is the iterative generation of the initial models of the suggested possible solutions that can be shared with the users. At the end of this stage, a strong

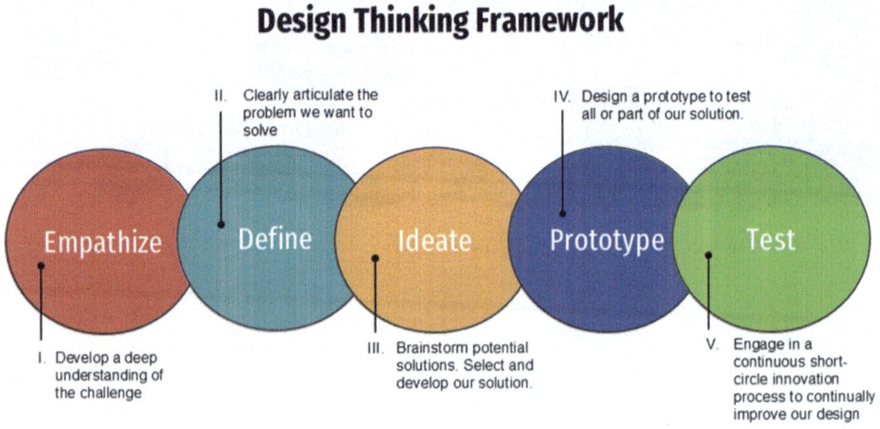

Fig. 7.3 The EDIPT model (D. School, 2010)

idea will be the final solution that would probably solve the problem. Finally, the 'Test' stage examines the best solution resulting from the prototype stage in the real environment setting. This process might lead to the next iterations of prototypes and provide a chance to refine the solutions and even learn more about the users. It also facilitates a more profound understanding of the design thinking process and aids in pinpointing areas for enhancement in future studies.

There are three rationales for using this EDIPT model. First, the process components of this model are similar to the scientific inquiry phases, emphasizing problem-solving and the production of actual products. Second, the model is largely consistent in its process with the sequences of engineering design processes used in this study. Third, this model suggests divergent thinking at the beginning of the process and ends with convergent thinking for assembling potential solutions (Rauth et al., 2010).

7.2.2 Engineering Design Process

One way to engage students in problem-solving to solve complex and real-world problems is through an Engineering Design Process (EDP). Engineering design challenges enable students to engage in science, technology, engineering, art, and mathematics (STEAM) instruction. The integrated STEAM instruction can take the form of project-based or problem-based tasks that involve solving complex and real-world problems (Garafolo et al., 2017; Honey et al., 2014).

Several types of structures of the EDP steps have also been proposed in the literature. However, the main idea of the different step models is still the same. Models with more stages only show a finer subdivision. For example, Kier et al. (2021) applied five phases of the engineering design process, and Rehmat and Owens (2020) used six stages, known as the Ask-Imagine-Plan-Create-Improve-Share model. In the other studies, Garafolo et al., (2017) implemented seven stages and Jolly (2017) proposed eight steps, which include Define, Research, Imagine, Plan, Create, Test and Evaluate, Redesign, and Communicate. In the following, we focus on Kier et al. (2021)'s model for simplicity and ease of use for seventh-grade junior high school students. The model consists of Unpack the problem, Imagine solutions, Build prototype, Test & Evaluate prototype, and Communicate.

7.3 Methods

This ongoing program aims to create an engaging and effective Ethno-STEAM education program for rural students using the Design Thinking framework while preserving endangered cultural heritage, *Klenang Sasak*. In our study, we adopted a comprehensive design thinking framework developed by the Hasso–Plattner Institution of Design at Stanford (D. School, 2010). This model consists of Empathize, Define, Ideate, Prototype, and Test (EDIPT model).

7.3.1 Case-Study Features

7.3.1.1 School Demography and Location

SMPN Satu Atap 2 Suela is located in Puncak Jeringo Village, within the Suela District of East Lombok Regency. This school is situated in a remote rural area, requiring significant travel to access urban centers. The socio-economic conditions of the students at SMPN Satu Atap 2 Suela are challenging; most students come from families where parents or guardians work as farm laborers with low and unstable incomes. This financial uncertainty often leads parents to seek better economic opportunities elsewhere, and some even require their children to work during the planting season, despite their young age.

The targeted school for this project, SMPN Satu Atap 2 Suela, shares a building with SDN 1 Puncak Jeringo, both managed by a single principal but with separate administrations. The school consists of two classrooms, but currently, only one is functional. The other classroom has been repurposed as a prayer room. For teaching and learning, the school relies on subject books, Chromebooks, Wi-Fi, and an LCD projector to support the educational process.

In the 2024/2025 academic year, the primary grades have the following enrolments: 13 students in grade 4, 18 students in grade 5, and 9 students in grade 6. In the secondary grades, there are 13 students in grade 7, 4 in grade 8, and 5 in grade 9. For this program, seven-grade students were selected using a purposive sampling method. They will be equipped with Ethno-STEAM learning to design, build, and play *Klenang Sasak* for a month. They have three science classes per week.

7.3.1.2 Ethno-STEAM Aspects

STEAM (traditionally known for "Science, Technology, Engineering, Art, and Mathematics) approach has emerged in educational research over the past decade as a new approach for improving students' thinking skills, practical skills, and motivation. It is more than just the collection of fields, it is an integration of disciplines used to find answers, solve problems, and create solutions. In this case study, *Klenang Sasak* is utilized as a form of cultural heritage and local wisdom, which is incorporated into STEAM learning, thus transforming it into Ethno-STEAM.

The design and manufacture of the *Klenang Sasak* can be viewed from all aspects, ethnography, science, technology, engineering, art, and mathematics. It provides useful assessment opportunities for designing *Klenang Sasak* activity that enables links to the science and mathematics assessments. Therefore, the idea of designing *Klenang Sasak* activity will also fully embody the aspects of the six E-STEAM disciplines, as described in detail below:

- ***Ethnography***—In this activity, Students will delve into the cultural history and traditional knowledge surrounding the *Klenang Sasak*.

- ***Science***—In this activity, students will explore the scientific principles behind the instrument's sound production through experiments and analysis.
- ***Technology***—The technology used in this idea will involve the use of bamboo, Desmos, TinkerCad, and Phyphox. Students leverage smartphone technology to conduct science experiments, potentially using sensor apps to record and analyze sound data. They also utilize Desmos to find the volume of the irregular-shaped bamboo and TinkerCad to design the *Klenang Sasak*.
- ***Engineering***—The engineering aspect is demonstrated with each group, using appropriate materials, designing and constructing different models of *Klenang Sasak*. Students could participate in redesigning or building improved versions of the *Klenang Sasak* using Engineering Design Process (EDP) principles.
- ***Art***—The art aspect is realized with each group creating the design of *Klenang Sasak* as the product that is going to market. Students need to consider how their design works so the customers are willing to pay more for products whose designs look and feel better than other group's products.
- ***Mathematics***—Each group needs to perform and validate calculations of the speed of sound, and instrument dimensions (volume and length) for design purposes, or to analyze data collected from smartphone sensors.

7.3.1.3 Content Coverage

The program incorporates four subjects alongside one extracurricular activity: Science, Mathematics, ICT, and Crafts. Table 7.1 provides a detailed overview of the curriculum topics.

Table 7.1 Topic maps of the curriculum contents

Subject	Topics
Science	Sound waves, including loudness and pitch; how sound travels; speed of sound; measuring volume of an irregular-shaped object using the displacement method; measuring frequency and speed of sound with smartphone applications and musical instruments
Mathematics	Calculating the volume of regular-shaped objects (3D space) using mathematical equations; calculating the volume of irregular-shaped objects (bamboo tubes) using Desmos (graphing software)
ICT	Learning to use Tinkercad for designing simple objects (musical instruments); using Desmos for calculating volumes of irregular-shaped objects; using Phyphox for conducting experiments on sound waves
Craft	Designing and creating traditional musical instruments using bamboo and technological toolkits
Extracurricular	Playing and performing with traditional musical instruments

7.3.2 Solutions Based on Design Thinking Framework

In this case study, the Design Thinking framework served as the basis for the development of the Ethno-STEAM with the *Klenang Sasak* program. The detailed implementation stages of this program can be seen in Table 7.2.

Table 7.2 The implementation stages of the entire program

Stage	DT component	Description
1	Empathize	Recognize the limitations of rural schools in providing STEAM education and the potential of utilizing local resources Observe the underutilization of technology and the need for innovative teaching approaches in rural schools Identify a gap in preserving local cultural heritage, specifically the *Klenang Sasak* musical instrument
2	Define	(a) User Needs: Teachers require practical strategies to integrate technology and interdisciplinary learning. Students need immersive learning experiences that connect with their cultural identity (b) Problem Statement: How can we leverage technology and local cultural resources to create an engaging and effective STEM education program for rural students, while simultaneously preserving endangered cultural heritage?
3	Ideate	Leverage previous experience integrating science and mathematics with musical instruments (flute and angklung) Recognize the *Klenang Sasak* as a valuable resource for an Ethno-STEM program due to its cultural significance and potential for scientific exploration Develop a framework for integrating technology, traditional instruments, and cultural heritage into the curriculum
4	Prototype	Create a detailed implementation plan for the Ethno-STEM program, incorporating lessons learned from previous projects Develop and validate worksheets and lesson plans that align with the program's objectives and build upon the success of previous interventions Conduct a preliminary implementation to refine the program and gather feedback, focusing on the effectiveness of integrating technology and cultural heritage
5	Test	Evaluate the program's effectiveness in enhancing student learning, student interest in science, teacher capacity, and cultural preservation Collect data on student and teacher experiences to inform program improvements and identify areas for scaling up Disseminate the program's findings and lessons learned to inspire similar initiatives in other rural communities

7.4 Introduction of Work

7.4.1 Implementation Process

The program will be implemented in a phased approach, ensuring a smooth transition for teachers and students. The implementation stages of this program are as follows.

7.4.1.1 Preliminary Phase

The preliminary phase is crucial in preparing resources, students, teachers, and the school community for the Ethno-STEAM program. Key activities include capacity-building workshops for teachers to integrate technology, introductory workshops on the *Klenang Sasak* for students and teachers, and the development and validation of Ethno-STEAM resources tailored to the school's specific needs and available technology. This phase ensures a smooth transition into the core activities of the program.

7.4.1.2 Main Phase

The program is initiated with an ICT introduction to Tinkercad, Desmos, and Phyphox, empowering students to embark on their *Klenang Sasak* Ethno-STEAM project. This foundational knowledge paves the way for the engineering design process. Furthermore, guided inquiry labs integrate technology into physics and mathematics classrooms. Students delve into the science of sound, measuring loudness, pitch, and speed of sound using Phypox which is installed on their smartphones. Phyphox (an acronym for physical phone experiments) is a free, open-source app that allows users to perform various physics experiments using a phone's sensors (Staacks et al., 2018). Students explore the relationship between frequency, the air column length, and the thickness of the bamboo plate in the *Klenang Sasak*. Mathematics takes center stage as students calculate bamboo volumes using Desmos, a mathematical software program. This integrated approach fosters a deeper understanding of scientific principles and mathematical concepts, while also preparing students for the engineering design process of their *Klenang Sasak* instruments.

The authors (SPS and MFGS) have also previously conducted practical works in physics and mathematics classes using traditional musical instruments like flutes, drums, gongs, and angklungs (Sriyansyah, 2023, 2024; Sriyansyah & Anwar, 2021). Drawing on this experience, the author developed an Ethno-STEAM program that incorporates *Klenang Sasak* as a form of local wisdom for rural students. Additionally, the flute, gong, and drums are also part of the instrument set used in the traditional musical art of *Klenang Sasak*. The documentation of students' previous activities can be seen in Fig. 7.4.

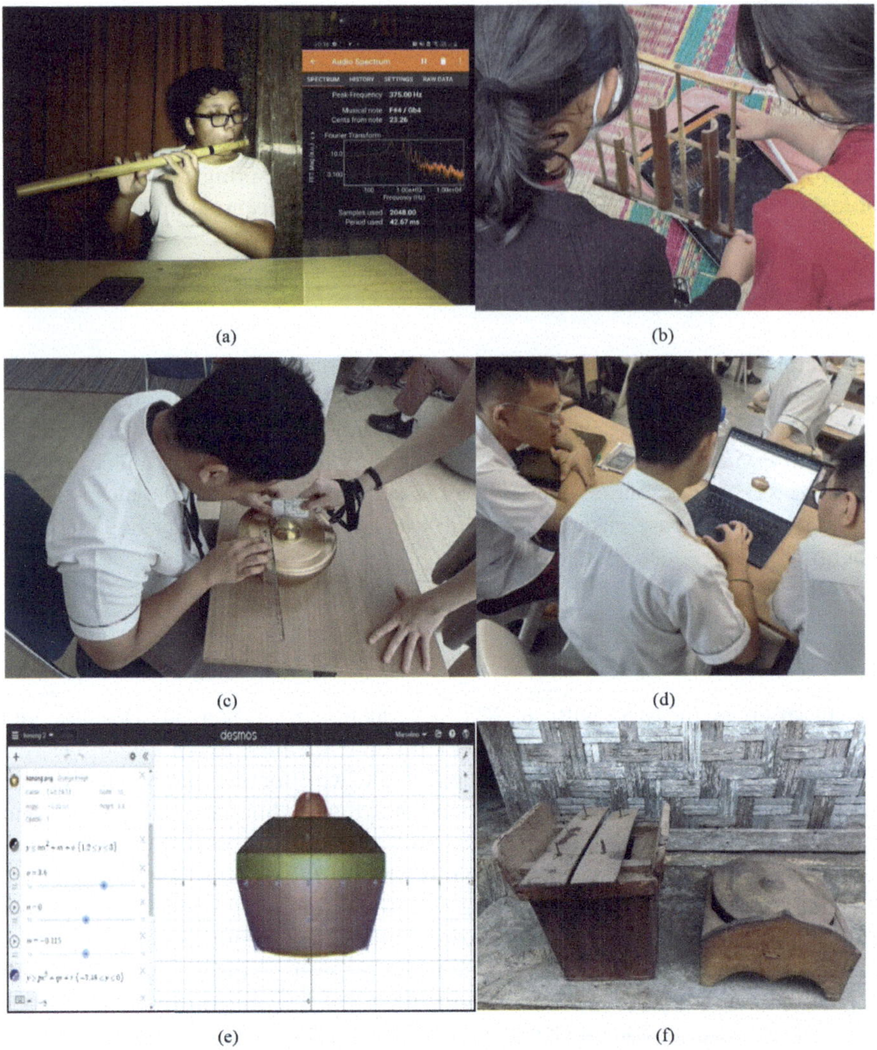

Fig. 7.4 Students measured the sound frequency of bamboo flute (**a**) and Angklung (**b**) using Phyphox, **c** students measured the dimensions of a gong, **d** then determined its volume using Desmos, **e** the sample of Desmos' worksheet when calculating volume, and **f** the sample of *Klenang Sasak* instruments

Once students have gained foundational knowledge in science and mathematics, as well as learning how to play the *Klenang Sasak*, they use this knowledge to design and create the instrument. This process follows the stages of engineering design: asking, imagining, creating, testing, evaluating, and communicating. During this process, students received guidance from teachers and skilled *Klenang Sasak* craftsmen, ensuring the authenticity and quality of their creations. Once the students had successfully made the *Klenang Sasak*, they also needed to learn how to play it. To facilitate this, experienced *Klenang Sasak* players will be invited to participate, promoting cultural exchange and the transfer of knowledge across generations. Ultimately, the activities of making and playing the *Klenang Sasak* are expected to create a collaborative learning environment and establish a sustainable extracurricular activity in the targeted schools.

7.4.1.3 Final Phase

The effectiveness of the program will be assessed using a quasi-experimental design. It will be closely monitored through evaluations of students' thinking skills, surveys on student interest in science, and interviews. This data will guide future iterations and the broader implementation of the Ethno-STEAM approach. Furthermore, a detailed report will document the program's implementation process, covering curriculum material development, teacher training, and student engagement. To promote the program's sustainability and encourage wider adoption, dissemination activities such as workshops and publications will be organized to share best practices and inspire other schools to adopt similar initiatives.

7.4.2 Progress of Project Implementation

7.4.2.1 Early Steps of the Ethno-STEAM Program

Initial efforts focused on curriculum development and feasibility testing.

- **Interdisciplinary Curriculum Mapping**: A thorough analysis identified key intersections between science, mathematics, ICT (Information and Communication Technology), and art. This ensures a holistic learning experience for students.
- **Development of Student Activity Worksheet**: The creation of comprehensive worksheets, including science, mathematics, and ICT components, is well underway. These guided activities will form the backbone of student learning during the program.
- **Preliminary Implementation and Evaluation**: A pilot test using the developed worksheets was conducted with Grade 8 students at BINUS SCHOOL Bekasi. This initial phase confirmed the feasibility and effectiveness of the materials for the target age group.

7.4.2.2 Adaptability and Collaboration

- **Versatility of the Program Design**: The successful application of the worksheets with various traditional instruments (Angklung, Flute, Drum, and coconut-leaf trumpet) demonstrates the program's adaptability (Sriyansyah, 2023, 2024; Sriyansyah & Anwar, 2021). This bodes well for the *Klenang Sasak*, as the underlying physics principles of sound production remain consistent.
- **Engaging Local Expertise**: Communication is ongoing with *Klenang Sasak* craftsmen and players. Their participation as resource persons and coaches is crucial for the program's success. Their insights will ensure authenticity and enrich the student experience.

7.4.2.3 Innovation and Exploration

Material Modifications: Promising results have been achieved through experimentation with *Klenang Sasak* component modifications. This includes replacing the metal flat with a bamboo flat as the sound source. This exploration demonstrates the program's capacity for fostering innovation in cultural preservation efforts. These initial steps lay a strong foundation for the full implementation of the Ethno-STEM program.

7.5 Application Outcomes

This innovative program is designed to seamlessly integrate ethnography with science, technology, engineering, and math to revive the almost forgotten *Klenang Sasak* musical instrument. By doing so, the project aims to achieve two primary objectives: to resuscitate the *Klenang Sasak* and equip students with the critical thinking, problem-solving, and technology skills necessary to excel in the twenty-first century. As we delve into the expected outcomes of this pioneering project, we will examine how it can revitalize cultural heritage, enhance student learning, and inspire a broader movement toward integrating culture into STEAM education. We will explore the potential of the *Klenang Sasak* project to:

- Bring back the *Klenang Sasak* musical instrument, thereby preserving cultural traditions and promoting intergenerational knowledge transfer.
- Boost student learning by providing students with critical thinking, problem-solving, and technology skills necessary to thrive in the twenty-first century.
- Inspire a wider movement towards increasing the appreciation of local culture and natural resources, integrate culture into STEAM education, and foster a deeper understanding of the interconnectedness of culture and science.

Through the *Klenang Sasak* project, we aim to demonstrate the transformative power of Ethno-STEAM, showcasing the potential for cultural revitalization, student empowerment, and innovation in education. By exploring the expected outcomes of this project, we hope to inspire educators, researchers, and policymakers to reimagine the role of culture in STEAM education and create a more inclusive and culturally responsive learning environment. The detailed desired outcomes are as follows.

7.5.1 Cultural Preservation

The project aims to revive the *Klenang Sasak*, a nearly forgotten instrument. Through student involvement in learning to play and creating new versions, the instrument will be reintroduced to the community. This can lead to performances, workshops, and increased cultural awareness. Students can document the traditional *Klenang Sasak* playing techniques and history through video recordings, interviews with local artisans, or creating educational materials. Sharing this knowledge ensures its preservation for future generations. Collaboration with local artists and cultural organizations can be fostered through joint workshops, performances, or exhibitions. Such involvement strengthens the project's impact and promotes a sense of community ownership over the cultural heritage.

7.5.2 Enhanced Student Learning and Skills Development

Students will gain valuable scientific knowledge through experiments using smartphone sensors and design software. This knowledge will be applied practically by creating new versions of the *Klenang Sasak*, fostering a deeper understanding of both scientific concepts and cultural heritage. The project promotes critical thinking, problem-solving, and creativity. Students will be challenged to design new instruments using technology, troubleshoot challenges, and collaborate effectively. These skills are crucial for success in the modern world. The hands-on, culturally relevant nature of the project is likely to boost student engagement and motivation. Learning becomes more meaningful as students connect it to their heritage and see the practical applications of scientific concepts.

7.5.3 Potential for Wider Impact

The project's success story, along with instructional materials and resources, can be shared with other schools. This can inspire the teachers of said schools to replicate the Ethno-STEAM model and integrate their cultural elements into science education. The use of readily available technology like smartphones and design software promotes resourcefulness and demonstrates that effective learning can occur without extensive lab facilities. This approach can be valuable for other resource-limited schools. The project empowers rural schools to offer innovative and culturally relevant STEAM education. This can help bridge the gap between rural and urban educational opportunities and create a more equitable learning landscape. By integrating cultural elements like the *Klenang Sasak* into science education, the project fosters a deeper appreciation for local heritage among students. This appreciation can extend to the natural resources used in creating instruments, such as bamboo. Students may become more aware of the importance of sustainable practices in preserving these resources, inspiring them to become stewards of their local environment and cultural traditions.

7.6 Promotion

The success of the *Klenang Sasak* project will present a compelling case study for promotion in similar rural schools. This initiative showcases a unique and effective approach to STEAM education by integrating intangible cultural heritage. Similar schools can adopt this model to preserve local cultural arts by creating projects that revive traditional instruments. This not only enhances cultural awareness but also fosters pride and connection to heritage.

The Ethno-STEAM approach is replicable in other rural areas with limited resources. Schools can leverage affordable technology like smartphones and design software to conduct science experiments and projects. The *Klenang Sasak* project demonstrates that even schools lacking traditional labs can effectively engage students in STEAM through creative solutions and existing technology.

The project's framework is highly adaptable and replicable. Schools can tailor it to their specific cultural context and resources. The use of design software and smartphones makes it feasible for schools with varying technological access. This adaptability ensures the model's wider implementation, promoting both cultural preservation and STEAM education in diverse settings.

By engaging students in hands-on, culturally relevant projects, schools can enhance critical thinking and creativity. This project exemplifies how practical, project-based learning stimulates problem-solving skills and innovation. Replicating this model can foster a similar environment in other schools, equipping students with crucial 21st-century skills.

In essence, promoting this case study offers a sustainable and innovative approach. It bridges cultural heritage with modern technology, providing a rich and engaging learning experience for students in diverse settings.

7.7 Summary

This ongoing case study presented an innovative Ethno-STEAM program designed to address the challenges faced by rural schools in Indonesia. The purpose of this case study is to create an engaging and effective Ethno-STEAM education for rural students using the Design Thinking (DT) framework, while simultaneously preserving the endangered cultural heritage of the *Klenang Sasak*. It also aimed to enhance students' 21st-century skills, including critical thinking, problem-solving, and creativity, utilizing the *Klenang Sasak*, a traditional Sasak musical instrument, as a central cultural resource. The DT framework, which included the stages of empathizing, defining, ideating, prototyping, and testing, was implemented across three phases: preliminary, main implementation, and final evaluation.

The expected outputs included the development and validation of research-based Ethno-STEAM curriculum materials and the preservation of cultural heritage through the reintroduction of the *Klenang Sasak* instrument. The significance of this case study lay in its demonstration of the potential of the Ethno-STEAM approach to transform rural education, offering a replicable model for enhancing learning outcomes and preserving cultural heritage.

Key findings of this case study up to this stage indicated that the integration of the Ethno-STEAM approach with available technology showed promise in bridging resource gaps in these environments. Furthermore, the innovative substitution of traditional metal components with bamboo in the *Klenang Sasak* demonstrated the program's potential for sustainable innovation rooted in local materials. This showed the application of educational theory by validating culturally responsive pedagogy and project-based learning in authentic contexts, alongside innovative practices in leveraging underutilized technology and fostering community-school partnerships for holistic development.

For practical recommendations, ongoing refinement of curriculum materials and implementation strategies was considered crucial. The project indicated the initial capacity building of teachers in interdisciplinary teaching and technology integration. Subsequent phases focused on guided inquiry-based learning, hands-on design through the Engineering Design Process with mentorship from local craftspeople, and consistent community involvement. For future research, further investigation into the long-term impact on students' thinking and inquiry skills, the potential for developing teacher training modules for wider dissemination, and evaluating the effectiveness of this case study as an extracurricular program within the school context would have provided valuable insights.

References

Baran, E., Bilici, S. C., Mesutoglu, C., & Ocak, C. (2019). The impact of an out-of-school STEM education program on students' attitudes toward STEM and STEM careers. *School Science and Mathematics, 119*(4), 223–235. https://doi.org/10.1111/ssm.12330

Barrett, B. S., Moran, A. L., & Woods, J. E. (2014). Meteorology meets engineering: An interdisciplinary STEM module for middle and early secondary school students. *International Journal of STEM Education, 1*(1), 1–7. https://doi.org/10.1186/2196-7822-1-6
Brown, T. (2008). Design thinking. *Harvard Business Review, 86*, 84–92. https://www.researchgate.net/publication/5248069_Design_Thinking
Chittum, J. R., Jones, B. D., Akalin, S., & Schram, Á. B. (2017). The effects of an afterschool STEM program on students' motivation and engagement. *International Journal of STEM Education, 4*(1), 1–16. https://doi.org/10.1186/s40594-017-0065-4
D. School. (2010). *Bootcamp bootleg*. Stanford University.
Dalimonte, C. (2013). Global STEM navigators: Developing project-based learning experiences helps students direct their learning. *Science and Children, 51*(2), 56–62.
Edelen, D., Bush, S. B., & Nickels, M. (2019). Crossing the Amazon by Lego: An interdisciplinary STEM adventure. *Science & Children, 56*(6), 30–36.
Garafolo, N., Makki, N., Halasa, K., Ahmed, W., Koskey, K., & Visco, D. (2017). Exploring the engineering design process through computer-aided design and 3-D printing. *Science Scope, 41*(1), 51–62. https://doi.org/10.2505/4/ss17_041_01_51
Honey, M., Pearson, G., & Schweingruber, H. (2014). *STEM integration in K–12 education: Status, prospects, and an agenda for research*. National Academies Press.
Jolly, A. (2017). *STEM by design: Strategies and activities for grades 4–8*. Routledge.
Kier, M. W., Grob, L. A., Leffel, K. G., Khalil, D., & Carroll, T. (2021). Culturally relevant partnerships in the engineering design process. *Science Scope, 45*(2), 20–28. https://doi.org/10.1080/08872376.2021.12291446
Liedtka, J. (2014). Perspective: Linking design thinking with innovation outcomes through cognitive bias reduction. *Journal of Product Innovation Management, 32*(6), 925–938. https://doi.org/10.1111/jpim.12163
Plattner, H., Meinel, C., & Weinberg, U. (2009). *Design-thinking*. Mi-Fachverlag.
Rauth, I., Köppen, E., Jobst, B., & Meinel, C. (2010). Design thinking: an educational model towards creative confidence. In *Proc. first international conference on design creativity (ICDC 2010)* (pp. 1–7). Kobe, Japan. https://www.designsociety.org/publication/30267/Design+Thinking%3A+An+Educational+Model+towards+Creative+Confidence
Rehmat, A. P., & Owens, M. C. (2020). Engineering clean water. *Science and Children, 57*(8), 42–48. https://doi.org/10.1080/19434812.2020.12291569
Sharobeam, M. H. (2016). The variation in spatial visualization abilities of college male and female students in STEM fields versus non-STEM fields. *Journal of College Science Teaching, 46*(2), 93–99.
Sriyansyah, S. P. (2023). *Teaching sound wave with sundanese traditional instruments*. UNESCO clearinghouse on living heritage and education. https://ich.unesco.org/en/documents/teaching-sound-wave-with-sundanese-traditional-instruments-62380
Sriyansyah, S. P. (2024). Science notes: Measuring the speed of sound with traditional instruments and a smartphone. *School Science Review in Depth, 105*(390), 16. https://www.ase.org.uk/resources/school-science-review/issue-390/science-notes-measuring-speed-of-sound-traditional
Sriyansyah, S. P., & Anwar, K. (2021). Pembelajaran Gelombang Bunyi Menggunakan Alat Musik Suling dan Gawai pada Pelajaran IPA SMP di Masa Pandemi Covid-19. *Journal of Natural Science and Integration, 4*(2), 175. https://doi.org/10.24014/jnsi.v4i2.13277
Sriyansyah, S. P., Sihole, T., & Nurhasanah, F. (2023). Gender differences in students' numerical literacy in the STEAM classroom. In *AIP conference proceedings* (Vol. 2724, p. 110003). https://doi.org/10.1063/5.0122549
Staacks, S., Hütz, S., Heinke, H., & Stampfer, C. (2018). Advanced tools for smartphone-based experiments: Phyphox. *Physics Education, 53*(4), Article 045009. https://doi.org/10.1088/1361-6552/aac05e
Tu, J. C., Liu, L. X., & Wu, K. Y. (2018). Study on the learning effectiveness of Stanford design thinking in integrated design education. *Sustainability, 10*(8), 2649. https://doi.org/10.3390/su10082649

Wahono, B., Lin, P., & Chang, C. (2020). Evidence of STEM enactment effectiveness in Asian student learning outcomes. *International Journal of STEM Education, 7*(1), 1–18. https://doi.org/10.1186/s40594-020-00236-1

Zakaria, I., & Cahyaningrum, S. Y. (2024). Menghadirkan Harmoni Klenang Nunggal yang Lama Hilang. *kompas.id*. https://www.kompas.id/baca/nusantara/2024/01/02/menghadirkan-kembali-harmoni-klenang-nunggal-pringgasela-selatan-yang-lama-hilang

Chapter 8
Unity in Diversity: Innovative Practices for Inclusive Education

Doina Otilia Filep, Szabolcs Vincze, and Péter Krajnyák

Abstract This case explores integrating students from diverse cultural and linguistic backgrounds through a teaching–learning approach rooted in common values. The project seeks to overcome communication barriers, foster trust, and encourage collaboration. Key aspects include inclusivity, cultural relevance, and student-centered learning, starting from shared experiences to build engagement and community. By using interdisciplinary methods and technology, the project redefines traditional education to promote equity and inclusivity. Digital tools—such as multimedia, online platforms, and virtual collaboration spaces—enhance accessibility, allowing students to interact in real-time across cultures. Teachers are vital to this process, creating and guiding tasks that support student participation. Positive outcomes include an improved campus atmosphere, enhanced academic achievement, and teacher development, resulting in stronger student relationships and confidence. To further impact, the project involves workshops, conferences, and resources for educators, encouraging inclusive practices in diverse communities. Finally, online platforms with adaptive learning and mobile apps gamify learning, offering cultural quizzes, interactive exercises, and virtual tours. These tools support immersive, culturally rich experiences that promote cross-cultural understanding, creativity, and collaboration.

Keywords Inclusive education · Student engagement · Cultural inclusivity

8.1 Background

In the increasingly globalized world, educational institutions face the challenge of effectively integrating students from diverse cultural and linguistic backgrounds (Markey et al., 2023). The initial lack of interaction and communication barriers between students and teachers can hinder the establishment of a trusting relationship, which is essential for successful integration and learning. Recognizing these

D. O. Filep (✉) · S. Vincze · P. Krajnyák
Megyaszói Mészáros Lőrinc Körzeti Általános Iskola, Megyaszó, Hungary
e-mail: lorincme2010@gmail.com

D. Liu et al. (eds.), *Empowering Educational Innovation Through Design: Volume One*, Smart Computing and Intelligence, https://doi.org/10.1007/978-981-95-4621-3_8

challenges, the project aims to address these issues by creating a teaching-learning interface grounded in common values and experiences.

Our goal is multifaceted: to alleviate the initial gap in communication, establish confidential communication between teachers and students, and to promote collaboration among students from different backgrounds. To achieve this, we propose the development of a task bank comprising interdisciplinary activities based on subject areas and skill levels. These tasks will be crafted using a universal language that spans natural sciences, arts, and folk art; they will also tap into the universal language of feelings and dreams, to ensure that all students, regardless of their social and cultural backgrounds, can relate to the starting points. We aim to facilitate the growth of students from different social, cultural, and linguistic backgrounds by engaging them in tasks that reflect shared values and experiences. By allowing students to see their own values reflected in the tasks' universal starting points, we seek to increase their engagement and involvement in the learning process.

Our project also addresses the STEAM needs of students from different cultural and linguistic backgrounds. In the teaching of natural sciences, for instance, we focus on language-independent experiments and artistic representations to engage students in learning.

Moreover, our project confronts the issues of migration, displacement, and relocation by strengthening the self-confidence of students with different disadvantages. By finding common starting points in history, literature, and folk tales, we aim to build relationships, enhance communication, and establish trust among students from diverse backgrounds.

8.2 Design Ideas

8.2.1 Theme

The central theme of our project focuses on fostering integration and learning among students from diverse cultural and linguistic backgrounds through the promotion of common values and experiences. By creating a teaching-learning environments grounded in shared cultural motifs and interdisciplinary collaboration, we aim to alleviate communication barriers and facilitate the development of inclusive learning environments. As shown in Fig. 8.1, interdisciplinary exploration plays a key role in bridging cultures by identifying common points of intersection and promoting mutual understanding among students from different backgrounds.

Fig. 8.1 Bridging cultures: interdisciplinary exploration

8.2.2 Method

Our method involves developing a bank of interdisciplinary tasks that address the needs of students with different cultural and linguistic backgrounds. These tasks are designed to incorporate common starting points familiar to students from various social and cultural backgrounds, thereby promoting engagement and fostering their sense of belonging. Teachers play a crucial role in the implementation process by collaboratively developing tasks and facilitating student participation.

8.2.3 Content

The content of our tasks spans multiple subject areas, including natural sciences, mathematics, history, literature, music, physical education, and technology. Each task is carefully crafted to align with common themes and values found across different cultures, ensuring relevance and accessibility for all students. For example, tasks in natural sciences may focus on language-independent experiments, while tasks in history may explore common events and narratives shared by multiple cultural groups.

8.2.4 Technology

While technology is not the primary focus of our project, it serves as a facilitator for learning and collaboration (Resta & Laferrière, 2007). We leverage various tech tools and resources to enhance the delivery of tasks and promote interactive learning experiences. For instance, online platforms may be utilized for task dissemination and student communication, while multimedia resources such as videos and interactive simulations may complement task content in certain subject areas.

8.2.5 Pedagogy

Our pedagogical approach emphasizes student-centred learning, collaborative inquiry, and experimental engagement. Tasks are designed to encourage active participation, critical thinking, and creativity, enabling students to construct meaning and develop essential skills in a supportive learning environment. Teachers adopt a facilitative role, providing guidance and support as students navigate the tasks and engage in collaborative problem-solving activities.

8.2.6 Implementation Process

The implementation process of our project involves several key steps. In the following, I present the steps of a lesson related to the topic of astronomy. As shown in Table 8.1, the process is divided into five steps of structure, including Research and Planning, Task Development, Teacher Training, Task Implementation, and Assessment and Evaluation.

8.2.7 Case Features

The case features several distinct elements that contribute to its effectiveness and relevance: By integrating multiple subject areas and disciplines, the project offers students a holistic learning experience that promotes interdisciplinary connections and critical thinking skills. The incorporation of common values and experiences from diverse cultural backgrounds ensures that the tasks resonate with all students,

Table 8.1 Steps of a lesson related to the topic of astronomy

Step	Activity	Description
Research and planning	Cultural exploration	Conduct research on students' cultural backgrounds and language skills. Identify common values, myths, and cultural experiences related to astronomy, such as star myths and constellation stories from various cultures. This helps develop inclusive lesson themes and content that resonate with all students
Task development	1. Exploring star myths 2. Constellation legend comparisons 3. Global astronomy presentations	Collaborate with educators to create tasks that merge astronomy with cultural stories, making lessons interdisciplinary and accessible. Tasks include exploring star myths from different cultures, comparing constellation legends, and creating multimedia presentations on global astronomical stories. These activities support cultural awareness and engagement across diverse backgrounds
Teacher training	1. Workshop on multicultural astronomy content 2. Collaborative learning development 3. Ongoing mentorship	Train teachers to facilitate culturally inclusive astronomy lessons. Workshops cover integrating multicultural content and collaborative learning methods, with ongoing mentorship to support adapting tasks to student needs. Teachers develop skills to guide students in exploring astronomy from a global perspective
Task implementation	1. Individual star myth assignment 2. Group constellation map project 3. Interactive cultural discussions	Teachers incorpoate tasks into their curriculum and adapt for varying learning styles. Students research and present star myths from their culture, work in groups to create a constellation map with cultural stories, and engage in discussions to share and compare cultural perspectives on astronomy. These activities foster cross-cultural understanding and collaboration
Assessment and evaluation	1. Quizzes on astronomy and cultural knowledge 2. Presentations and projects evaluation 3. Peer and self-assessment	Use a mix of formative and summative assessments to track student progress, including quizzes on lesson content, project evaluations, and reflective assessments. Peer assessments encourage cross-cultural learning, while continuous evaluation helps adjust tasks to ensure inclusivity, engagement, and achievement of learning goals

fostering a sense of cultural inclusivity and belonging. Tasks are designed to be student-centred, allowing for individual autonomy, collaboration, and creativity. This promotes active engagement in the learning process and encourages them to take charge of what they are learning. Teachers play a central role in the development and implementation of tasks, fostering a sense of ownership and investment in the project's success. Collaboration among teachers enables the sharing of best practices and the development of innovative instructional strategies (Mahat et al., 2023).

Technology plays a pivotal role in enhancing learning experiences. We integrate digital tools and resources that cater to diverse linguistic and cultural backgrounds. This may include interactive simulations of astronomical phenomena, virtual field trips to culturally significant astronomical sites, multilingual educational apps, and collaborative online platforms for global storytelling.

Overall, the project represents a comprehensive approach to addressing the integration and learning needs of students with diverse cultural and linguistic backgrounds. By leveraging common values, interdisciplinary collaboration, and innovative pedagogy, we aim to create inclusive learning environments where all students can thrive and succeed.

8.3 Application Outcomes

The outcomes of the project are multifaceted:

- **Cultural awareness and appreciation**: students develop a deeper understanding and appreciation of cultural perspectives related to astronomical knowledge and global stories about stars.
- **Language proficiency**: language barriers are minimized as students engage with multilingual resources and collaborative activities, enhancing their language skills.
- **Critical thinking and problem-solving**: through interdisciplinary tasks and technology-enhanced learning experiences, students hone their critical thinking abilities and develop solutions to real-world challenges.
- **Digital literacy**: students gain proficiency in utilizing digital tools and resources, preparing them for future academic and professional endeavors.
- **Collaborative skills**: collaborative projects foster teamwork, communication, and empathy among students from diverse backgrounds, promoting a supportive learning community.

By focusing on these application outcomes, we aim to empower students to become globally minded individuals who appreciate *diversity*, leverage technology effectively, and excel academically in the field of astronomy and beyond.

8.4 Innovations

The innovation of this project lies in its comprehensive integration of cultural sensitivity, advanced technology, and collaborative learning strategies into a science lesson plan focused on astronomy. It uniquely combines global stories, legends of stars, and culturally relevant content to ensure inclusivity and engagement across diverse linguistic backgrounds. By leveraging digital tools such as interactive simulations, virtual field trips, and multilingual educational apps, the project enhances accessibility and caters to varied learning styles. Through interdisciplinary tasks, teachers foster critical thinking, language proficiency, and collaborative skills among students. This holistic approach not only enriches scientific education but also prepares students to succeed in a globalized, technologically driven world.

The project innovates by recognizing the importance of establishing meaningful connections between students and their learning experiences, regardless of cultural or linguistic differences. By incorporating common values and experiences into interdisciplinary tasks, we create a bridge that spans across cultural divides, fostering a sense of belonging, confidence, and collaboration among students.

This innovative approach challenges traditional notions of education by prioritising inclusivity, cultural relevance, and student-centred learning. Rather than focusing solely on academic content, the project places equal emphasis on promoting communication, understanding, and empathy among students, thereby creating a more supportive and cohesive learning community.

Furthermore, the project innovates by empowering teachers to play a central role in the development and implementation of tasks, promoting collaboration and professional growth. By providing opportunities for ongoing training and support, we equip teachers with the tools and resources needed to effectively address the diverse needs of their students and create inclusive learning environments.

8.5 Promotion

The success and innovative approach of the case study offer promising opportunities for promotion and implementation in similar schools seeking to address the integration and learning needs of students from diverse cultural and linguistic backgrounds. By showcasing the positive outcomes and benefits experienced within our project, we can inspire and encourage other educational institutions to adopt similar strategies and practices.

One avenue for promoting the case-study involves disseminating our findings and methodologies through workshops, conferences, and professional development sessions targeted towards educators and school administrators. These events can serve as platforms for sharing best practices, facilitating discussions, and providing hands-on training on how to implement interdisciplinary tasks grounded in common values and experiences.

Additionally, we can collaborate with educational organizations, government agencies, and non-profit institutions to develop resources and materials that support the implementation of our approach in similar schools. This may include the creation of task banks, lesson plans, and instructional guides tailored to different subject areas and grade levels, as well as the development of online platforms and communities where educators can share ideas and collaborate.

Furthermore, we can leverage technology and digital platforms to reach a wider audience and promote our case-study to schools beyond our immediate geographical area. This may involve creating online courses, webinars, and instructional videos that provide step-by-step guidance on how to implement our approach effectively.

By highlighting the potential benefits and impact of our approach on campus environment construction, students' academic achievement, and teachers' professional development, we can attract interest and support from schools looking to create more inclusive and supportive learning environments. Through strategic promotion and collaboration, we can inspire positive change and contribute to the advancement of inclusive education practices in schools across diverse communities.

8.6 Summary

This case study explored the challenge of integrating students from diverse cultural and linguistic backgrounds into a cohesive and inclusive learning environment. The project aimed to reduce communication barriers, build trust, and foster collaboration through a shared teaching-learning interface based on common values and lived experiences. It redefined traditional educational structures by emphasizing inclusivity, cultural relevance, and student-centered methodologies.

8.6.1 Core Findings

Key findings highlighted that learning tasks anchored in universal experiences, effectively cultivated students' sense of belonging and supported active engagement. Interdisciplinary approaches, combining elements from various subjects, served as a bridge between differing cultural perspectives. The team incorporated digital tools to improve access and encourage participation, including collaborative platforms, multimedia resources, and interactive modules. Real-time collaboration through online tools facilitated meaningful cross-cultural dialogue, improving students' interpersonal relationships and broadening their global awareness.

8.6.2 Theoretical and Practical Significance

The significance of this project demonstrated how educational theory could be translated into innovative classroom practices. It showed that culturally responsive pedagogy and technology-enhanced learning can work synergistically to address equity and inclusion. Teachers, acting as both facilitators and co-creators of knowledge, reported professional growth and increased confidence. Students, in turn, showed improved academic outcomes and deeper engagement, contributing to a more supportive school climate.

8.6.3 Recommendations

- **Task Design**: Develop learning activities rooted in common values and everyday experiences to promote cultural neutrality and relevance.
- **Technology Integration**: Utilize online platforms with adaptive learning features to personalize content and pacing. Develop mobile applications that gamify cultural learning through quizzes, interactive exercises, and virtual cultural tours.
- **Teacher Training**: Provide targeted professional development to equip teachers with skills in culturally responsive teaching and digital facilitation.
- **Collaboration Tools**: Foster continuous cross-cultural student exchange through shared digital environments, such as virtual whiteboards, co-edited documents, and video conferencing.
- **Dissemination Strategy**: Organize workshops and seminars, and partner with educational institutions to share resources and promote broader adoption of inclusive practices.

Future research should focus on long-term impact culturally inclusive digital pedagogy on student identity development, academic pathways, and social integration. Studies could also explore the scalability of this model across different educational systems and cultural settings, and the role of artificial intelligence in enhancing adaptive and culturally responsive learning environments.

In sum, this project underscores the transformative potential of inclusive, tech-enabled pedagogy in fostering cross-cultural understanding, educational equity, and collaborative learning in diverse classrooms.

References

Mahat, M., Plenty, S., Jackson, H., Greene, J., Allen, M., & Wong-Ratcliff, L. (2023). The impact of teacher collaborative pedagogies on student learning. In J. E. Morris & W. Imms (Eds.), *Teachers as researchers in innovative learning environments: Case studies from Australia and New Zealand schools* (pp. 107–122). Springer Nature Singapore. https://doi.org/10.1007/978-981-99-7367-5_8

Markey, K., Graham, M., Tuohy, D., McCarthy, J., O'Donnell, C., Hennessy, T., Fahy, A., & O'Brien, B. (2023). Navigating learning and teaching in expanding culturally diverse higher education settings. *Higher Education Pedagogies, 8*(1), 1–15. https://doi.org/10.1080/23752696.2023.2165527

Resta, P., & Laferrière, T. (2007). Technology in support of collaborative learning. *Educational Psychology Review, 19*, 65–83. https://doi.org/10.1007/s10648-007-9042-7

Chapter 9
AIGC-Integrated Art Course: Pathway to Cultivate High-Quality Artistic Talents

Cheng Zang and Muyao Li

Abstract China's basic education has entered a new era focused on core competencies. Art education now increasingly emphasizes innovative thinking and artistic expression. Artificial Intelligence-Generated Content (AIGC), is transforming industries worldwide, including art and design (Fathoni in Leveraging generative AI solutions in art and design education: bridging sustainable creativity and fostering academic integrity for innovative society, 2023), by lowering creative barriers and significantly enhancing efficiency. Recognizing this potential, our school has integrated AIGC into art education, using thematic creation to guide teaching and address challenges in conceptualizing themes during the early stages of creation. This approach goes beyond mere technological integration, aiming to cultivate students' artistic sensibility and critical thinking skills. Through interaction with AI during the creative process, students engage in material research and conceptual development early on, fostering empathy and imbuing their artworks with a sense of "life." It illustrates the application of AIGC in art education through the case, "*AI Design: Future City*." This human–computer interaction mode has changed the traditional teaching mode, greatly enhancing students' initiative and efficiency in learning. Practical research confirms that this integration enriches teaching resources, provides intelligent support for art creation, and nurtures innovative talent. The study's findings offer new perspectives for art education and hold practical, innovative, and scalable implications.

Keywords AIGC · Art education · Interdisciplinary teaching · Technology-driven curriculum · Creativity

---

C. Zang (✉)
Beijing No. 109 Middle School, Beijing, China
e-mail: zangcheng_75@163.com

M. Li
Beijing Huiwen Middle School, Beijing, China
e-mail: limuyao100@sina.com

D. Liu et al. (eds.), *Empowering Educational Innovation Through Design: Volume One*, Smart Computing and Intelligence, https://doi.org/10.1007/978-981-95-4621-3_9

9.1 Background

9.1.1 Policy Support

The *Compulsory Education Curriculum Plan (2022 Edition)*[1] emphasizes the importance of linking curriculum content with students' experiences and social life, while enhancing the integration of knowledge across disciplines. It advocates for the coordinated design of comprehensive courses and interdisciplinary thematic learning. Interdisciplinary project-based learning, as an emerging form of classroom instruction, must continually adapt to stay relevant, reflecting the fundamental mission and responsibilities of education in the new era.

The *New Generation Artificial Intelligence Development Plan*,[2] issued by the State Council, explicitly calls for utilizing intelligent technologies to accelerate reforms in talent cultivation models and teaching methods. President Xi Jinping has emphasized China's strong focus on the significant impact of artificial intelligence on education, actively promoting deep integration of AI and education to drive educational innovation. During his visit to Tsinghua University, President Xi Jinping further underscored the interdependence and mutual reinforcement between art, aesthetics, science, and technology, advocating for their complementary and synergistic development.

9.1.2 Deepening Curriculum Reform

The author is from Beijing No.109 Middle School,[3] which has a 38-year history of operating a high school art program and has become Beijing's only exemplary general high school with a specialized art program. Over the years, it has nurtured thousands of outstanding art talents for universities and the nation.

In response to contemporary demands, the school has deepened teaching reforms. Guided by the Beijing research project "*Development and Implementation of Specialized High School Art Curriculum*",[4] the interdisciplinary course "*Art and Technology*" (Fig. 9.1), collaboratively developed by art and information technology teachers, was officially launched in September 2022. This course introduces AIGC into teaching and can support multiple other art courses and activities.

This new course closely aligns with the objectives of the recently introduced art college entrance exam reforms, integrating core competencies into classroom teaching to cultivate top-tier art talents. It reflects the school's commitment to preparing students for future development through thoughtful and innovative teaching practices.

---

[1] https://www.ictr.edu.cn/download_center/ywjy.html.

[2] https://www.gov.cn/zhengce/zhengceku/2020-08/09/content_5533454.htm.

[3] https://baike.sogou.com/v459877.htm.

[4] https://mp.weixin.qq.com/s/4uXQSVZOibLuCTAvOtVaXw.

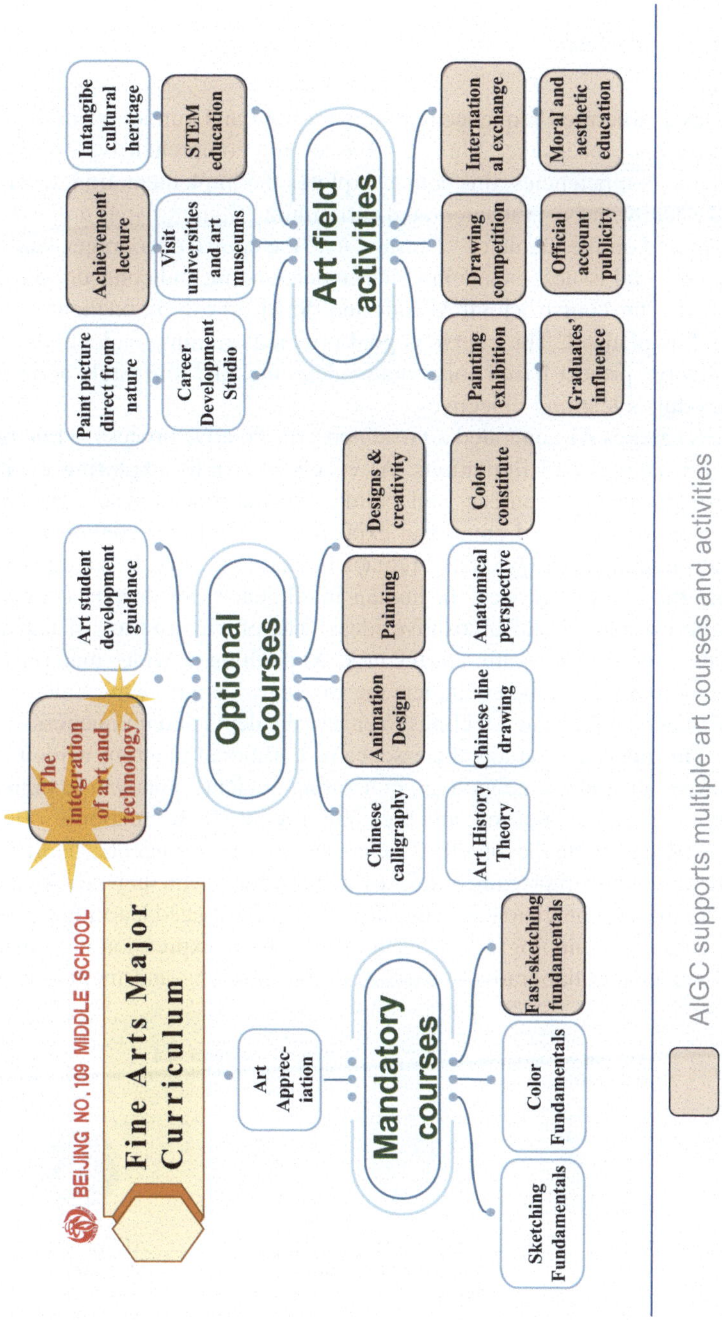

Fig. 9.1 New system of art curriculum in Beijing No.109 Middle School

9.2 Design Process

9.2.1 Project Analysis

This course uses a structured sequence that integrates AI and human creativity into the artistic education process (see Fig. 9.2). Based on a comprehensive evaluation of artistic creation competencies, this course explores the implementation, teaching strategies, and effectiveness of the integrated curriculum "*Generative AI and Artistic Creation.*" Centered on the theme of "*Future City,*" the course encourages students to engage with societal issues, enhancing their understanding and connection to the world. At the end of the course, a joint AI and hand-drawn art exhibition themed "*The Future Is Here*" is planned. The course is goal-oriented, guiding students through artistic appreciation, project-based collaborative learning, and ultimately achieving the painting module's learning objectives.

The course leverages AI-generated art to support the creative process, while fully recognize its advantages and limitations. AI excels in rapidly exploring creative directions, providing rich resources, and building inspiration libraries. However, its lack of human aesthetic and emotional expression capabilities cannot be overlooked. Furthermore, technology cannot replace human creativity. It is emphasized that grounding the creative process in human intelligence can deepen creativity, leading to more diversified and innovative ideas. Therefore, in terms of teaching content, we take proposition creation as the task, AI painting provides materials for secondary creation. Students then complete the works by hand drawing, Integrating their personal understandings and feelings to ensure the depth and uniqueness of the work's ideas. This approach retains the essence of traditional painting while introducing the power of modern technology, cultivating students' innovative thinking and the ability of artistic expression, and breathing life into their creations.

This course aligns with the practical needs of art exam students, addressing teaching challenges while improving efficiency without compromising the essence of education. It emphasizes precision and stability, even in fast-paced learning environments. By integrating technology, new avenues for artistic expression are opened. Teachers and students collaboratively explore design thinking and methodologies

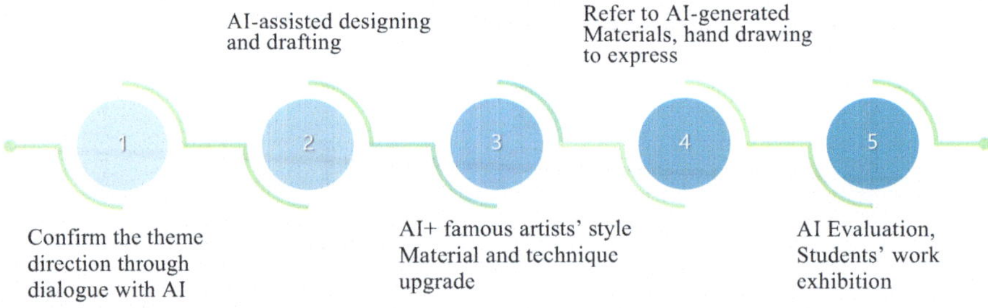

Fig. 9.2 Course process

within the AIGC era, gradually forming a personalized learning support system. This system serves as a "learning companion" for students and a "lesson-planning assistant" for teachers, showcasing the vision of a future school. It represents a forward-looking approach to cultivating high-quality art talents in a digitalized world.

9.2.2 Technical Resources

9.2.2.1 Teaching Hardware Resources

Ensuring "stability and control" in applying AI to classroom teaching is the top priority. For this course, open-source AI projects were selected, adopting an "Open-Source AI + Domestic VPS Cloud Server" DIY solution. This approach effectively addresses the technical barrier of external network access while reducing costs and improving efficiency.

The solution includes flexible server GPU configurations, allowing teachers to easily adjust and redevelop the setup as needed. For AI-generated art, tablets were selected instead of desktop computers typically found in computer labs. Tablets are better suited for use in art studios and are highly compatible with blended online and offline courses.

9.2.2.2 Teaching Software Resources

The course utilizes Stable Diffusion for AI-generated art. This model is open-source, user-friendly, and popular, capable of running smoothly on consumer-grade CPUs while delivering powerful results. Its simplicity and effectiveness make it an ideal for classroom integration.

Additionally, the course incorporates resources such as the visual inspiration platform Pinterest, AI drawing software Photoshop, and prompt exploration tools like OpenArt. ChatGPT is also used in a limited capacity to support teaching activities. The course is designed to stay up to date with the latest technologies, ensuring continuous updates and iterations to enhance the learning experience.

9.2.2.3 Addressing Integration Challenges

During the implementation of the course, several challenges emerged. For instance, AI-assisted artistic creation requires technical support, and the software involved is often quite specialized, with a steep learning curve. Additionally, the diverse types and styles of AI painting tools demand that students possess a certain level of aesthetic literacy and creativity to adjust parameters and guide the output effectively. Beginners often need considerable time to generate satisfactory materials using AI painting tools.

To address these issues, targeted strategies were implemented in teaching. These strategies included enhancing the analysis of thematic concepts and optimizing prompts, which significantly improved teaching outcomes. Furthermore, repeated testing revealed that incorporating the artistic styles of masters plays a critical role in improving the quality of AI-generated works. Therefore, teachers need to deepen their learning and preparation to enhance the quality and efficiency of AI-generated art, making this a key aspect of the teaching process.

9.2.3 Teaching Overview

The course follows the *General High School Art Curriculum Standards* [5]*(2022)* (hereinafter referred to as Standards). These Standards define the teaching content for modern media arts as the use of multimedia tools and technologies to express concepts, ideas, and emotions. They emphasize the importance of understanding the characteristics of modern media arts, which blend technology, art, and humanistic concepts. Students are expected to master modern digital media techniques, develop artistic sensitivity, modeling, and design skills, and engage in profound humanistic reflection and social awareness.

Additionally, the course adheres to the Standards for introductory artificial intelligence content. These Standards require students to understand the basic processes and principles behind intelligent technology applications. They encourage the use of open-source AI application frameworks to build simple AI application modules. Students are also expected to stay informed about new developments and applications in AI and apply them appropriately in their studies and daily lives.

9.2.4 Student Analysis

In China, art college entrance exams are divided into the general unified examination and selective school-specific exams. Over the past 4 years, more than 95% of our students have achieved undergraduate admission through the general art exam, with approximately 50% participating in the school-specific exams. Among these, the acceptance rate for school-specific exams is about 15% of the total number of students.

This project targets students enrolled in the elective *Creative Design* course for first-year high school art majors. Notably, 90% of these students intend to take prestigious school-specific exams in their senior year. These exams demand not only high-level technical drawing and modeling skills but also increasingly emphasize creative thinking, with a focus on testing through creative composition and production tasks.

[5] https://www.pep.com.cn/ebook/2022yjkcbz/ys/mobile/index.html.

Survey results reveal that students in this elective course have a basic foundation in modeling skills, but their levels vary significantly. Professional scores, out of a total of 300 points, show a gap of 76 points between the highest and lowest scorers. Most students focus on general exam preparation with little exposure to design studies. They tend to feel overwhelmed by creative tasks, with a confidence level averaging only 6.7. Specifically, conceptualization is perceived as the most challenging aspect, with a creativity score averaging just 5.2.

To address these issues, the course prioritizes methods to effectively inspire students' enthusiasm for creation. By integrating AI painting tools into the artistic conceptualization process, students can rapidly expand their thinking and overcome traditional teaching challenges within a short period. The curriculum uses a thematic approach with interconnected tasks, designed as a cohesive unit completed over four sessions. Creativity is emphasized in the second and third lessons, focusing on breaking through conceptual barriers.

Themes are closely aligned with students' daily lives and personal interests, enhancing emotional resonance and deeper understanding of the world. This approach maximizes students' enjoyment of the creative process, making artistic exploration both engaging and accessible.

9.2.5 Teaching Process

Using the unit "*AI Design: Future City*" as an example, the teaching process is guided by constructivist theory. The teacher designed a progressive chain of questions and corresponding task clusters to lead students toward independent thinking and precise design direction. Through a structured process of pre-class, in-class, and post-class learning and evaluation, students' performance is assessed using a multifaceted evaluation system that includes self-assessment, teacher evaluation, and peer evaluation. This comprehensive approach helps students balance external expression and internal understanding in their artistic learning.

In AI-assisted drawing, the core focus is on refining high-quality prompts (creative keywords) to establish the best design concepts (Fig. 9.3). These teaching methods and practices enable students to quickly grasp effective learning strategies for artistic creation.

By leveraging AI painting tools, students are better equipped to engage in creative practice and master the complete creative process, including establishing themes, gathering materials, conceptualizing compositions, and expressing through painting (see Table 9.1). Through thematic guidance, the course fosters students' sense of social responsibility and humanistic spirit, encouraging them to boldly explore and embrace the scientific spirit of advanced technology. This approach enhances their interest and awareness in creative expression, improving their artistic abilities. It also aims to comprehensively elevate students' core competencies in art—encompassing aesthetic perception, artistic expression, cultural understanding, and creative practice—thereby strengthening their overall artistic literacy.

Fig. 9.3 Explanation of prompting techniques

9.3 Implementation Outcomes

This course tracks the world's most advanced technologies, using technical empowerment to optimize teaching and effectively address challenges. It has become an innovative highlight in our school's art curriculum system and has expanded to influence several other art courses and interdisciplinary activities.

The course's advanced concepts, correct direction, and effective implementation have yielded significant results. Within just one year, it has won 15 awards, including 3 publications in core national journals and 5 first-place awards at the national and municipal levels. It is hoped that the course will not only benefit the school but also generate valuable results that can be promoted more broadly, advancing the integration of AIGC in education and contributing to policy development.

9.3.1 Expert Evaluation

In November 2022, the course participated in a municipal-level research session. Experts in the field of art education, including Zhao Mingru and others, evaluated the course positively. They noted that the new curriculum standards emphasize interdisciplinary thematic teaching aimed at solving real-world problems. The incorporation of AI painting technology in this course is designed to enhance students' creative abilities, focusing on thematic creation while also emphasizing humanistic education. This approach aligns with the goal of fostering students' core competencies. Additionally, the course attracted the attention of technology expert Liang Senshan, Director of the New Technology Department at the Ministry of Education's

Table 9.1 Key stages of the teaching design process

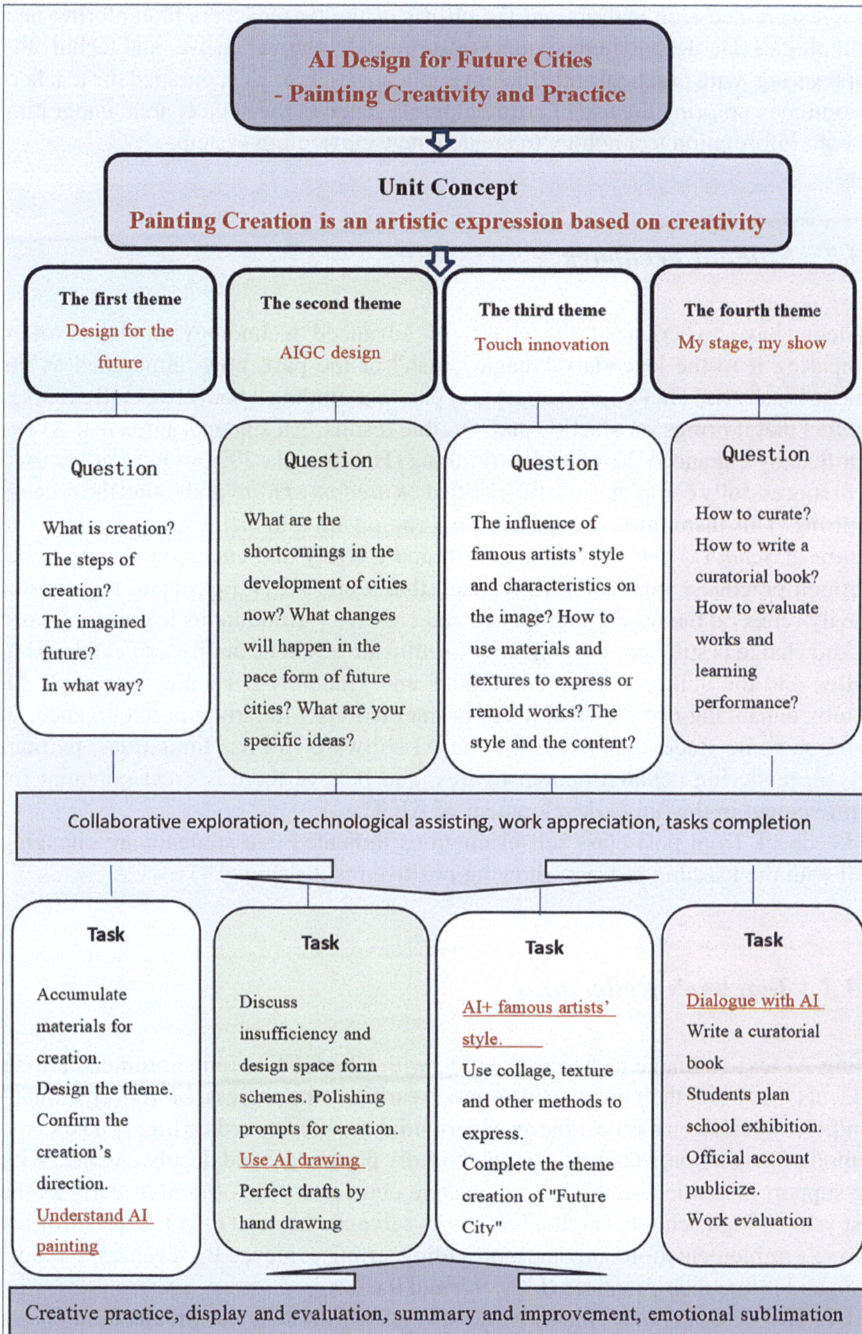

Educational Technology and Resource Development Center. He visited the school for a research session and praised the efforts of the two teachers in exploring new technologies. He described the course as profound, comprehensive, and technically empowering, with practical and efficient applications. Liang encouraged the teachers to continue exploring the use of artificial intelligence in the subject area, integrating art with information technology to create a new curriculum system.

9.3.2 Student Feedback

Students have experienced the charm of advanced technology firsthand, often comparing it to the legendary "magic brush" of the past, now reimagined as the "magic brush AI." They enjoy using AI for painting, finding it both novel and exciting, and feel that it brings satisfaction and tangible results. All students agree that AI has significantly enhanced their creative thinking (Habib et al., 2024). One student used AI to successfully complete an artwork titled "*Kingdom of Kun*" and found the process inspiring. This inspiration carried over to a composition in their Chinese class, "*The Future of Schools*," where they demonstrated a lively and energetic writing style. Through practices, students have realized that while AI is a powerful tool, it still heavily relies on the user's intelligence and creativity. Continuous learning of basic art knowledge is still necessary in order to cultivate aesthetic ability, critical thinking ability, and the ability to have continuous and gradually deepening talk with AI, thereby enhancing one's own abilities in interactions with artificial intelligence. In addition, some students pointed out that AI software still has limitations, particularly in rendering detailed human figures, and believe there is great potential for improvement in the future development of AIGC.

Feedback from post-class self-evaluations indicated that students are engaging well with the learning process, showing positive results (Fig. 9.4).

9.3.3 Teacher's Reflections

Using AI text-to-image technology, along with ChatGPT to refine prompts, allows students to solidify their creative ideas much earlier in the process. Moreover, human-computer interaction has become an experiential classroom method highly advocated in modern education, enabling students to fully participate and deeply engage. With the support of artificial intelligence, creative courses can be offered as early as the first year of high school. The final artworks serve as the most effective proof of the course's implementation outcomes, providing strong evidence for teachers to reflect upon and refine their practices (Figs. 9.5 and 9.6).

Creative design courses have always struggled with the conceptualization phase, as teachers used to prepare a large amount of image materials to guide students. However, student engagement was typically low, and the results were only average,

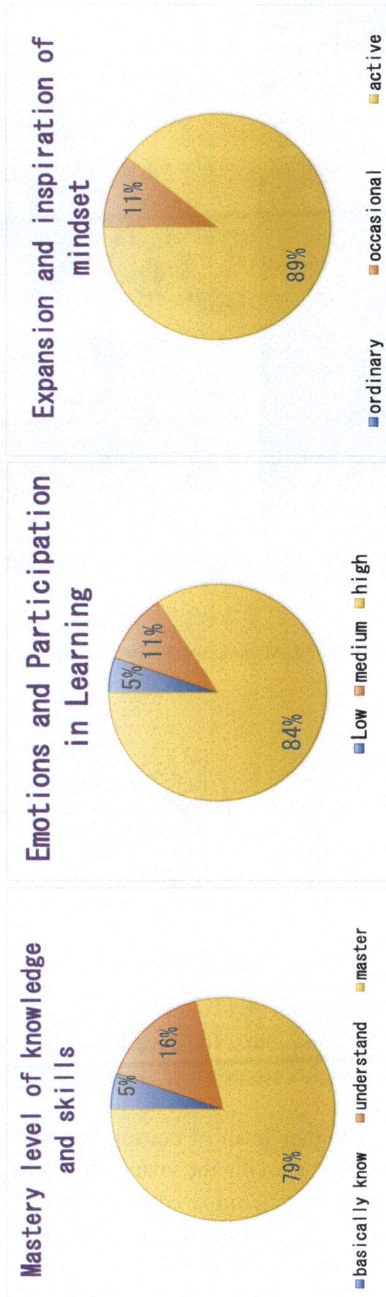

Fig. 9.4 Student post-class self-assessment feedback

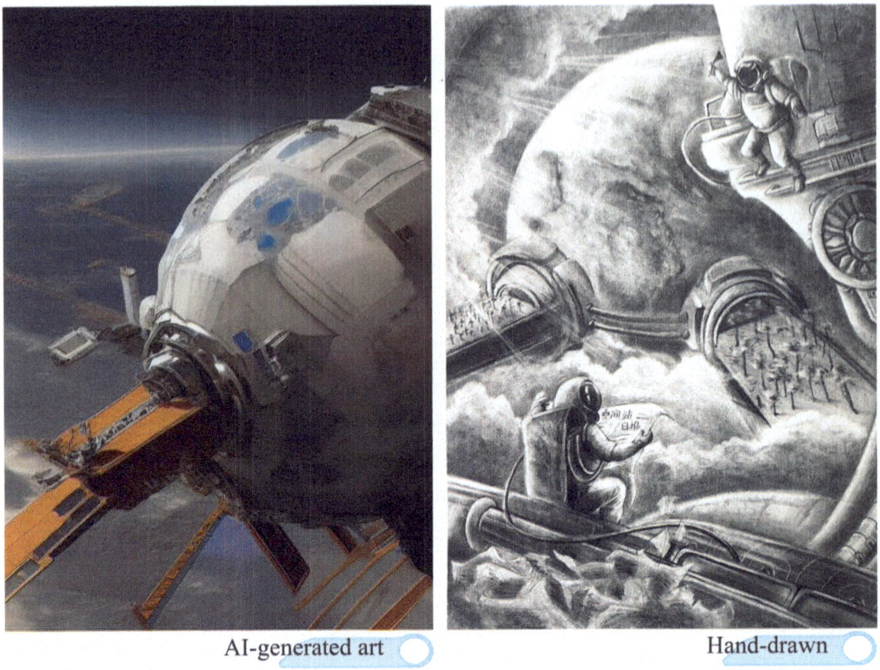

Fig. 9.5 Hand-drawn reinterpretation using AI-generated artwork as source material

Fig. 9.6 Student AI-designed artwork featured on public account

often requiring 8 class hours to complete. Now, we have successfully reduced the course duration to 4 h, helping students create more creative design drafts with greater ease.

AI-generated images have a higher level of completion and can assist students with weaker foundational skills in improving the visual quality of their work, making a significant contribution to the course. Compared to traditional methods, there has been a qualitative leap in terms of the completeness of the shapes, richness of the details, and the creativity of the concepts (Fig. 9.7).

Fig. 9.7 Comparison of traditional teaching and AI-enhanced artwork

9.4 Innovations

Stable Diffusion was open sourced in August 2022 and was immediately applied to teaching practice in September. This approach revolutionizes traditional teaching models by continuously stimulating students' intellectual development through human-computer interaction. This approach effectively fosters students' creative thinking, accelerates the creative process. The program is dedicated to nurturing character and virtue, combining technology, art, and humanistic ideas to cultivate art talent that is proficient in advanced techniques, forward-thinking, and solid foundational skills.

9.5 Promotion

This study offers a pioneering case and provides new ideas for innovative practice. Teachers have fully utilized the advantages of both advanced technology and traditional teaching methods to create a deep integration, making valuable explorations in the field of "Artificial Intelligence + Education." This innovation offers new directions and solutions for educational theory.

By bringing cutting-edge technology into basic education, we conduct research on methods, pathways, processes, and mechanisms to maximize the promotion of students' sustained development. This approach not only offers a seamless connection to higher education but also lays a solid foundation for future careers.

This research integrates AIGC with art education, helping to break down disciplinary boundaries, broaden students' perspectives, and develop their comprehensive abilities, thereby accelerating the cultivation of artistic talent. The course aligns with the goals of the art entrance exam reform, which will include creative assessments with a difficulty level similar to that of school-based exams. Consequently, this course is no longer limited to students intending to take school-specific exams but is now open to all high school art students. According to incomplete statistics, over 600,000 students take the art entrance exams each year, with approximately 4,000 art exam candidates annually in Beijing alone. Therefore, this research holds significant promotional value for educational authorities, decision-makers, relevant schools, and individuals.

In future educational practice, we will deeply explore AIGC technology, leveraging the power of technology to broaden the boundaries of education and ignite the spark of innovation and practice in students. We firmly believe that technology empowerment will bring unprecedented learning experiences to students, guiding them to achieve more remarkable accomplishments in art and across multiple disciplines.

9.6 Summary

9.6.1 Research Objective

This project explored the in-depth integration of AIGC technology into art education, addressing challenges in traditional teaching, such as difficulty in stimulating creativity, slow skill improvement, and insufficient interdisciplinary integration, and promote the digital transformation of art education.

9.6.2 Core Value Discoveries

Teaching Model Innovation: Constructed a "human–machine co-creation" model that significantly enhances students' innovative capabilities and substantially increases the proportion of innovative works.

- **Curriculum System Optimization**: Developed a three-dimensional curriculum of "traditional-digital-interdisciplinary", which integrated elements of technology and humanities, effectively breaking down disciplinary boundaries to break disciplinary barriers.
- **Evaluation System Upgrade**: A multiple-intelligences evaluation system was established to cover the full learning process, providing accurate and individualized feedback through AI.
- **Significant Practical Achievements**: Teachers' and students' works were selected for national textbooks, the proportion of high-achieving students in studio assessments increased, and the school emerged as a distinctive benchmark in the field.

9.6.3 Significance for Educational Theory and Practice

- **Theoretical Level**: The study validated the relevance of constructivist theory in the context of digital and intelligent learning environment and expanded the evaluation dimensions of the multiple intelligence theory.
- **Practical Level**: The project provided a replicable paradigm for integrating AIGC with aesthetic education to support the transformation of teachers into creative facilitators, and contribute to a more balanced distribution of educational resources. It helped to promote the transformation of teachers' roles into creative guides, and facilitate the balance of educational resources.

9.6.4 Recommendations

Practical Application: Strengthen teachers' AIGC technology training, establish interdisciplinary collaboration mechanisms, and formulate data security and ethical norms.

9.6.5 Future Research

Emerging Technology Integration: With the continuous evolution of AIGC (e.g., AI-generated video/music), future studies should explore multimodal generative models to enable cross-disciplinary artistic fusion and AI-powered “image-to-text” evaluation systems.

- **Dynamic Content Frameworks**: Develop synchronized update mechanisms for teaching materials to adapt to rapid technological advancements, ensuring pedagogical relevance.
- **Extended Applications**: Conduct long-term tracking on AIGC’s impact on professional literacy, and expand integrated applications with VR/AR and other immersive technologies.

References

Fathoni, A. (2023). Leveraging generative AI solutions in art and design education: Bridging sustainable creativity and fostering academic integrity for innovative society. In *E3S web of conferences* (Vol. 426). https://doi.org/10.1051/e3sconf/202342601102

Habib, S., Vogel, T., Anli, X., & Thorne, E. (2024). How does generative artificial intelligence impact student creativity? *Journal of Creativity*, *34*(1), 100072. https://doi.org/10.1016/j.yjoc.2023.100072

Chapter 10
Food Museum: Curriculum and Venue Development for Food Education in Kindergartens

Yanling Wang, Cong Qian, Tingting Ye, Kaka Xiang, Xiaoqing Zhang, and Tingting Lin

Abstract Building on years of practical research from previous kindergarten nutrition and health education pilot project, this project aims to create a space for children's active exploration and diversified growth through a comprehensive food museum curriculum. This initiative relies on the development of the "Food Museum" software and hardware, as well as other digital resources, to address typical challenges in advanced kindergarten nutrition and health education, such as general objectives, fragmented content, and limited resources. The approach of food museum curriculum addresses these challenges by advanced nutrition and health education in kindergartens of the advanced stage, aiming to meet the needs of cultivating children's nutrition and health literacy. Specifically, this project constructs a comprehensive space, where exhibits are knowledge and children can learn while visiting. The food museum is where food education happens. In essence, this study has pioneered a novel method to enhance children's nutrition and health literacy, using "Food Museum" as the primary activity carrier.

Keywords Kindergarten learning · Food education · Food museum curriculum · Venue development · Nutrition and health literacy

10.1 Background

Healthy body, nutrition first. Since Binhai No.1 Kindergarten became one of the first pilot schools for students' nutrition and health education in Wenzhou in 2015, it has been devoted to building a "Food Museum" and has conducted plentiful research in pilot projects for creating "theme venues" focused on nutrition and health. Ideal results have been achieved through the long-term experience of the Food Nurturing

Y. Wang (✉) · C. Qian · T. Ye · K. Xiang · X. Zhang · T. Lin
Wenzhou Economic and Technological Development Zone Binhai No.1 Kindergarten, Wenzhou, Zhejiang Province, China
e-mail: 171305388@qq.com

D. Liu et al. (eds.), *Empowering Educational Innovation Through Design: Volume One*, Smart Computing and Intelligence, https://doi.org/10.1007/978-981-95-4621-3_10

Education Research Institute and the concept of children's museum as an educational form. Focusing on the overall improvement of children's nutrition and health literacy and aiming to promote curriculum reform, our kindergarten began constructing the Food Museum, a theme venue for food education in 2019. In the early stages, by developing a set of systematic food museum courses and relying on the software and hardware of the Food Museum and the construction of digital resources, we created a brand-new strategy for promoting children's food education. This initiative provides a space for children to actively explore and grow in multiple ways. Through themed activities and experiential programs inside and outside the museum, children can acquire nutrition and health knowledge via rich curriculum resources and independent venue activities and games. By interacting with the environment, curriculum, teachers, and peers, the "Food Museum" can serve as a significant medium for improving children's nutrition and health literacy.

"Food Museum" refers to utilizing the kindergarten's site resources and leveraging the achievements from years of research and development of relevant courses and supporting teaching resources. We created spaces with various functions such as food display, food cooking, and food culture experiences. In the free-to-play space, children can initiate learning behaviors through interaction with the objects. Learning by doing and learning through play can help children enrich their life experiences, acquire nutrition and health knowledge, establish nutritious diet attitudes, and develop good eating habits. Through the optimization of the existing curriculum system of Food Museum, and the revision of the development guideline for children's nutrition and health literacy, we have enhanced the three major systems: education goal, content, and evaluation. The Food Museum is established as the main body for curriculum implementation. Based on the optimized curriculum content, we extend the themes and exhibition content of the theme exhibition area and discuss the education methods used by the Food Museum, forming a replicable model for food museum construction, as shown in Fig. 10.1.

10.2 Design Process

10.2.1 Implementation of a Good Curriculum Content System

The "Dietary guidelines for Chinese residents" points out that "school-age children should actively learn nutrition and health knowledge, inherit China's unique dietary culture and etiquette, improve nutrition and health literacy, recognize food, participate in food selection and cooking, and develop healthy eating behaviors" (Chinese Society of Nutrition, 2016). Therefore, according to the food pyramid, the kindergarten divided ingredients into seven categories: fruits and vegetables, cereals, milk and dairy products, potatoes, nuts, beans, and fish and eggs. After experts' discussions and reviews, the kindergarten developed eight stories with topics such as "Happy

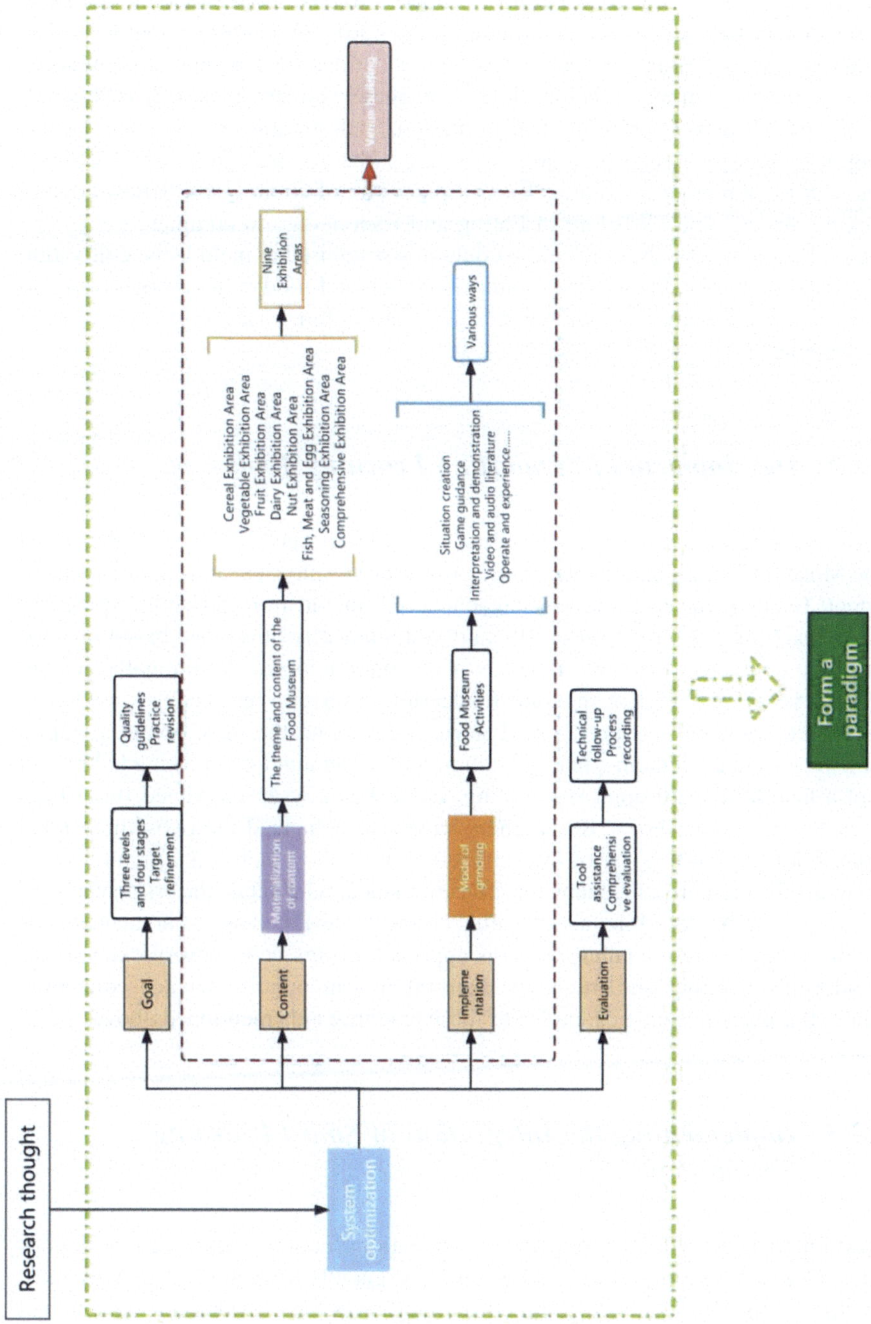

Fig. 10.1 Food museum construction model

Fruit," "Come on, Millet," "Exploration of Potato Kingdom," "World of the Meng Bean Family," and "Vegetable Family," respectively. We hope students can build healthy dietary habits and correct attitude toward eating by learning about food. The Dietary Guide for China provides six recommendations for the general population. However, the previous eight themes did not adequately cover topics such as "less salt and less oil," "sugar control and salt limitation," "eliminating waste," and "developing new dietary habits." To address these gaps, the research group has developed three additional themes, including "Little Spices Are of Great Use," "Knowledge of Drinking Water," and "Wisdom in Eating and Exercising" to complete the course content. Based on this, the curriculum content is structured to include the cultivation of good eating habits, knowledge of nutritious diet, and food culture, organized into eleven classified themes. A number of micro-curriculum cases were derived from this structure.

10.2.2 Development of Supporting Learning Tools

"Nutrition and Health" self-made teaching aids are important resources in the "Food Expo Museum," significantly contributing to children's physical and cognitive development. To tackle the issue of limited teaching aids in nutrition and health education, custom teaching aids are purposefully and systematically developed based on children's age characteristics and the needs of thematic courses. Additionally, picture books serve as an excellent medium for learning nutrition and health knowledge. The kindergarten will create targeted stories addressing the most common eating problems within each theme, helping children to understand good eating habits and adopt a healthy diet through storytelling. Each theme is accompanied by at least one matching picture book. Researchers have also enhanced the two-dimensional picture books by adding various designs creating three-dimensional picture books with multiple interactive mechanisms. These enhancements retain the original advantages of picture books while incorporating carefully designed operable mechanisms to address key content challenges. This approach brings new sensory experiences to children's learning and increases their interest in popular science education. Figure 10.2 presents some examples of these teaching aids mentioned above.

10.2.3 Implementing the Integration of Smart Learning Technology

This study focuses on the concretization of educational content within the nine themes of the "Cereal Museum," "Potato Museum," "Vegetable Museum," "Nut Museum," and "Meat Museum," as well as the educational objectives, evaluation system, and implementation subjects of the "Food Museum" course. This involves integrating the

Fig. 10.2 Examples of teaching aids

exhibition themes, content, and methods across the nine exhibition areas of the "Food Museum." The challenge lies in enhancing the role of the "Food Museum" in course learning through activities such as context creation, game guidance, explanations and demonstrations, audio-visual literature, and operational experience. This represents the most critical and challenging aspect of integrating smart learning technology.

10.2.4 Implementing Cloud Sharing of Curriculum Resources

We developed the "Food Museum" app to make the resources of the physical museum available on the cloud, so that children, parents, and teachers can enjoy a mobile version of the "Food Museum." See Fig. 10.3 for the Cloud Food Museum resources. Users can access the resources through three sections: "Teachers' World," "Parents' World," and "Children's World." Take "Children's World" as an example, after parents install the app, they can choose a virtual character image for their children based on prompts, and the app will provide age-appropriate themed picture books and teaching aids according to the user's initial age.

Children can browse independently in a safe and healthy environment. Each time they browse some content, they earn gold coins, which can be exchanged for virtual medals after accumulation. In addition to the default display of each medal, children can add stickers from the material library to decorate their medals according to their preferences and make each medal unique. The medals can be used as decorations for the character image, enhancing the fun. The gamified format allows children to learn nutrition knowledge in a pleasant and relaxed atmosphere. Additionally, in the "Content" section, clicking on each concept map of the venue links to the thematic curriculum content for that topic.

Teachers can access complete teaching plans and relevant resources, while parents can obtain parenting books and teaching materials tailored to their children's age groups. The "community" feature facilitates communication among app users through messages, questions, and answers, so as to enable ubiquitous learning anytime and anywhere.

10.3 Implementation Outcomes

The "Food Museum" has become a core resource and a pivotal medium for kindergarten nutrition and health education, which has brought far-reaching effects to our school. As direct beneficiaries, children have not only improved their nutritional literacy and eating habits but also developed a strong interest in learning about nutrition and health and a desire for active investigation. More importantly, children's physical fitness has been improving year by year.

Fig. 10.3 Cloud food museum resources

After 5-year practical research, we have developed a specialized kindergarten curriculum focused on "nutrition and health" education. We have implemented teaching activities of nutrition and health education, involving all the teachers and students to participate in creating a specialized environment for nutrition and health education, and established a nutrition education practice theme library. As a result, we have achieved two significant outcomes.

10.3.1 Development of the Teacher Education

Throughout the project research, teachers' scientific research abilities and educational innovation levels have been significantly improved. Thirteen teachers have shared and exchanged ideas in municipal and district-level lectures, six teachers have presented public lessons at these levels, and two teachers have created picture books published by Zhejiang Education Publishing House. Three projects have been established in the city and have received awards, while one project has been recognized at the provincial level. Additionally, seven teaching aids have successfully applied national copyright patents, including two design patents. Furthermore, 15 individuals received awards in the Wenzhou "Little Nutritionist" mini-course collection. These outcomes are shown in Fig. 10.4.

10.3.2 Promotion of the Brand Characteristics

Since becoming a pilot school in 2015, the construction of the "Food Museum" has been a key project in the kindergarten's development planning over two cycles. Through years of persistence and practice, the kindergarten has remained true to its original aspiration, steadfast in research, and vigorous in innovation. After years of accumulation, the kindergarten's food education course has been recognized as an excellent course by the Wenzhou Development Zone and has been invited to showcase its achievements at the 2nd, 3rd, 4th and 5th Student Nutrition Festivals in Wenzhou for four consecutive years.

In 2021, the kindergarten was invited by Wenzhou University to showcase its achievements at the first Service Basic Education Achievements Exhibition. Recently, the achievements of this project were also presented at the 6th China Education Innovation Achievements Public Welfare Expo. The kindergarten has developed distinct brand characteristics. With its influence continually expanding, nutrition and health education has gradually gained attention. Along with recognition and praise, it has also garnered motivation and confidence for further innovation.

Fig. 10.4 Examples of food museum project outcome

10.4 Promotion

In 2018, the "Paper" Museum of Nanjing No.1 Kindergarten was opened, which has attracted widespread attention and praise, and has become a model for venue construction (Ge, 2018). The Children's Museum creates a space for active exploration and diverse growth. Here, children learn by starting from their life experiences and exploring the science, nature, and humanities of life together, so as to deepen their understanding and become active learners. It functions both as a venue and as a course.

Recognizing the advantages of children's museums in terms of freedom, autonomy and diversity compared to traditional teaching mode, Bin Yiren, who is full of creativity, envisioned creating a tailored "Food Museum" for children. She actively sought breakthrough points and footholds for nutrition and health education in kindergarten. We have not only developed a comprehensive system of food and education courses but also created various peripheral products such as picture books, teaching toys, resources for co-education at home, and materials for teachers' education and instruction.

While integrating advanced information technology into the construction of physical venues, we conceptualized the "Food Museum" and transformed it from being confined to a specific site into a comprehensive resource package, presented in an electronic virtual form. The focus is not only on the museum's style and furnishings but also on the collection of resources within it and the evolution of children's learning styles. This approach enhances food education at kindergarten and expands the educational venue to a broader audience.

This new educational model of constructing courses, developing teaching aids resources, building special venues, and sharing resources in the cloud, can promote the educational reform and provide a set of excellent templates that can be used for reference, imitation and replication in future educational developments.

10.5 Conclusion

As a core resource and significant medium for nutrition and health education in kindergarten, the construction of "Food Museum" has had a far-reaching influence on children's nutrition and health education. Through years of practice in kindergartens, we have developed nutrition literacy guidance that emphasizes knowledge and practice. We have integrated different fields to construct a comprehensive curriculum system for food and education. By prioritizing research and learning, we have formed an effective curriculum implementation strategy. Additionally, we have incorporated technology to create a model for venue construction. Based on this new educational model of constructing courses, developing teaching aid resources, building themed venues, and sharing resources in the cloud, we have promoted educational reform and provided a set of excellent templates that can be referenced, imitated, and replicated for future educational development.

Based on the practical research achievements of the previous case "Nutrition and Health Education Based on Thematic Venues" in kindergartens, this study identified the typical challenges faced at the advanced stage of kindergarten nutrition and health education, such as general goals, scattered contents, and limited resources, as well as the growing need to cultivate children's nutrition and health literacy. It explored how to construct a comprehensive "Food Museum" curriculum through the development of the "Food Museum" itself, aiming to achieve the research goals of standardizing evaluation, systematizing the curriculum, and materializing and transforming resource achievements. The case developed a Guide for Children's Nutrition and Health Literacy, constructed a curriculum system for the "Food Museum," identified curriculum implementation strategies, and built the "Food Museum" as a learning space. It also further optimized the content of kindergarten nutrition and health education, introduced an innovative model, and improved the nutrition and health awareness of parents and teachers, as well as children's nutrition and health literacy. Moreover, it cultivated teachers with strong capabilities for educational innovation and helped establish a distinctive educational identity for the kindergarten. In the next step, it would aim to develop a "Mobile Food Museum," extend the food education resources beyond the kindergarten settings, implement innovative practice of borderless teaching, and let more audiences to benefit from the outcomes of food education.

References

Chinese Society of Nutrition. (2016). *Dietary guidelines for school-age children in China*. People's Health Publishing House.

Ge, L. D. (2018). Make "paper" by yourself Nanjing built the first municipal children's museum. *Jiao Hui Dian*. https://jnews.xhby.net/waparticles/59/DMSuDVn50l7XdZGb/1

Chapter 11
Crafting and Deploying a Metaverse Curriculum Leveraging the WorldWide Telescope

Lei Liang, Yingchen Mao, Chenzhou Cui, Shanshan Li, Yuexin Wang, and Xuemin Zhang

Abstract With the gradual maturity of virtual reality and mixed reality technology, online education is transitioning from a two-dimensional plane to a three-dimensional space, from a third-person one-way perspective outside the screen to a first-person surround-view inside the screen, and from sequential guidance to comprehensive independent exploration. The educational metaverse will cater to this development and provide immersive online learning scenarios for both students and teachers. The metaverse curriculum needs to take into account the scientific nature, interactivity, and experiential learning, putting forward new requirements and challenges for its development and application.

Keywords Educational metaverse · WorldWide Telescope · Curriculum design and development · Virtual reality · Mixed reality · Augmented reality

L. Liang (✉)
Dalian No. 20 High School, Dalian, China
e-mail: ray13jing@126.com

Y. Mao
School of Physics and Electronic Technology, Liaoning Normal University, Dalian, China
e-mail: myc@lnnu.edu.cn

C. Cui · S. Li · X. Zhang
National Astronomical Observatories, Chinese Academy of Sciences, Beijing, China
e-mail: ccz@nao.cas.cn

S. Li
e-mail: lishanshan@nao.cas.cn

Y. Wang (✉)
Lingyuan No. 2 High School, Lingyuan, China
e-mail: lnnu64@126.com

D. Liu et al. (eds.), *Empowering Educational Innovation Through Design: Volume One*, Smart Computing and Intelligence, https://doi.org/10.1007/978-981-95-4621-3_11

11.1 Introduction

The Metaverse represents the latest stage in the development of visual immersion technology and is essentially an online digital space parallel to the real world (Patel, 2017). The educational metaverse can be viewed as the application of the metaverse in education. It propels online education into the innovative development stage of experiential learning and immersive interaction. According to the Department of Basic Education of the South African Government (DBE, 2015), this will break the bottleneck of online education development, which has been dominated by a two-dimensional plane. Although domestic Internet giants have ventured into the metaverse, their focus remains on commercial applications rather than the educational Metaverse. Therefore, in order to realize the educational metaverse, we should actively choose a well-developed, open-source big data platform that can create virtual reality environments and is maintained by professional institutions.

This case explores the development and application of a metaverse course using the scientific big data platform, WorldWide Telescope (WWT). We hope this case will encourage and inspire more teams to use WWT to develop and create more courses in astronomy, physics, geography and other subjects, finally realizing online teaching in a virtual reality environment through external devices (see Fig. 11.1).

11.2 WWT—Developing Platform with Scientific Big Data

WWT provides extensive real astronomical data from global observatories, telescopes, and observation projects based on advanced Internet, data visualization, and virtual reality technologies. Additionally, it has powerful data visualization capabilities to ensure that users can easily browse the latest astronomical data and related information. Managed by the American Astronomical Society, WWT is a free, fully open-source scientific big data software. The Chinese Virtual Observatory (China-VO) team conducted in-depth research on it and developed a version more suitable for the application of science education.

After more than ten years of refinement and updates, WWT now supports full Chinese operation mode, adheres to the latest data standards, integrates extensive astronomical and geographical data, and offers high scalability. It can connect various equipment, such as ball-screen projection, large arc screens, virtual reality/augmented reality devices, and gesture-interaction capture devices. This enables seamless connection of multiple operation modes and methods, creating diverse immersive universe environments (see Fig. 11.2). It can also provide users with immersive experiences in different scenes and meet the basic needs of the educational metaverse platform.

Fig. 11.1 User interface of the WWT

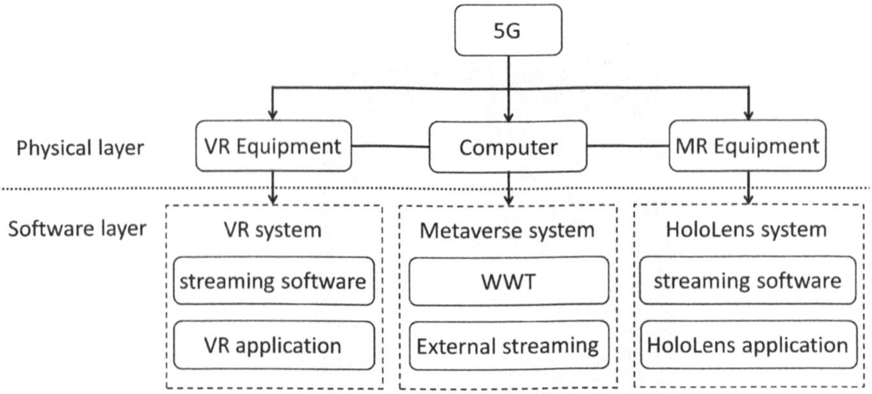

Fig. 11.2 Construction of educational metaverse platform with WWT

11.3 Design of Metaverse Curriculums Using WWT

Relying on the unique "tour" production function of WWT, teachers can easily obtain data through simple operations and then freely develop and design activities to produce courseware and develop curriculum. Besides that, the WWT has the advantage of integrating platform data through external equipment, which allows teachers to acquire images, model, and create a realistic picture of the universe by "perceiving" the user's surrounding environment and terrain.

Teachers can use virtual reality (VR), augmented reality (AR), and other technologies to superimpose projected virtual images and 3D models onto the real environment. This helps create a multi-dimensional interactive environment based on real data and prepares interactive experience with gestures. Furthermore, teachers can also combine sound, video, and other interactive behaviors to design and develop educational metaverse curricula, create situational learning environments from different perspectives, and provide learners with an all-around immersive learning experience. Next, teachers can release learning tasks through the resource sharing function and students can complete online learning using external devices, such as VR headsets and HoloLens wearable devices that support mixed reality. Using WWT, two kinds of metaverse curriculums can be realized.

- **The "Otherness" Curriculum**: Based on the view of the spectator, this type of course allows learners to engage in experiential learning and deepen their impressions through a "tour" approach.
- **The "Ontology" Curriculum**: Based on the view of the learner, this type of course enables learners to engage in experiential learning and explore independently through an "immersion" approach (see Fig. 11.3).

Next, let's take "Space Voyage" as an example, as shown in Table 11.1, which is an important context in the course of "Interactive Teaching of High School Physics." The first step is to produce the main interface of knowledge points using WWT.

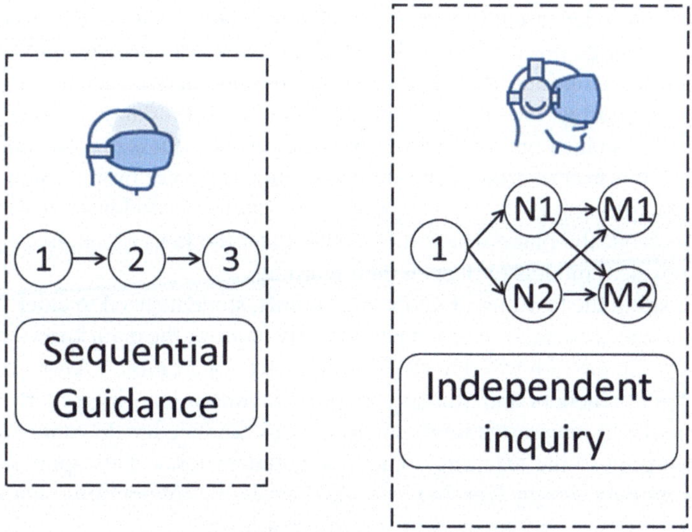

Fig. 11.3 Course form of Metaverse based on WWT

Table 11.1 The knowledge points, teaching formats, contents and learning methods for "Space Voyage", respectively

Knowledge	Format	Content	Method
Satellite orbit	Independent inquiry	Observe the motion of the artificial earth satellite and preliminarily understand the motion characteristics and orbit characteristics of the satellite	Knowledge detection
Newton's conjecture	Sequential guidance	Understand Newton's establishment of artificial satellite model	Knowledge detection
Gravity	Sequential guidance	Establish the formula of universal gravitation and centripetal force, deduce and master the expression of linear velocity, angular velocity, centripetal acceleration and period of artificial satellite motion	Knowledge detection
Satellite orbit works	Independent inquiry	Understand the law of the orbit of the satellite, and extend to conclude that the orbit law of the planet is the same as the orbit law of the satellite	Knowledge detection
Gold substitution and cosmic velocity	Independent inquiry	Deduce and master the gold substitution, understand the cosmic velocity	Knowledge detection
Near earth satellite and geostationary satellite	Independent inquiry	Understand the basic parameters of near-earth satellite and artificial satellite operation	Knowledge detection

Students can independently select knowledge points for learning according to their own knowledge mastery and learning ability. For example, by clicking "satellite orbit" in the main interface, they can enter the interactive tour interface where artificial satellite orbits are created using real satellite orbit data, providing immersive visual effects (see Figs. 11.4 and 11.5). After the content is played, students can operate the WWT user interface according to subtitles or voice prompts and carry out independent exploration. Through the interactive platform, students can observe the shape of the orbit, the plane where it is located, and the laws of satellite operation, using real physical models to deepen their learning.

After finishing the learning of knowledge points, students need to enter the corresponding knowledge detection link. They can only return to the main knowledge interface and start the next knowledge point if all answers are correct. After completing all the learning, teachers and students jointly generate a learning report. Teachers can adjust and edit interactive tours, micro-classes, knowledge detection, and other content according to the teaching content and student level. This approach allows them to accurately assign learning tasks and promote collaborative learning and synchronous progress among students of different levels.

11.4 Metaverse Curriculum Development Based on WWT

At present, the developed metaverse curriculums cover astronomy, physics, geography, and other disciplines. More courses are being developed using WWT to create an educational metaverse. The following three types of curriculum systems have been established.

11.4.1 Spacecraft in the Digital Universe

In this type of curriculum, there are three kinds of tours, including space telescopes, space exploration, and Chinese aerospace (see Table 11.2).

11.4.2 Interactive Curriculum in a High School Physics Course

This course is developed with reference to the teaching content of the high school physics textbook published by PEP and can be applied to high school physics teaching. The content focuses on chapters related to gravity and aerospace (see Table 11.3).

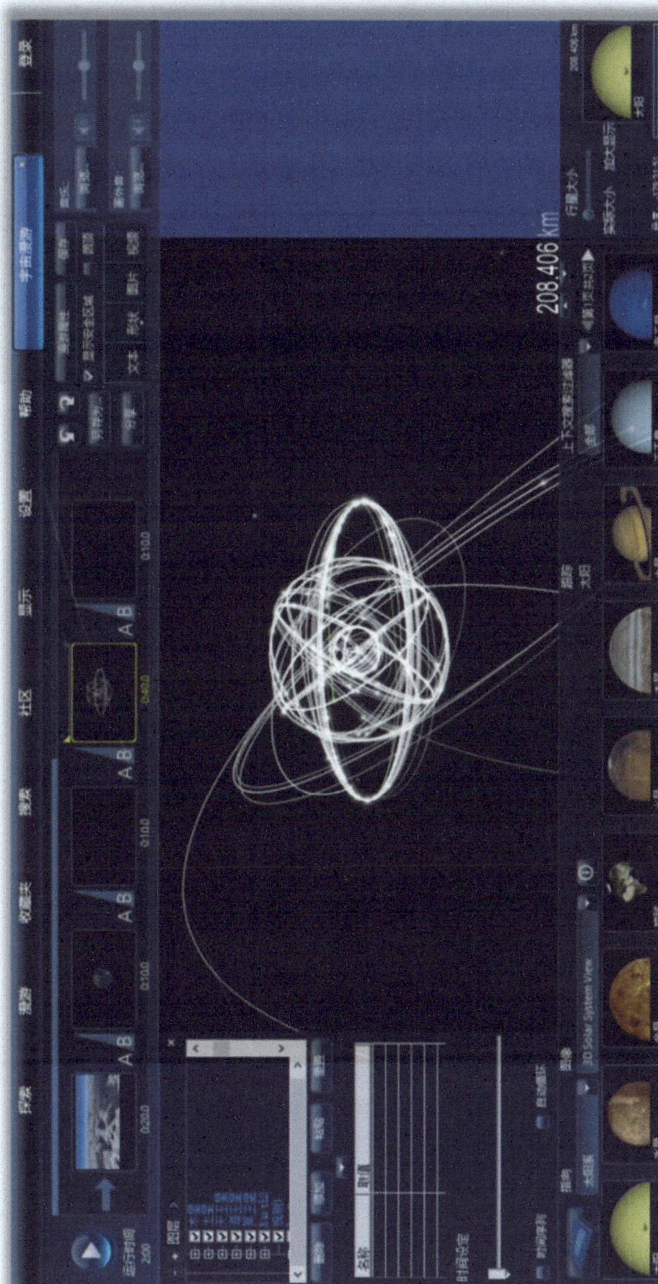

Fig. 11.4 Course Catalogue of "Space Voyage"

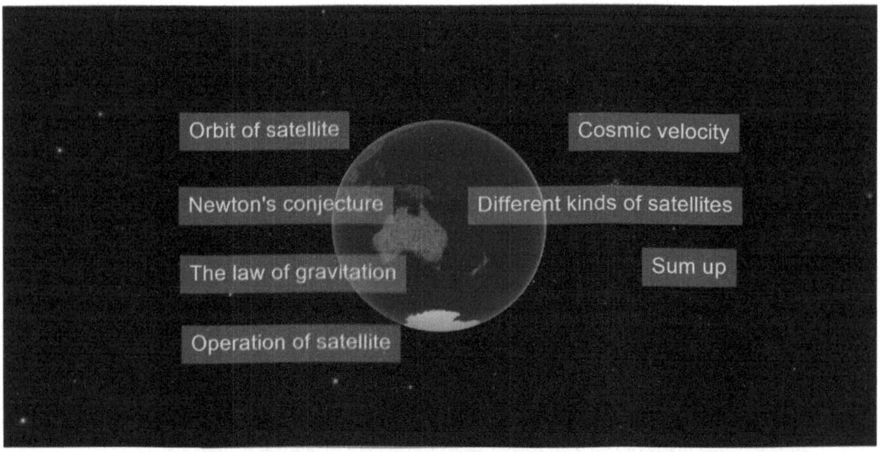

Fig. 11.5 The metaverse curriculum of "Space Voyage" using the WWT

Table 11.2 The curriculum structure of "spacecraft in the digital universe"

Space telescope	Space exploration	Chinese aerospace
Hubble space telescope	Voyager	Dongfanghong-1
Spitzer space telescope	Children of planet earth	Tiangong-2
	Bye cassini	Chang'e-4
	Starlink	

11.4.3 FAST in Sight: Chinese Major Scientific Device Demonstration Works

Relying on the panoramic resources of WWT and combined with hybrid reality technology, the artwork "FAST in Sight" can transport the FAST radio telescope from Qiannan Autonomous Prefecture, Guizhou Province, directly to students. This allows them to personally experience the magnificence of the world's largest single aperture radio telescope, self-designed and built by China. This artwork showcases

Table 11.3 The curriculum structure of "high school physics interactive curriculum"

High school physics interactive curriculum	
Planetary motion	Completed
Space Voyage	Completed
Orbit change of satellite	Completed
Cassini	Completed
Binary star	To develop
Black hole	To develop
Limitations of classical mechanics	To develop

the miracles created by the combination of technology, engineering, and nature. It guides students to understand the science and technology innovations contained in the FAST project, as well as its scientific objectives, through various interactive methods (see Figs. 11.6 and 11.7).

11.5 Application of the Metaverse Curriculums

The only worldwide astronomy classroom was built in Lingyuan No. 2 high school in Liaoning Province. It was established jointly by the National Astronomical Observations of China (NAOC) and the Chinese Astronomical Society (CAS) as the WWT astronomy teaching base and the popularization education base for the China-VO and CAS's Information Working Committee (IWC-CAS). At the front of the classroom, a curved hard screen and three special projectors are fused to provide a 120° visual angle. A dual-screen display can be achieved, with the "big screen" showing the images from WWT and the "small screen" (Oculus or HoloLens) providing students with VR or MR experiences. Data can be exchanged between the two screens, and users can synchronously control them through gesture operation (see Fig. 11.8).

Lingyuan No. 2 high school has collaborated with scientific research institutions and universities to build a joint-action mechanism for science and education. It has also established a science popularization studio named Star Horizon. Based on the scientific data released by the NAOC and the National Science Data Center, the

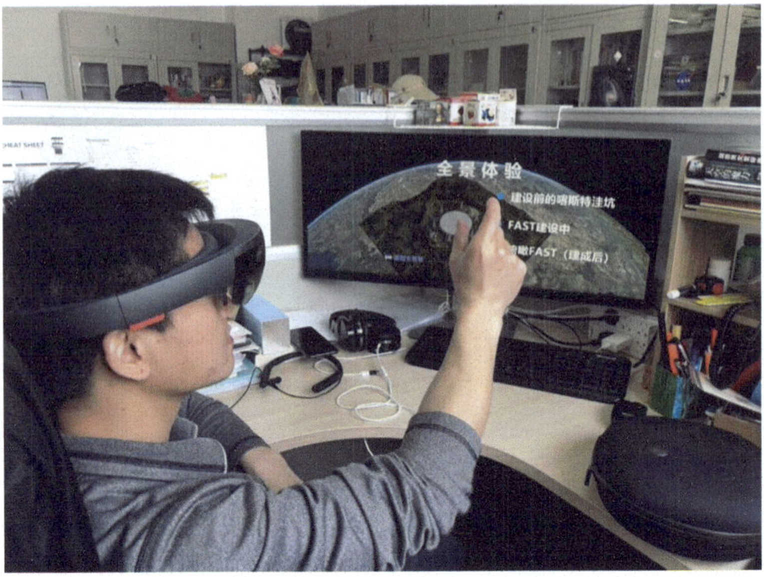

Fig. 11.6 The third-person perspective of Demo screen of "Fast in front"

Fig. 11.7 The first-person perspective of Demo screen of Fast in front

studio adopted suggestions from relevant experts at the IWC-CAS and Liaoning Normal University. It implemented specific resource transformation and application, providing "one-stop" services for the collection, collation, transformation, and application of integrated scientific data (see Fig. 11.9).

Fig. 11.8 Worldwide astronomy classroom of Lingyuan No. 2 High School

Fig. 11.9 The joint-action mechanism of science and education

11.6 Promotion

Since WWT was introduced into China in 2008, the China-VO team has been continuously developing, updating and maintaining the platform for more than 10 years. To promote WWT, the China-VO has carried out numerous activities such as teacher training and the Universe Tour Contest and has also created an astronomy curriculum system. The WorldWide Astronomical Hall, the WorldWide Astronomical Classroom, and other infrastructure facilities have also been built in universities, primary and secondary schools, and science and technology museums across the country. This work has provided a solid foundation to design and develop metaverse curriculums using WWT, as well as strong support for the promotion of these curriculums. At present, the data-driven astronomy interactive teaching based on WWT has been recognized both internationally and domestically, is widely known in the astronomical community, and has gained good popularity (see Tables 11.4 and 11.5).

The metaverse curriculums developed using WWT can be directly shared with schools and science and technology museums equipped with the Worldwide Astronomical Hall or Classroom, as well as supporting interactive equipment. This

Table 11.4 Worldwide astronomy classroom and worldwide astronomical hall in China

List of worldwide astronomy classroom	List of worldwide astronomical hall
Daqing Experimental Middle School	Shixinlu Primary School
Lingyuan No. 2 High School	Beijing Normal University
Lijiang Teachers College	Middle School Affiliated to South China Normal University
Sanya Middle School affiliated to Renmin University of China	Shahe Station, NAOC
Bishan Middle School	Museum of Hebei Normal University
Lize Primary school	Zunyi Aerospace High School
Xiamen University	Sheshan Astronomy Museum, Shanghai
Shanghai Minhang District No. 3 Middle School	Zhangjiawan Primary School
	China West Normal University
	Bishan Middle School
	Hangzhou High School
	Lhasa Astronomical Science Museum
	Hangzhou Jingzhou Primary School
	Middle School Affiliated to Southwest University
	Hangzhou Fengtan Middle School

Table 11.5 The WWT national teacher training and the WWT tour contest

WWT national teacher training		WWT touring creation competition	
Time	Place	Sequence	Co-organizer
2010	Beijing	1st	Beijing Planetarium
2011	Hohhot		
2011	Beijing		
2012	Xinjiang		
2012	Beijing		
2014	Chongqing		
2015	Beijing	2nd	Beijing Normal University
2016	Daqing		
2017	Shijiazhuang	3rd	Beijing Science and Technology Museum
2018	Qingdao		
2019	Hangzhou	4th	Guangdong Experimental Middle School
2021	Chongqing	5th	Shanghai Planetarium

approach helps build a teaching network under multiple application scenarios, and construct a metaverse space. Some applications have received extensive attention from society and the media, such as the signing of a memorandum of cooperation between the International Virtual Observatory Alliance (IVOA) and the Office of Astronomy for Development (OAD) of the International Astronomical Union (IAU) in 2021.

The bilingual promotion of data-driven popularization of astronomical science has been reported by Xinhua News Agency and other media. The publicity and reports from central media and other major outlets, together with regular activities such as continuous teacher Training and Tour Contest, are set to further expand the scope of the joint action between science and education. This is expected to attract more universities, primary and secondary schools across the country to participate in the development of the metaverse curriculum, thereby accelerating the promotion of these curriculums (see Figs. 11.10 and 11.11).

Fig. 11.10 WWT national teacher training and WWT tour contest

Fig. 11.11 Data-driven astronomy science education advocated by IAU

11.7 Summary

11.7.1 Objective of the Case

This case explored the development and application of an educational metaverse curriculum based on the scientific big data platform WorldWide Telescope (WWT). By integrating VR/MR technologies with global astronomical/geographical datasets, the case aimed to construct an immersive and interactive online learning environment. It showed the limitations of traditional two-dimensional instructional formats and provided new teaching approaches for disciplines such as astronomy, physics, and geography.

11.7.2 Core Value Discoveries

- **Technical Platform Adaptability**: As an open-source platform, WWT integrated data from global observatories and supported full Chinese-language across multiple devices. Its "tour" function enabled teachers to develop two types of courses using real data: the "Otherness" curriculum, which showed spectator-view experiential learning, and the "Ontology" curriculum, which supported learner-view independent exploration).
- **Innovative Teaching Model**: Shifting from one-way instruction to two-way "operation-feedback" interaction model, the "Space Voyage" course served as an example where students constructed knowledge by observing real data and conducting virtual experiments. This approach strengthened deep learning.
- **Educational Ecosystem Collaboration**: Schools like Lingyuan No. 2 High School built "Worldwide Astronomy Classrooms" in collaboration with research institutions, forming a closed loop of "data collection-course development-teaching practice." They promoted resource sharing through teacher training and academic competitions.

11.7.3 Significance for Educational Theory and Practice

- **Theoretical Level**: It validated the Constructivist Learning Theory and Embodied Cognition Theory, expanding the interdisciplinary application boundaries of mixed reality education.
- **Practical Level**: It narrowed the urban–rural digital divide through the open-source WWT, achieved the rapid transformation of scientific research data into classroom resources, and provided a replicable model for the development of interdisciplinary courses.

11.7.4 Practical Suggestions and Future Research

- **Practical Applications**: It expanded horizontally to include disciplines such as biology and history, and vertically through the design of K-12 curricula with clear progression across grade levels.
- **Future Research**: It would help introduce AI to analyze learning data for personalized teaching and explored blockchain technology for learning outcome certification. It would help establish a "knowledge-skill-emotion" three-dimensional evaluation system, and address data privacy and virtual identity ethics in the metaverse.

References

Department of Basic Education (DBE), South African Government. (2015). *Action plan to 2019: Towards the realization of schooling 2030.* https://www.education.gov.za/Portals/0/Documents/Publications/Action%20Plan%202019.pdf

Patel, P., & Patel, N. (2017). ICT pedagogy for effective learning, education and quality evaluation. *International Journal of Computer & Mathematical Sciences IJCMS, 6*(5), 105–106.

Chapter 12
Magic Space: The Construction and Practical Study of a Comprehensive Learning Space in Schools

Yanping Lu, Chong Chen, Jing Yang, Sijia Li, Yishun Mo, and Yueyue Nie

Abstract Seven-Color Flower Elementary School, located in the heart of Shanghai's central district, makes the most of its highly valuable and limited space. As education reform continues, expanding the school's educational offerings through creative use of space has become essential for schools, especially in a global metropolis, with a strong focus on arts education. This case explores innovative transformations in learning spaces in the digital age. By redesigning the school's 400 m^2 indoor gymnasium, it showcases the many possibilities for small urban schools, often described as "making the most of tight spaces." The project effectively integrates space technology, curriculum, and people, aligning the physical, social, and mental dimensions of space. It also creates a future-forward model by linking the virtual and real worlds, developing a data-driven, open-resource ecosystem that meets the diverse needs of all learners.

Keywords Comprehensive learning space · Multidimensional integration · Technology empowerment · Interconnectedness · Educational innovation

12.1 Background

Seven-Color Flower Elementary School is located in the Huaihai Road commercial district in Huangpu District, Shanghai, with a building area of over 7,700 m^2. The school currently has 15 classes, accommodating more than 550 students. It has around 30 classrooms, including specialized classrooms, along with a 600 m^2 outdoor playground and a 400 m^2 indoor gymnasium. Although the average space per person is limited, substantial educational investment from the central district ensures that both hardware and software in various learning spaces are well-equipped. This enables effective curriculum implementation.

Y. Lu · C. Chen · J. Yang (✉) · S. Li · Y. Mo · Y. Nie
Seven-Color Flower Elementary School, Shanghai, China
e-mail: 39863077@qq.com

D. Liu et al. (eds.), *Empowering Educational Innovation Through Design: Volume One*, Smart Computing and Intelligence, https://doi.org/10.1007/978-981-95-4621-3_12

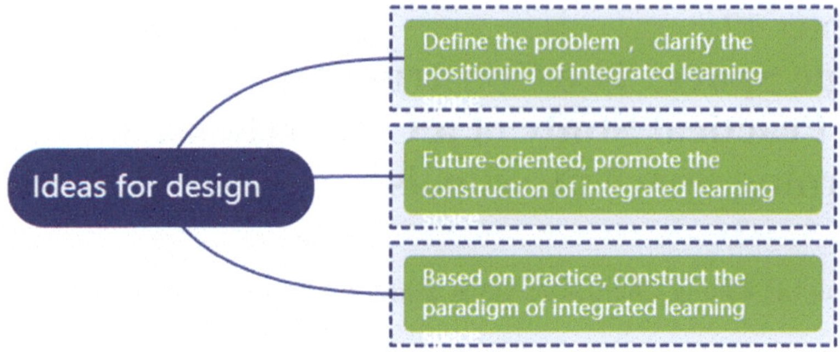

Fig. 12.1 Conceptual framework for building a comprehensive learning space

As education reforms like the "New Curriculum Standards and New Teaching Materials" "Innovative Education" and the "Integration of the Five Educations"[1] take hold, and as we move closer to an era of smart collaboration between humans and machines, interdisciplinary learning, and shared creativity, it becomes more apparent that our current learning spaces limit the development of future education and talent.

Looking ahead to 2050, how can we take proactive steps to prepare for the future? A few key questions are worth considering:

- Can we design a comprehensive learning space that meets the diverse needs of learners, moving beyond limited functionality and monotonous environments?
- Can we harness information technology to create a multidimensional learning environment that nurtures an intelligent social space for education?
- Can we optimize these spaces to fully support students' mental development while seamlessly integrating the five core educational areas?

12.2 Conceptual Framework

This case aims to develop a replicable, adaptable, and scalable model for constructing comprehensive learning spaces through action research that maximizes the potential of small spaces. As shown in Fig. 12.1, the conceptual framework follows three key principles: problem-oriented, innovative thinking, and sustainable development. The main focus of the design concept highlights the following:

- Identifying the issues and defining the role of the comprehensive learning space.
- Building a comprehensive learning space with a future-oriented approach.
- Grounding the design in practical application to establish a model for comprehensive learning spaces.

[1] Five Education Integration refers to the blending of moral education, intellectual education, physical education, aesthetic education, and labor education.

12.2.1 Understand and Define the Comprehensive Learning Space

The comprehensive learning space should be distinct from regular classrooms, specialized classrooms, playgrounds, or gymnasiums. Its primary goal is to maximize the learning experience for teachers and students, creating an immersive learning atmosphere and fostering shared growth (Kuhail et al., 2022). The space typically exhibits the following characteristics.

Teachers can rearrange the classroom tables and chairs based on learning needs, allowing students to sit on the floor in a circle or divide the space flexibly. For example, the front area can have tables and chairs for regular lessons, while the back area can remain open for subject-based learning activities, enabling versatile teaching methods.

This space not only supports students' learning of national curriculum subjects but also accommodates activities related to the school's four major aesthetic education projects, which are based on the "Integration of the Five Educations." Additionally, it serves as a venue for teacher development, promoting professional growth and moral education, and can host important school events.

Immersive technologies create a seamless blend between the physical world and a digital or virtual environment (Wagner & Liu, 2021). In this space, the walls and floor come to life, allowing both teachers and students to experience the joy of educational activities in an immersive environment that beautifully combines dynamic and static elements. The setting transforms single-sensory learning into multisensory experiences, shifts from one-dimensional to multi-dimensional understanding, and makes abstract concepts concrete, simplifying complexity.

The space should be enhanced with projection systems, lighting, and AI smart devices integrated into the ceiling, walls, and floors. Whether for learning or activities, the beauty of the environment, interaction, design, and layout should be amplified by technology, making the entire setting feel like a "Magic Space."

12.2.2 Establish the Path for Building a Comprehensive Learning Space

The construction of a comprehensive learning space is not something that can be achieved overnight. Whether it involves selecting a location, defining its functions, designing creatively, or implementing renovations, it is a long-term project aimed at future education. Its construction should be based on the school's specific conditions and should adapt to the flow of developments, representing a transformation from thought to action. The building of the school's "Magic Space" has gone through the following three stages.

The initial construction of the space was modest, considering the average class size of around 20 students at the time. We installed two projectors on the ceiling of the indoor gymnasium and laid a white mat, about 100 m^2 in size, on the floor as a projection area. The larger immersive environment, moving from wall-based projection to the floor, created a magical experience for the primary school students.

As class sizes gradually increased, we added four more projectors to the ceiling and expanded the mat to three times its original size, calling it the "Magic Carpet." The boundaries of the space were opened up, allowing for a wide variety of learning activities for different group sizes and formats. This change reflected an evolving concept of space that catered to the multifunctional needs of users, achieving gradual internal growth.

We pushed beyond traditional thinking to truly integrate curriculum learning, technology, and human interaction within the space. With the advanced technological equipment, we upgraded the space further by introducing two AI interactive magic walls. This allowed for two-dimensional interaction between the floor and walls, resulting in the successful creation of a panoramic and visualized comprehensive learning space.

12.2.3 Refine the Implementation Model for the Comprehensive Learning Space

The construction of the "Magic Space" is not only an action research project but also a sustainable development strategy grounded in educational philosophy and integrated though, aimed at transforming teaching and learning. The challenge lies in ensuring that this comprehensive learning space deeply embodies its core values in future education. This includes meeting curriculum needs, transforming teaching and learning methods, encouraging the joint development of teachers and students, and supporting the school's sustainable and distinctive growth. Our case provides strong validation of this approach.

12.2.3.1 Multidimensional Integration Environment Model

Learning spaces should be multimodal. Traditional classrooms or specialized areas support static learning, while the "Magic Space," with its intelligent walls and interactive floor, facilitates the integration of learning activities. It blends physical spaces with virtual ones, combining scenarios and data. Implementing multidimensional integration between formal and informal learning spaces on campus is the environmental model for future learning.

For example, in the English class "1B M1 In the Park." In the first period, the teacher uses the interactive walls to introduce language learning objectives, while the interactive floor creates real-life situations based on the lesson content. The

Fig. 12.2 Rhythmic golf exercise—scenario-based teaching

floor's partitioning feature divides the space into sections, allowing students to complete language practice tasks in groups. In the second period, formal language application and communication activities are held in the regular classroom, making English learning more effective with the support of the multidimensional learning environment (see Fig. 12.2).

12.2.3.2 Technology-Enabled Teaching Model

The Ministry of Education, in its Guiding Opinions on Promoting the Construction of New Educational Infrastructure to Build a High-Quality Education Support System, clearly states the goal to develop AI-based teaching applications, such as intelligent teaching assistants and learning companions, to achieve human-machine co-teaching and co-education and improve the quality of teaching. So, what should the teaching model in a comprehensive learning space look like?

We believe it should provide an immersive learning experience through the "Magic Carpet," where sound, light, and visuals continuously engage students in the learning process. At the same time, the "Magic Wall" should enable interactive projections, data feedback, learning assessments, and adjustments to teaching strategies, all driving improvements in teaching and learning methods.

Fig. 12.3 Physical education class with intelligent human–machine interaction system

For example, in the physical education lesson "Endurance Running," the teacher makes full use of the "Magic Carpet" to create immersive scenarios where students can enjoy running through lively alleyways with market sounds or strolling on a beach alongside sea turtles. The "Magic Wall" adds an interactive element, allowing students to use specially designed modules to time their runs while engaging with the wall during shuttle sprints. This setup encourages students to observe and interpret data before, during, and after their practice, using scientific methods to improve their pacing and running technique, helping them overcome key challenges in the lesson. Moreover, the "Magic Wall" includes a sports load monitoring system that provides real-time, comprehensive heart rate feedback throughout the activity. This not only helps teachers ensure the safety of students during physical exercises but also enhances the scientific and effective aspects of fitness training. This technology-empowered model exemplifies high-quality teaching that is not possible in a traditional classroom (see Fig. 12.3).

12.2.3.3 A Curriculum Model of Multidimensional Integration

An ideal learning space should be a platform that encourages students to explore, innovate, collaborate, and engage in hands-on practice. It should also serve as a comprehensive educational environment that integrates across all subjects, courses, and activities. For example, we effectively utilize a blend of virtual and physical spaces, fostering continuous interaction between space and curriculum to support the school's mission of "developing all education and teaching toward beauty."

Events such as the "End-of-Term Comprehensive Assessment" "Subject Teaching Extension Activities" "Third-Grade 10th Birthday Celebration" (see Fig. 12.5)

Fig. 12.4 "Commands on Campus" comprehensive practical activity

"Technology Festival Robot Battle Challenge" and the "Children's Day Carnival" are all vibrant examples of what takes place within this comprehensive learning space. These remarkable showcases bring even more life and brilliance to our small yet dynamic learning environment, often described as creating greatness in a small space (see Fig. 12.4).

12.3 Implementation Outcomes

As an innovative teaching space designed within China's curriculum reform context, this case holds great potential for broad implementation across the region. It has played a key role in transforming learning approaches and advancing classroom teaching reforms, achieving significant results.

12.3.1 Catalyzing the Enhancement of Curriculum Leadership

Enhancing curriculum leadership is essential for teachers' self-fulfillment and is a crucial pathway for advancing curriculum reform under the Core Competencies and Values framework (Harris et al., 2020). The construction and practice of the

Fig. 12.5 "Third-Grade 10th Birthday" themed educational activity

comprehensive learning space—referred to as the "Magic Space"—focuses on each individual student. It leverages new opportunities provided by modern information technology to break through spatial limitations and address challenges encountered in curriculum teaching at the school level.

By effectively utilizing a blended virtual and physical learning environment, the school has continually reorganized and optimized its resources. It has established an open framework for its unique aesthetic education programs. This framework spans four key domains—arts, physical education, moral education, and technology—and integrates 20 courses across five grade levels into four major practical projects: "Arts Stage" "Rhythm Club" "Mind Station" and "Playful Beauty Workshop." These programs aim to nurture students' innovative competencies. The interaction between the learning space and curriculum has significantly promoted the development of curriculum leadership.

12.3.2 Nurturing an Innovative Team of Teachers

As curriculum and learning spaces continue to evolve and integrate, teachers are exploring resources, collaborating, and expanding interdisciplinary teaching concepts. The methods of teaching and learning are continually refined through ongoing interaction with the learning space, leading to the continuous improvement of teachers' diverse skills.

Whether it's the physical education, English, and science classes mentioned earlier, various school-based courses and activities, or lessons conducted in the "Magic Space," such as the Chinese class "The Four Suns," the music appreciation class "The Sun Holds My Hand" and the moral education class "The Gentle Breeze Blows." Each successful case demonstrates the transformation and innovation in teaching content, methods, and educational practice. These cases highlight the creative application of technology, enhancing students' access to and use of resources while adhering to the basic principles of subject-specific teaching.

As a result, teachers have not only earned national, municipal, and district-level awards for teaching achievements, but have also nurtured a group of interdisciplinary teachers proficient in both physical education and information technology, as well as in arts and design.

12.3.3 Promoting Diverse Student Development

The "Magic Space" extends the learning environment beyond the classroom, encompassing the entire campus. Through vivid scenario creation, subject-specific scene displays, problem-driven contexts, and real-life re-creations, it sparks students' interest in learning and enhances their genuine emotional experiences. In this immersive, interactive, and engaging learning environment, students face challenges, actively explore, and gradually strengthen their creative thinking and comprehensive learning abilities.

Meanwhile, within the integrated curriculum framework guided by the "Five Educations" (moral, intellectual, physical, aesthetic, and labor education), students enjoy the learning process, unlocking their potential for diverse development. In this warm, safe, and open curriculum structure, they achieve both comprehensive growth and individual development, fully embodying the "holistic education involving everyone, throughout the entire process, and in all dimensions" approach through the integration of curriculum and space. One example includes students engaging in traditional culture activities on the "Clouds of Auspiciousness" platform (see Fig. 12.6), where cultural heritage and modern learning environments come together to support comprehensive development.

12.3.4 Driving the School's Distinctive Development

The construction and practice of the comprehensive learning space provide a conducive environment for fostering the subject competencies emphasized by the

Fig. 12.6 Students inherit and develop Chinese traditional culture on the "Clouds of Auspiciousness"

Curriculum Standards 2022.[2] It also serves as a powerful engine for improving the quality of subject teaching and enhancing the school's distinctive educational offerings.

The "Magic Space" highlights the school's unique focus on aesthetic education, not only in terms of content and form but also from its core values to its external expression. This approach has inspired the school to adopt a mindset centered around the development of such spaces, integrating aesthetics, technology, curriculum, and collaboration between teachers and students across all artistic spaces, thereby establishing a shared vision for growth and practice.

12.4 Innovations

This case represents an innovative study and practical exploration of learning space transformation in the digital age. Throughout the process of constructing and implementing the comprehensive learning space, we have identified the following key innovations:

The "Magic Space" integrates the light, shadows, and scenarios projected by the "Magic Wall" and "Magic Carpet," seamlessly linking the limited physical space with the virtual world. This not only extends the learning environment for students but also allows them to experience the joy of learning in a boundless, immersive space.

[2] The *Curriculum Standards 2022* or *New Curriculum Standards* in China are the fundamental guidelines for national education. They serve as the official framework that outlines the basic norms, expectations, and quality standards for primary and secondary education in the country. These standards define what should be taught, the goals of each subject, and the competencies that students are expected to develop at each educational level. They are designed to ensure consistency across schools and provide a benchmark for measuring educational quality nationwide.

The deep integration of the school's aesthetic curriculum with the design of beautiful learning spaces has created an environment where aesthetic education and teaching activities can flourish. This establishes both an environmental and teaching model for conducting educational activities within a comprehensive aesthetic learning space.

12.5 Promotion

The design of the "Magic Space" is grounded in the fundamental principles of education and the overarching goals of educational reform and development. It is replicable and adaptable, making it highly scalable and suitable for broader implementation.

In international metropolitan areas, such as Beijing's hutongs or Shanghai's alleyways, where schools are often small and space is limited, using information technology to design smart, multidimensional learning spaces offers an efficient way to expand and enhance the functionality of such environments. This approach is particularly suited for schools in historic or densely populated areas.

The "Magic Space" uses sound, light, and visual effects to create diverse learning environments, enabling students to explore both the macro and micro worlds. This smart space allows students to experience different scenarios, regions, and even countries' astronomical and geographical features without ever leaving the school, making the teaching methods adaptable for wider use.

As an effective platform for integrating information technology into teaching, the "Magic Space" drives continuous improvement in teaching and learning methods. This promotes innovative teaching in the new era and serves as a forward-looking tool for teacher development.

Mechanisms such as "principle-setting, smart reservations, and reasonable adjustments" effectively increase the efficiency of the "Magic Space," ensuring it meets the educational and teaching needs of various subjects.

12.6 Summary

The "Magic Space" case aimed to explore the innovative transformation of learning spaces in the digital age, and to develop a comprehensive learning space model that was replicable, adaptable, and scalable. It integrated physical and virtual spaces using smart walls, interactive floors, and combines formal and informal learning spaces to provide students with a diverse learning experience. By incorporating immersive technologies and smart devices, it created an interactive and richly experiential learning environment that stimulates student interest and enhances the quality of teaching. The integration of space design with the curriculum broke down disciplinary boundaries, promoted changes in teaching methods and facilitated the comprehensive development of students, forming a unique model of aesthetic education.

"Magic Space" broke through the limitations of traditional classrooms by offering more possibilities for teaching and learning. It encouraged teachers to explore new teaching methods and promoted the integration of digital technology into everyday educational practice. Further recommendations included developing curriculum resources for cross-disciplinary learning, developing curriculum resources that are compatible with cross-disciplinary learning, strengthening teacher training to improve the effective use of learning spaces, evaluating the effectiveness of applications, and improving the approach to give scientific guidance for future educational development.

References

Harris, A., Jones, M., & Crick, T. (2020). Curriculum leadership: A critical contributor to school and system improvement. *School Leadership & Management, 40*(1), 1–4. https://doi.org/10.1080/13632434.2020.1704470

Kuhail, M. A., Elsayary, A., Farooq, S., & Alghamdi, A. (2022). Exploring immersive learning experiences: A survey. *Informatics, 9*, 75. https://doi.org/10.3390/informatics9040075

Wagner, C., & Liu, L. (2021). Creating immersive learning immersive learning experiences: A pedagogical design perspective. In A. Hui & C. Wagner (Eds.), *Creative and Collaborative Learning through Immersion: Interdisciplinary and International Perspectives* (pp. 71–87). Springer International Publishing. https://doi.org/10.1007/978-3-030-72216-6_5

Chapter 13
Teaching Accuracy in English Writing Through Process-Oriented Data Feedback Technology

Weiwei Wang and Qian Yao

Abstract With the development of educational technology, educators increasingly focus on using data feedback to improve teaching effectiveness. *Process-Oriented Data Feedback Technology* refers to installing a high-speed camera on a traditional Bluetooth pen tip, which collects and tracks student progress in real time without changing conventional writing habits and provides immediate feedback (Tian and Zhang in An intelligent handwriting teaching system supported by dot matrix recognition technology. Artif Intell 03:64–70, 2019). This case aims to explore the effectiveness of this technology in improving instructional precision in English teaching.

Keywords Data feedback technology · English writing · Precision education · Instructional evaluation

13.1 Background

In the *China Education Modernization 2035* plan,[1] emphasis is placed on making education more inclusive and adapting instruction to individual student needs. By leveraging technology, differentiated teaching and personalized learning can be achieved, effectively reducing students' academic burdens. However, the reduction in teaching hours and homework assignments has inadvertently increased challenges in English instruction, particularly in English writing. Traditional English writing instruction faces four key issues that hinder the development of students' writing skills.

[1] http://www.moe.gov.cn/jyb_xwfb/s6052/moe_838/201902/t20190223_370857.html.

W. Wang (✉) · Q. Yao
The High School affiliated to Beijing Normal University, Beijing, China

Q. Yao
e-mail: yaoqian@bjsdfz.com

D. Liu et al. (eds.), *Empowering Educational Innovation Through Design: Volume One*, Smart Computing and Intelligence, https://doi.org/10.1007/978-981-95-4621-3_13

Firstly, excessive emphasis is placed on fragmented grammatical knowledge. Students are required to memorize and apply numerous grammatical rules and structures, causing them to focus too much on grammatical accuracy while neglecting content expression and organization (Fu, 2012). Secondly, there is a lack of sustained skill development. To ease the workload for students and teachers, writing tasks are often assigned at the end of a unit, depriving students of sufficient opportunities for continuous practice and improvement. Thirdly, real-time, personalized feedback is scarce. Students typically have to wait for teachers to grade their assignments before receiving feedback, delaying opportunities for progress and revision. Moreover, teachers' feedback is often limited to overall scores and grammar corrections, lacking targeted guidance on thought design and structural organization. Fourthly, traditional data collection methods, such as scanners and document cameras, are inefficient and cumbersome (Wu, 2023). These methods burden teachers with additional tasks and fail to capture students' writing processes effectively, making it difficult to provide timely, individualized support.

Although various technologies have been employed to assist English writing instruction, they come with limitations that hinder their effectiveness. In traditional English writing instruction, teachers often rely on scanners and document cameras, which have various limitations in the data collection phase. In contrast, *Process-Oriented Data Feedback Technology* uses a Bluetooth pen with a high-speed camera to collect and track students' writing progress in real time without altering conventional writing habits. This technology records students' writing strokes and thought processes, generating process-oriented data. The application of *Process-Oriented Data Feedback Technology* in English writing instruction aims to align with the fundamental principles of student writing development and serve as a technological aid in the following ways.

- **Planning (Organization):** By tracking students' writing progress and decision-making processes in real time, this technology helps teachers understand students' thought patterns and content expression, thereby enhancing language organization.
- **Structure (Coherence):** This technology records students' writing processes, including planning, organizing, and revising. Students can review their writing journeys to identify issues and areas for improvement, ensuring the logical coherence of their work.
- **Details (Grammar and Vocabulary):** Students receive efficient, targeted feedback on grammar and vocabulary. They can communicate with teachers in real time, ask questions, and seek advice. Teachers can annotate and provide specific feedback directly on students' compositions, addressing particular challenges as they arise.

This case is grounded in constructivist theory, social learning theory, cognitive load theory, and multiple intelligences theory. It aims to explore the effectiveness and impact of *process-oriented data feedback technology* in achieving precision teaching for high school English writing, providing insights and references for the advancement of educational technology.

13.2 Design Concept and Teaching Process

13.2.1 Traditional Lesson Design—A "Reading-to-Writing" Lesson Following Writing Principles

Paragraph writing is a critical component of English writing. In paragraph writing, the topic sentence, supporting sentences, explanatory sentences, and linking sentences are the fundamental elements. These four elements are interconnected, and their structure and organization (the PEEL structure) significantly affect the quality and clarity of the writing. This lesson plan is a traditional "reading-to-writing" class, where the teaching objective is to use the four elements of the PEEL structure to complete paragraph writing. The diagram below illustrates the relationship and hierarchy of the four PEEL elements (see Fig. 13.1).

This case uses *Unit 5: Education, Lesson 1: Enlightening a Mind*, from the optional compulsory textbook published by Beijing Normal University. The reading material focuses on the story of Helen Keller's childhood and her first encounter with her teacher, Anne Sullivan, describing how she learned four words. This lesson is the second teaching session in this unit. After reviewing the first session, where students learned about Helen Keller's experience learning the four words, students reflect on Anne Sullivan's qualities as a teacher. The rich and moving content of this text provides ample examples to support writing, making it an excellent resource for writing instruction. The compelling nature of the story inspires students to write. Based on the text's characteristics and fundamental writing principles, the lesson's design process is outlined in Fig. 13.2.

13.2.2 Enhanced Technology Design—Using "Process-Oriented Data Feedback Technology" to Support Instruction

To help students improve their writing in terms of planning, structure, and details, and to effectively record their learning process and provide feedback, this high school English writing lesson demonstrates how to use *Process-Oriented Data Feedback Technology* while adhering to traditional English writing principles. Incorporating this technology into the traditional "reading-to-writing" lesson can follow these steps:

Assign the Writing Tasks: Choose a specific writing task that includes examples and explanations, ensuring that the task aligns with the students' English proficiency levels and learning objectives.

Students Use Technology (Bluetooth Pen) to Record the Writing Process: Provide students with a Bluetooth pen equipped with process-oriented data feedback technology. As they write, technology tracks their progress in real time and records their decisions and actions.

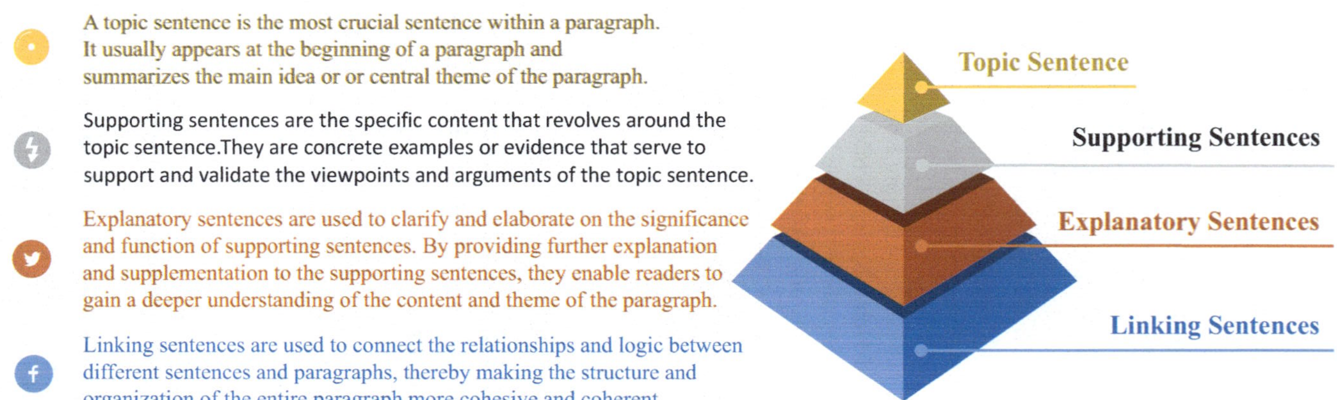

Fig. 13.1 Internal structure of paragraph writing (PEEL structure)

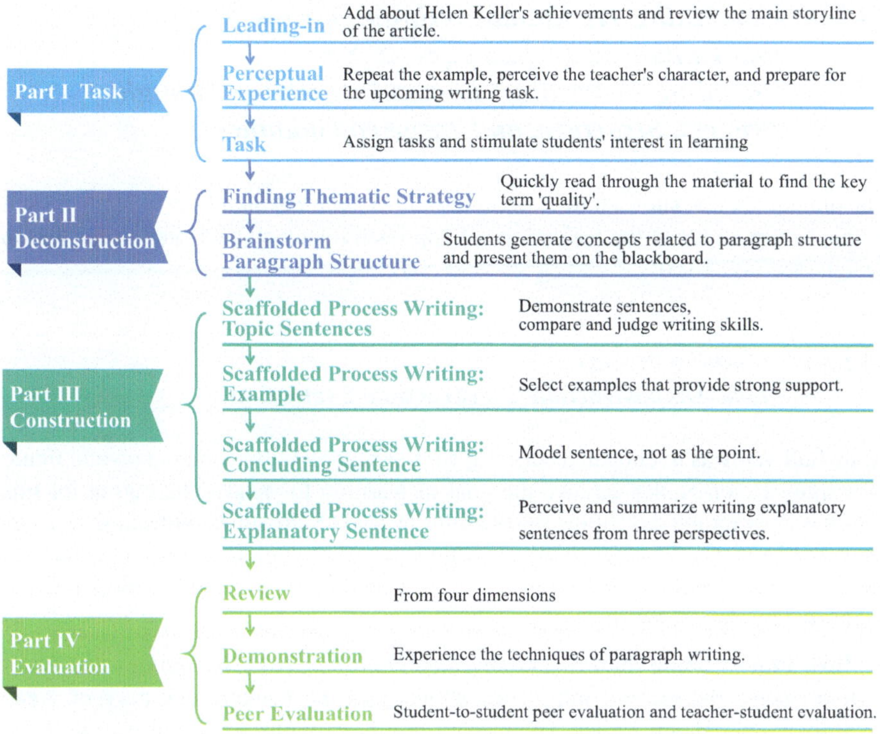

Fig. 13.2 Teaching process and steps

Collect Full-Process Data: Gather process-oriented data generated during the writing process, including information on writing speed, word count, paragraph structure, and grammar usage. Additionally, collect students' annotations and notes to understand their thought processes and writing strategies. (This step is completed by the teacher after class).

Analyze Data and Provide Feedback: Perform statistical and visual analysis on the collected data. Based on the analysis results, provide personalized, immediate feedback and guidance to students, highlighting their strengths and suggesting areas for improvement.

Personalize Instructional Adjustments: Adjust teaching strategies and content based on the process-oriented data feedback to meet the specific needs and writing abilities of each student.

13.2.3 Combining Traditional Writing Lessons and Enhanced Technology—A New "Reading-to-Writing" Lesson Integrating Thought Process, Structure, and Targeted Guidance

The diagram below shows how *Process-Oriented Data Feedback Technology* is used in this lesson and how the technological approach relates to each step of the lesson design (see Fig. 13.3).

13.2.3.1 Teaching Process: Task–Deconstruction–Construction–Evaluation

Anne Sullivan was a renowned educator for the deaf and blind and the pivotal figure in helping Helen Keller achieve the goal of learning to speak. The output for this lesson is a paragraph describing the qualities of Anne Sullivan, taught using *Process-Oriented Data Feedback Technology*. Students will engage in writing practice and receive teacher feedback and guidance to improve their writing skills, with evaluation supported by this technology.

- **Task Phase**: Using the *Process-Oriented Data Feedback Technology* and Bluetooth pens, the teacher introduces photos and the background story of Anne Sullivan and Helen Keller. The teacher briefly explains Anne Sullivan's educational philosophy and values, and the technology is used to compile data on the qualities that left the strongest impression on students (see Fig. 13.4).
- **Deconstruction Phase**: During the knowledge explanation phase, the teacher uses *Process-Oriented Data Feedback Technology* to create a mind map and instructional strategy chart for paragraph writing, focusing on one of Anne Sullivan's qualities. This helps students better understand her teaching methods and philosophy (see Fig. 13.5).
- **Construction Phase**: In the practice phase, the teacher uses *Process-Oriented Data Feedback Technology* to record students' writing processes and feedback. This includes tracking students' thoughts, ideas, and providing specific feedback and guidance during the writing process (see Fig. 13.6).
- **Evaluation Phase**: In the summary phase, the teacher uses *Process-Oriented Data Feedback Technology* to facilitate peer and teacher evaluations of the best-written paragraphs. Students also evaluate the qualities demonstrated by their teachers, helping them summarize their learning gains and reflections (see Fig. 13.7).

13.2.3.2 Integrating *Process-Oriented Data Feedback Technology* with Process Writing Instruction

Process writing instruction views the writing process as an essential learning phase, emphasizing reflection and revision. Unlike traditional "product-oriented" writing

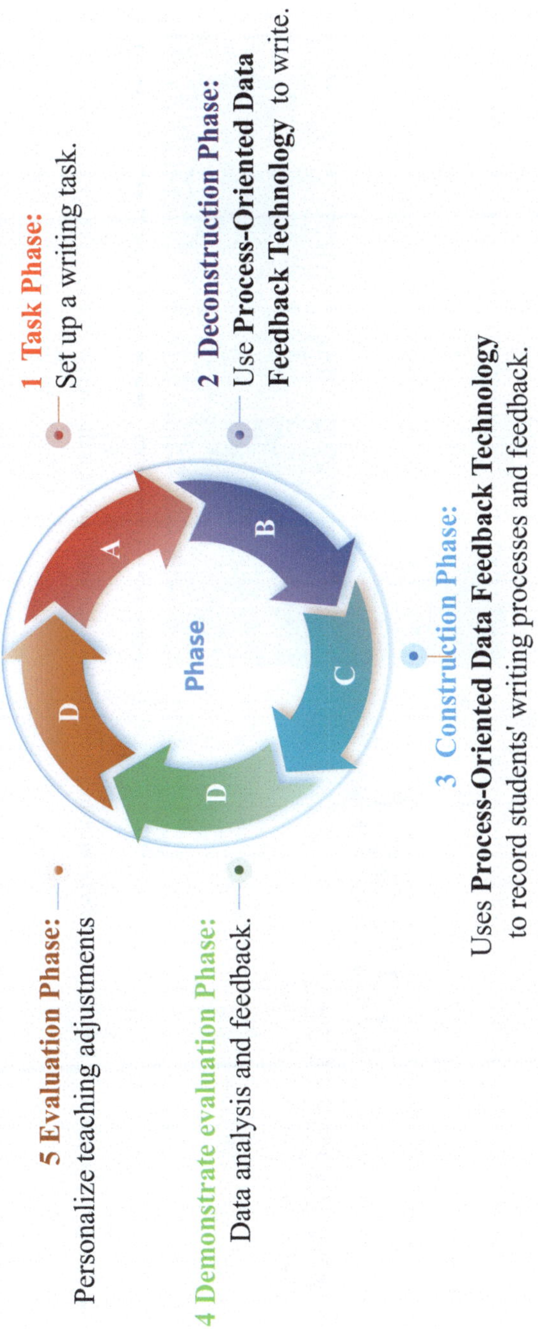

Fig. 13.3 Using *Process-Oriented Data Feedback Technology*

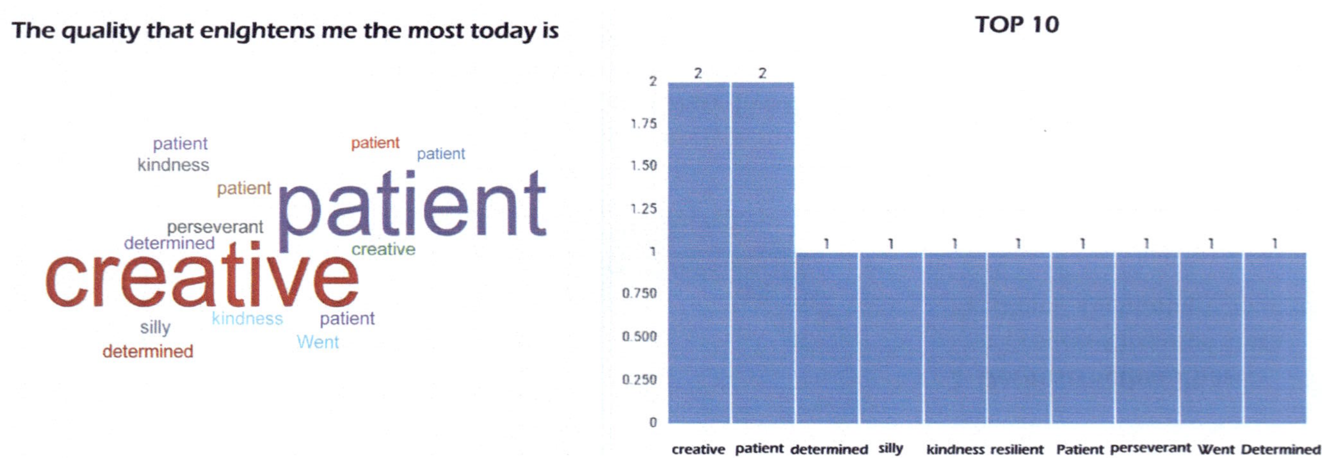

Fig. 13.4 Using *Process-Oriented Data Feedback Technology* to identify qualities that impressed students most

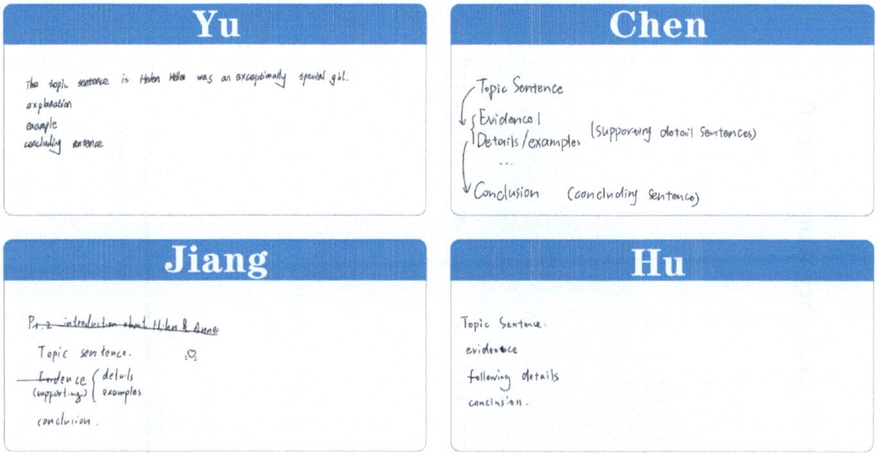

Fig. 13.5 Using *Process-Oriented Data Feedback Technology* to create a paragraph writing mind map

instruction, process writing focuses on students' active participation and self-regulation. This approach enhances students' flexibility and creativity in thinking, fostering better writing skills and innovative thought processes.

Process-Oriented Data Feedback Technology supports process writing by recording every stroke and action during writing. This data is fed back to teachers and students, enabling teachers to better understand students' writing processes, learning habits, and challenges. It also helps students recognize and analyze their writing behaviors, allowing them to adjust and improve their strategies in real time.

13.2.3.3 The Role of *Process-Oriented Data Feedback Technology* in Writing Instruction

Process-Oriented Data Feedback Technology offers several irreplaceable benefits for high school English writing instruction:

- **Real-Time Feedback:** The Bluetooth pen records students' writing strokes in real time and transmits data to the teacher's computer or tablet. Teachers can quickly identify students' issues and provide precise guidance.
- **Accurate Assessment:** The pen converts writing strokes into digital data. Through data analysis, teachers can determine students' writing scores, error rates, and language expression abilities, allowing for more objective assessment of writing skills.
- **Interactive Collaboration:** The Bluetooth pen supports simultaneous use by multiple students. This enables collaborative writing activities, while teachers can monitor and guide several students in real time.

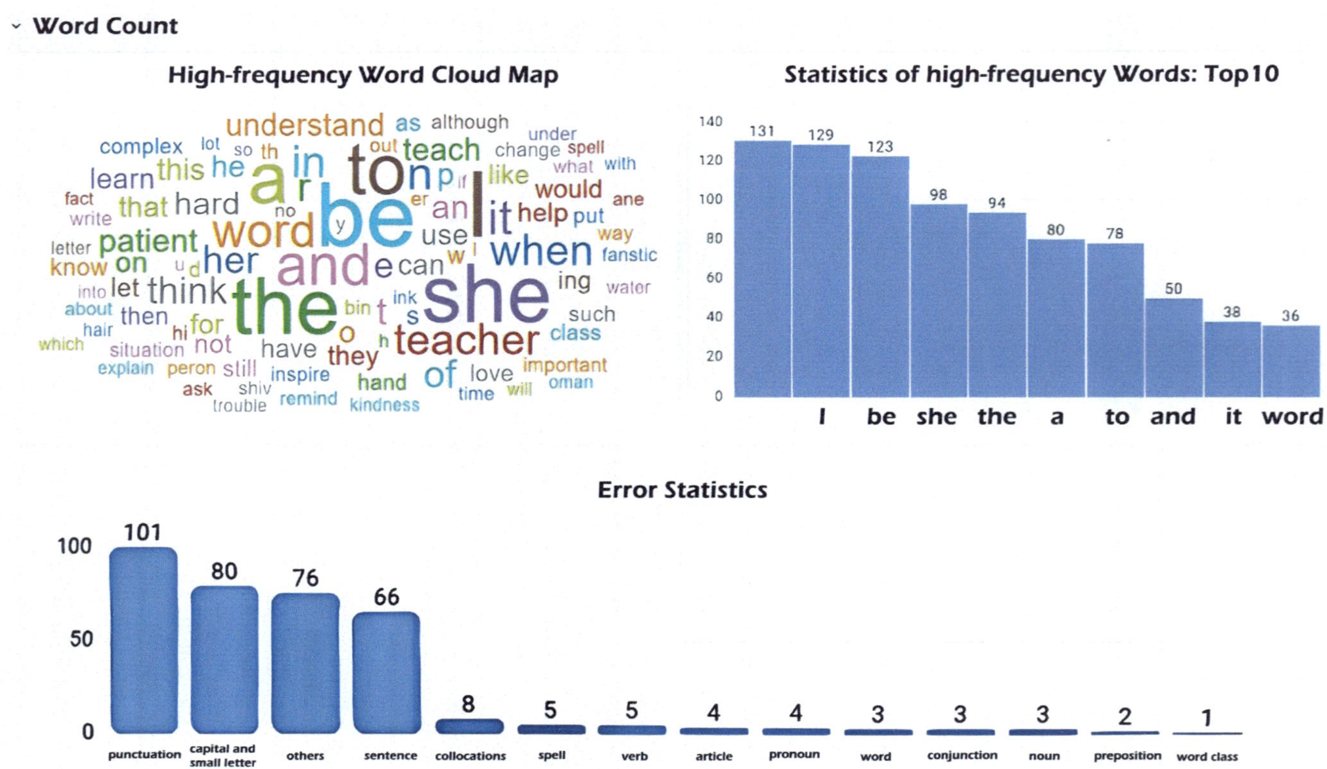

Fig. 13.6 Using *Process-Oriented Data Feedback Technology* to track students' writing data in real time

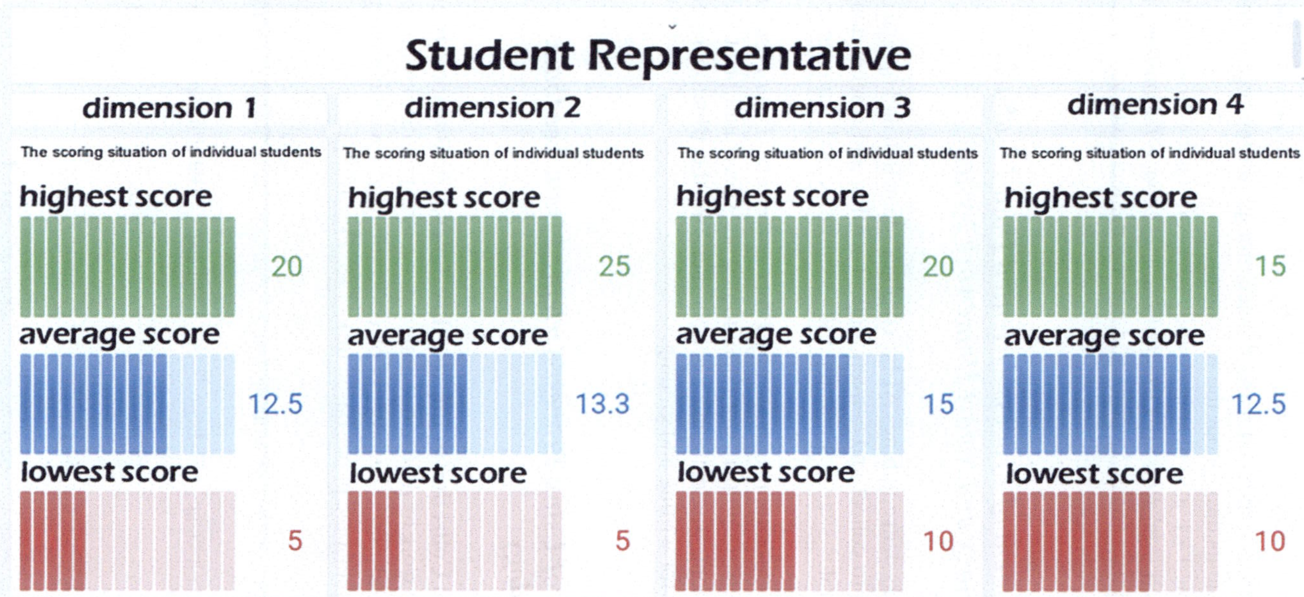

Fig. 13.7 Using process-oriented data feedback technology for peer and teacher evaluations

13.3 Implementation Outcomes

This case applies *Process-Oriented Data Feedback Technology* in the lesson *Enlightening a Mind*. Teachers continued using this technology during regular instruction and weekly assessments throughout the semester.

As shown in Table 13.1, a comparative study between classes shows that Class B1 and Class B2, which used *Process-Oriented Data Feedback Technology*, achieved faster improvement than Class A, which did not use the technology.

13.3.1 Deployment of Process-Oriented Data Feedback Technology and Smart Homework System

The Bluetooth pen system combines AI-based object detection and optical character recognition (OCR) to collect and convert student writing and teacher annotations. The backend generates statistics and reports to optimize and innovate evaluation systems.

To implement a classroom supported by *Process-Oriented Data Feedback Technology*, teachers need to prepare dot-code homework sheets and print them using specialized equipment. Students and teachers then use the Bluetooth pen with a high-speed camera for writing and grading.

13.3.2 Effects of Using Process-Oriented Data Feedback Technology

- **Real-Time Data Collection to Improve Efficiency:** The Bluetooth pen replaces traditional board work by digitizing handwritten content for display on a computer screen. This streamlines the process of collecting and reviewing assignments, eliminating cumbersome paper-based tasks (see Fig. 13.8).

Table 13.1 Comparative analysis of utilizing *Process-Oriented Data Feedback Technology* and non-utilization

Class	Entry average score	Midterm average score	Entry gap	Midterm gap	Improvement
Class A (no technology)	72.29	77.7	—	—	5.41
Class B1 (with technology)	44.7	54.1	27.59	23.6	9.4
Class B2 (with technology)	44.22	62.6	28.07	15.1	18.38

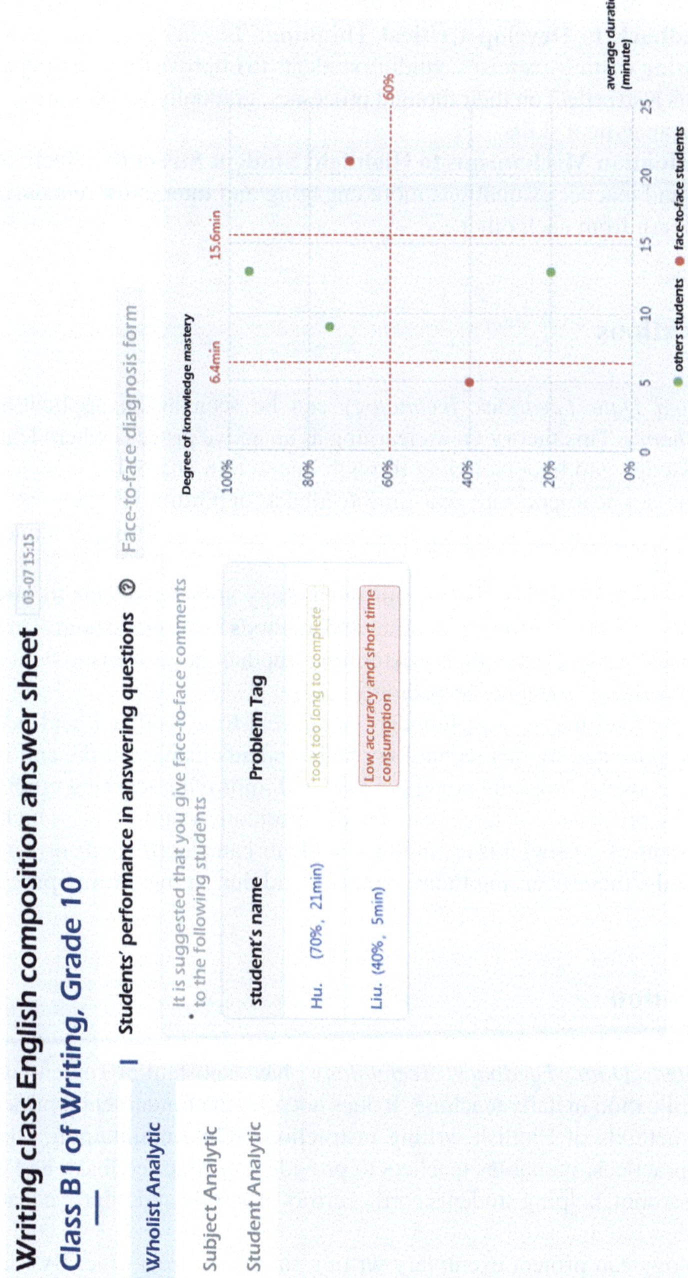

Fig. 13.8 Real-time generation of student responses and analysis in the backend

- **Data Analysis to Help Teachers Understand Student Learning:** Teachers use technology to track student responses and compile data, such as word clouds, to identify overall trends and insights into students' ideas.
- **Timely Feedback to Develop Critical Thinking:** Teachers provide real-time feedback during writing exercises, guiding students to improve their writing structure. Students also reflect on their thought processes, gradually developing critical thinking and analytical skills.
- **Diverse Evaluation Mechanisms to Highlight Student Strengths:** Technology makes peer and teacher evaluations more engaging and interactive, encouraging students to learn from each other.

13.4 Innovations

Process-Oriented Data Feedback Technology can be seen as an application of *constructivist theory*. This theory views learning as an active process where learners construct knowledge and understanding through interaction and collaboration. The technology provides learners with real-time feedback opportunities, helping them develop knowledge and understanding of writing skills while offering guidance for skill improvement.

Supported by social learning theory, this technology allows students to observe and learn from their peers' writing techniques and methods in the classroom. Through real-time feedback, students can quickly determine whether they are on the right track and adjust their writing strategies accordingly.

With cognitive load theory as a foundation, the real-time feedback and personalized guidance provided by this technology help students manage tasks and information more effectively, reducing cognitive load and improving learning efficiency. Writing involves processing a large amount of information and tasks, which can overwhelm students. By using this technology, students can identify and correct their mistakes promptly, thereby easing their cognitive load during the writing process.

13.5 Promotion

Process-Oriented Data Feedback Technology has substantial potential for widespread application in daily teaching. It does not alter the fundamental principles or traditional methods of English writing instruction. While maintaining conventional writing practices, it enables teachers to provide real-time feedback and guidance in the classroom, helping students correct errors promptly and improve learning outcomes.

The technology can project exemplary writing samples in real time. By reading and analyzing these examples, students can learn and adopt effective structures, logical flow, and language expression techniques.

The technology focuses on teaching objectives, allowing students to repeatedly practice paragraph writing. Students are encouraged to define clear topic sentences, provide supporting arguments and examples, and develop logical and coherent paragraphs.

With integrated optical character recognition (OCR) technology, the pen offers intelligent error correction, facilitating vocabulary and language exercises. This includes expanding vocabulary, learning common expressions, and familiarizing students with sentence structures, thereby improving the accuracy and richness of language expression.

Technology helps teachers innovate their teaching methods, creating more engaging and dynamic classroom environments. It also fosters reflective teaching practices, allowing teachers to quickly and accurately record students' learning progress and feedback, improving overall teaching efficiency.

Given considerations of cost and student privacy, *Process-Oriented Data Feedback Technology* may not yet be applicable to all students. However, it is poised to become an integral part of digital education, promoting its development and application. Through real-time data feedback, this technology enhances education quality and efficiency, truly achieving the goals of reducing burdens and improving effectiveness while supporting personalized teaching.

13.6 Summary

13.6.1 Purpose, Core Value, and Educational Implications

This case examined the application of process-based data feedback technology to improve the accuracy of English writing instruction in Chinese high schools. The primary objective was to investigate how real-time data collection, enabled by smart digital pens with high-speed cameras, can personalize feedback, reduce cognitive load, and support individual learning needs without disrupting traditional writing habits.

The core value of this study lays in its innovative integration of educational theory and technological practice. Built on constructivist, social learning, cognitive load, and multiple intelligences theories, the study demonstrated how data-driven insights can make students' thinking visible, enabling more targeted teacher intervention. Real-time tracking of writing processes, including planning, organization, and revision, enabled teachers to provide precise, individualized feedback while empowering students to reflect on their own learning strategies. Comparative data across classes confirmed measurable improvements in academic performance when the technology was applied consistently.

From a theoretical perspective, the case validated that timely, individualized feedback, which is an often-missing link in large classrooms, can be effectively scaled through the use of smart tools. By externalizing cognitive processes and enabling immediate adjustments, the approach reduces learners' working memory load, supported formative assessment, and reinforced both peer and teacher scaffolding.

Practically, the study provided a replicable instructional model by combining traditional "reading-to-write" pedagogy with process-tracking tools to enhance both writing quality and student engagement. The technology enabled multiple forms of assessment, including self-assessment, peer assessment, and teacher assessment, which contribute to a more interactive and responsive classroom environment. Moreover, the process data supports instructional decision-making and teaching design improvement.

13.6.2 Recommendations for Practice and Future Research

To ensure broader impact, future implementations should focus on three aspects:

- **Scalability and Affordability**: Develop cost-effective, privacy-conscious versions of the technology for wider access.
- **Teacher Training**: Equip educators with data literacy and instructional design skills to interpret and act on feedback effectively.
- **Cross-Disciplinary Application**: Extend the model beyond English writing to other subjects emphasizing process-based learning.

Future research may explore long-term effects on student autonomy, metacognition, and creativity, as well as integration with AI-driven adaptive learning platforms for real-time instructional optimization.

References

Fu, C. (2012). Challenges and strategies in teaching English writing in secondary schools. *Secondary School English Garden (Teaching Guidance), 32*, 78–79. CNKI:SUN:ZYCC.0.2012-32-039.

Tian, X., & Zhang, X. (2019). An intelligent handwriting teaching system supported by dot matrix recognition technology. *Artificial Intelligence, 03*, 64–70. https://doi.org/10.16453/j.cnki.issn2096-5036.2019.03.009

Wu, R. (2023). A study on attitudes toward teacher feedback and online feedback methods in English writing instruction. *Overseas English, 23*, 115–117. CNKI:SUN:HWYY.0.2023-23-036.

Chapter 14
Designing a Creative "Smart+" P.E. Class Break Using Educational Neuroscience

Ke Li and Guoliang Zhao

Abstract Physical education during long breaks is a crucial component of school physical education, ensuring one hour of daily physical activity within school. It plays an irreplaceable role in promoting students' physical and mental health, enhancing their physical fitness, and cultivating good exercise habits. Relevant research in educational neuroscience suggests that exercise can maintain brain function and promote brain plasticity, making it the best investment for adolescent brains. However, traditional physical education during long breaks lacks fun and innovation, making it difficult for students to gain rich exercise experiences and cognitive benefits. Moreover, there are no clear standards for exercise load, and the effectiveness of these activities is often questioned. This case leverages educational neuroscience theories to analyze students' learning situations, considering brain and physical development characteristics of junior high school students. It proposes some strategies such as optimizing activity content, creating exercise scenarios, and arranging exercise intensity appropriately. These strategies provide theoretical guidance and practical references for the design of junior high school physical education during long breaks, contributing to the innovation and development of school physical education programs.

Keywords Educational neuroscience · Physical education · Cognitive development · Exercise monitoring · Smart sports platform

---

K. Li (✉)
Boshan Middle School, Zaozhuang, Shandong, China
e-mail: 690936871@qq.com

G. Zhao
Zibo Boshan Middle School, Zibo, Shandong, China

D. Liu et al. (eds.), *Empowering Educational Innovation Through Design: Volume One*, Smart Computing and Intelligence, https://doi.org/10.1007/978-981-95-4621-3_14

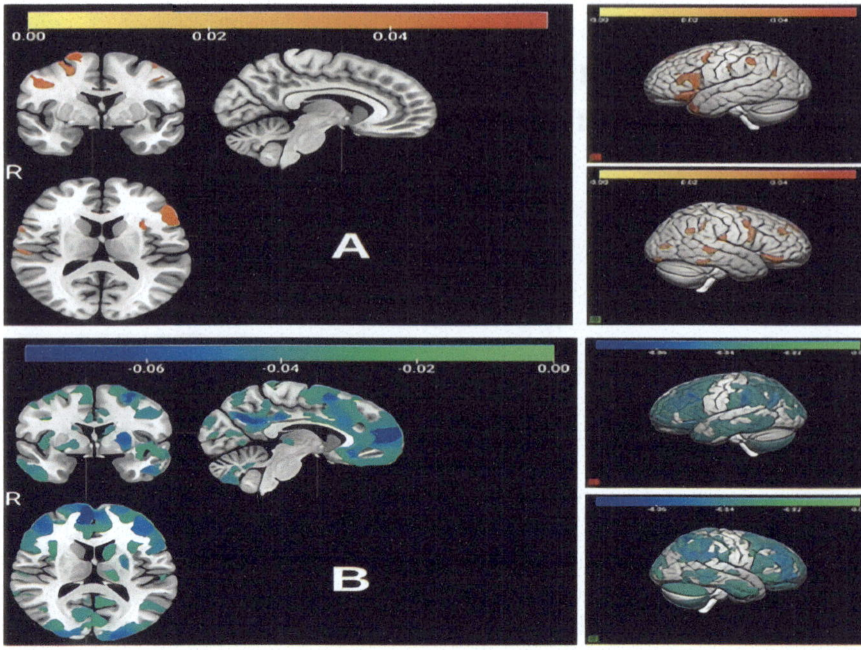

Fig. 14.1 **a** Positive effect of physical activity (orange) on the aggregation of nerve units throughout the brain, **b** significant negative effect of BMI (green) on the brain (Brooks' et al. 2021)

14.1 Background

The Opinions on Comprehensively Strengthening and Improving School Physical Education in the New Era, issued by the General Office of the Central Committee of the CPC and the General Office of the State Council, puts forward that schools should reasonably arrange the time for off-campus sports activities and strive to ensure that students have one hour of physical activity both in and out of school every day, so as to promote the development of lifelong exercise habits.[1]

Currently, due to high academic pressure and addiction to electronic devices, junior high school students are experiencing prolonged sedentary behavior and a lack of effective physical exercise, leading to a decline in physical fitness and Body Mass Index (BMI). Research in educational neuroscience has shown that these factors play crucial roles in neural information processing and cognitive function. Unhealthy weight (higher BMI) negatively impacts motor function, executive control, memory processes, attention, overall brain flexibility, and emotional processing (Fig. 14.1). Therefore, it is urgent to address how to guide teenagers to actively engage in physical exercise, reduce widespread obesity, and scientifically instruct students to comprehensively improve their physical fitness.

[1] http://www.moe.gov.cn/jyb_xxgk/moe_1777/moe_1778/202010/t20201015_494794.html.

14.2 Design Process

Taking Boshan Middle School in Zibo City as an example, this case deeply explores and applies the principles of educational neuroscience. By relying on a smart sports platform and analyzing students’ learning situations, it optimizes activity content to stimulate students’ interest in learning, creates exercise scenarios to enhance students’ participation, and scientifically and reasonably arranges exercise loads to ensure students’ physical and mental health. This approach has effectively improved the quality of physical education during long breaks, creating a new “smart + ” model for physical education based on educational neuroscience.

The design of this case is based on the requirements of the *Compulsory Education Physical Education and Health Curriculum Standards (2022 Edition),* adhering to the guiding principle of “Health First.”[2] Grounded in the principles of educational neuroscience, it takes students’ learning situations as the starting point and considers the school’s existing faculty, facilities, equipment, and digital resources, emphasizing scientific and practical design. Through a cycle of “practice—validation—reflection—improvement,” a typical, replicable and scalable case of reform for physical education during long breaks has been formed.

14.2.1 Investigation and Analysis

To gain deeper insights into students’ perceptions and expectations regarding the physical education (PE) class long break activities, the school conducted a survey using an electronic questionnaire. The analysis of the questionnaire data revealed that students have urgent needs regarding the content of PE class long break activities, organizational methods, construction of sports scenarios, and adjustments to exercise intensity

14.2.2 Teaching Method Design

This case summarizes research findings in educational neuroscience and current investigation results on physical education during long breaks and proposes the following optimization plan for designing physical education activities during these long breaks to reasonably improve students’ physical fitness.

[2] http://www.moe.gov.cn/srcsite/A26/s8001/202204/W020220420582362336303.pdf.

14.2.2.1 Optimize Activity Content

The *Compulsory Education Physical Education and Health Curriculum Standards (2022 Edition)* emphasizes the need to ensure the foundational, diverse, and systematic nature of teaching and guiding students to master specific athletic skills while experiencing the charm of different sports. Different types of sports have varying impacts on brain regions. During long breaks in winter and spring, a "3-km daily" aerobic endurance running program is arranged. According to educational neuroscience research related to endurance exercises, adolescents who consistently engage in 12 weeks of endurance training demonstrate significant improvements in cognitive flexibility, work capacity, and executive function. A structural magnetic resonance imaging (MRI) study further reveals that aerobic endurance levels are closely correlated with cortical surface area and volume in adolescent brains. Those with higher levels of aerobic exercise show greater increases in thickness and volume in the frontal lobe, parietal lobe, and motor areas (Fig. 14.2).

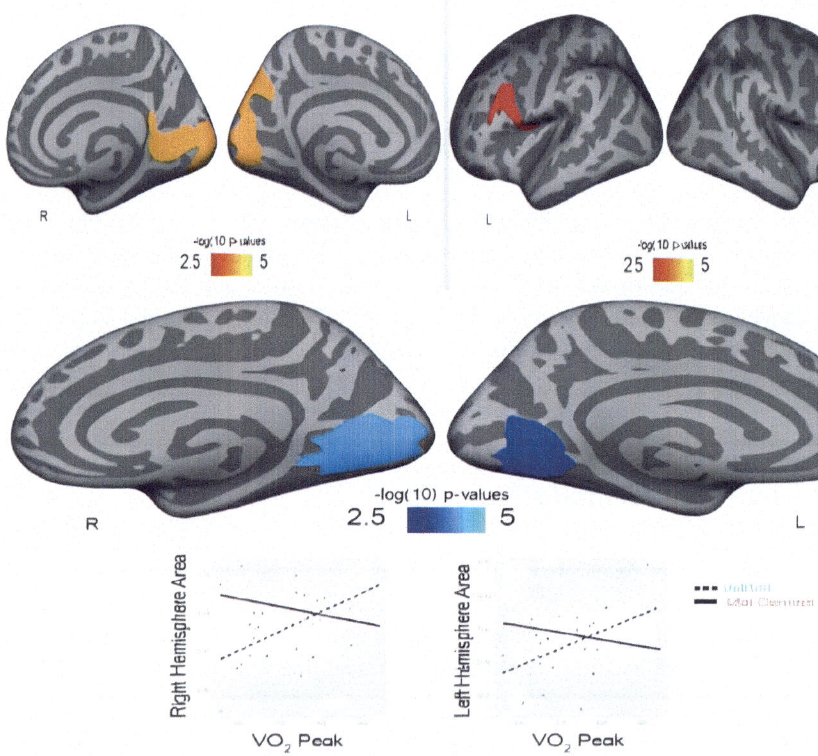

Fig. 14.2 Aerobic endurance level was positively correlated with cortical thickness and volume in motor areas of the frontal and parietal lobes (Herting et al., 2016)

Fig. 14.3 "3 km a day" aerobic endurance running program

By maintaining a consistent pace to complete the daily 3-km endurance run, students accumulate a total of 15 kms per week and 66 kms per month. Throughout their junior high school years, the total distance covered by their running can reach 1,200 kms. This long-term, moderate-to-vigorous aerobic endurance exercise not only effectively improves students' physical fitness and class cohesion but also promotes the development of brain structure and function, enhancing cognitive and executive functions of the brain (Fig. 14.3).

During the long breaks in summer and autumn, an "album-based" activity mode is adopted, focusing on the learning content of motor skills. According to a study on rope skipping, the power of the electroencephalogram spectrum in the parietal $\alpha 1$ and temporal $\alpha 2$ regions increases after rope skipping. An increase in α waves may indicate an increase in the excitability of the cerebral cortex, and rope skipping can affect cognitive function by activating the noradrenaline and serotonin systems and their interactions. This proves that just 3 min of rope skipping can bring the brain into a state conducive to cognitive learning (Fig. 14.4).

Additionally, a study on martial arts suggests that the skill acquisition in martial arts follows a "slow-then-fast" pattern. This learning mode enhances changes in brain activity in the left temporal lobe and precuneus, contributing to the regulation of emotions and memory, as well as the improvement of high-level cognitive functions. The unique routines of martial arts provide learners with opportunities to optimize brain function (Fig. 14.5).

The design of the "album-based" activity mode fully considers the comprehensive development of physical abilities during sensitive periods, including agility and

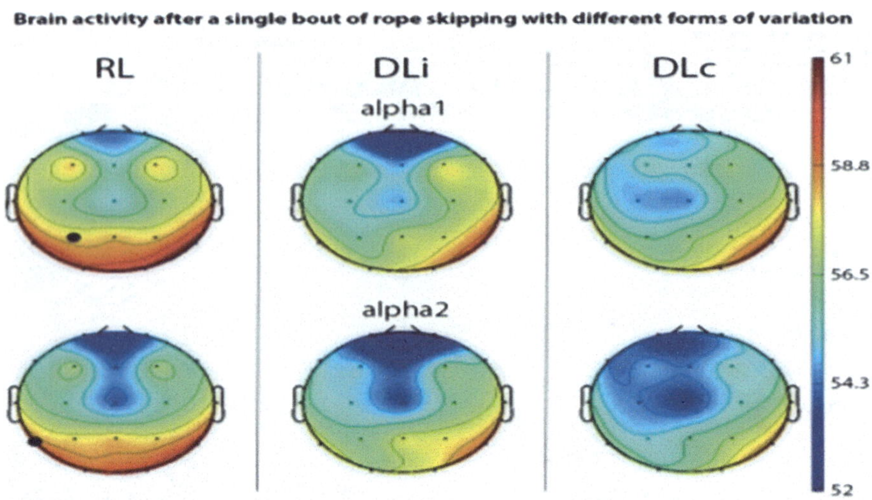

Fig. 14.4 Performance of electroencephalogram spectrum power when applying specific learning methods within the first 5 min of recovery after rope skipping (John et al., 2018)

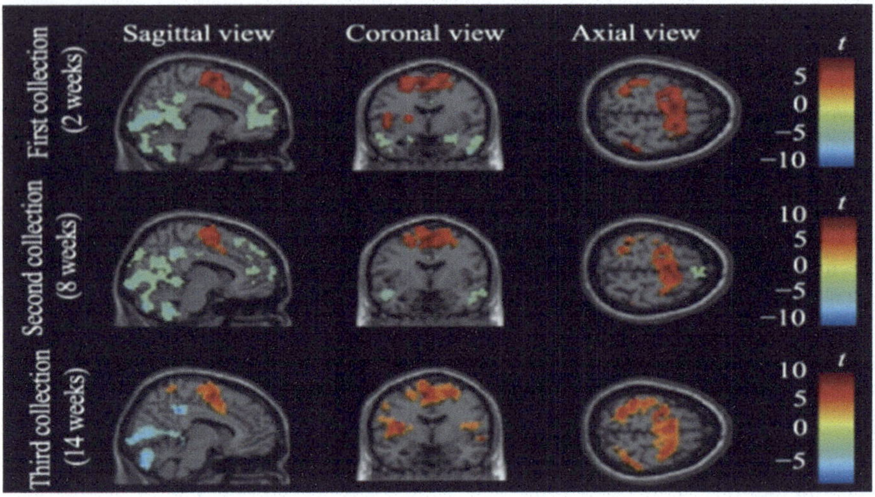

Fig. 14.5 Brain regions activated by movement imagery in martial arts at different time points (Li et al., 2023)

coordination, speed, endurance, strength, and flexibility, helping students master 1-2 sports skills. The activity content is structured as: 1 + 1 + 1 + X, which includes colorful rope exercises + martial arts exercises + fitness running + independent activities. The "X" content, which refers to the type of exercise, is adjusted once every half semester (Figs. 14.6 and 14.7).

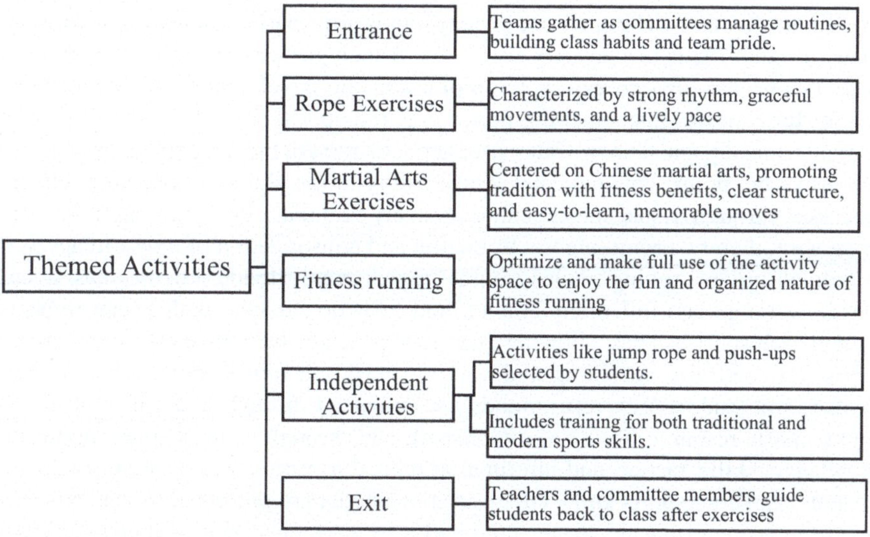

Fig. 14.6 Content of the "album-based" physical education activities during long break

Fig. 14.7 "Album-based" activity mode during long breaks

14.2.2.2 Create Sports Scenarios

The *Compulsory Education Physical Education and Health Curriculum Standards (2022 Edition)* suggests creating a variety of complex sports scenarios to provide students with rich athletic experiences and cognitive development. A diverse environment can increase the frequency of neuronal connections in adolescents and open up memory channels. The environmental stimulus information embedded in different types of motor skills is a primary factor in enhancing cognitive abilities. Similarly, these factors are also necessary for promoting the development of motor skills. Studies have shown that watching image materials positively affects the formation of motor imagery and demonstrates stronger activation in the left parietal and insular regions of the brain. This indicates that learning motor skills or observing athletic processes through visual means can effectively stimulate brain regions related to movement, thereby promoting the formation and consolidation of motor imagery.

Schools utilize existing venue resources to reasonably divide activity areas, scientifically arrange and utilize equipment, and provide students with visual impacts through rich sports scenes, allowing them to appreciate the unique charm of various sports. By skillfully switching and varying music, different sports scenarios are created, and students' exercise moods are enhanced through musical rhythms. A digital sports resource library is established, and through a smart sports platform, knowledge, skills, tactics, and information related to various sports are provided to students. Abstract sports knowledge and movements are presented to students in a three-dimensional manner using multimedia technologies such as images, micro-lecture videos, and animations, which stimulates students' enthusiasm for learning and helps them master movement methods more quickly and comprehensively.

14.2.2.3 Regulate Exercise Load reasonably

Various intensities of exercise can have positive effects on physical health. However, in the field of neuroscience, not all intensities of exercise yield positive results for the brain. A controlled experiment showed that compared to low-intensity and high-intensity exercise, the experimental group engaged in moderate-intensity exercise demonstrated the highest motivation and learning efficiency. Long-term moderate exercise can improve brain structure and cognitive function. Furthermore, a structural magnetic resonance imaging (MRI) study revealed that short-term moderate-intensity aerobic exercise can induce positive changes in the activation patterns of brain regions related to executive function in adolescents. Therefore, appropriate exercise intensity plays a crucial role in the development of the adolescent brain.

To ensure effective physical exercise during long breaks, schools utilize intelligent sports bracelets, exercise load monitoring systems, and smart sports platforms (Fig. 14.8). These intelligent devices precisely monitor students' exercise intensity through heart rate evaluation and feedback. Through multiple rounds of exercise data monitoring and analysis of the "3-km daily" aerobic endurance running mode, it is ensured that when students run at a moderate pace in a group to complete 3 km, the effective exercise time reaches over 15 min; the exercise intensity reaches an average heart rate of 140–160 beats per minute; and the effective exercise density is around 50%. At the same time, the "album-based" activity schedule is scientifically divided to ensure that the 30-min-long break falls within the effective exercise load range. For each student, the system automatically calculates personalized maximum and warning heart rates based on factors such as age and provides timely feedback

Fig. 14.8 Intelligent sports bracelet and exercise load monitoring system

during long breaks activities. Relevant statistical data are also provided for physical education teachers to analyze and improve activities afterward, safely enhancing students' exercise effects.

14.3 Implementation Outcomes

Based on educational neuroscience, the "Smart+" physical education during long breaks is designed to connect and arrange content according to students' physical development and motor learning patterns. The diverse activity arrangements can improve adolescents' cognitive flexibility, work capacity, and executive function. The various activity components are coherent and interconnected, with appropriate exercise load and the flexible use of different teaching methods, promoting students to explore, create, and attempt new things during the activities, as well as apply what they have learned and mastered outside the classroom. This design helps stimulate students' interest in sports, enhances their sense of participation, provides different stimuli and influences the brain, and promotes coordinated brain development.

By utilizing multimedia technology, the barriers of time and space in learning are broken, achieving a perfect combination of sports knowledge and physical activities, allowing students to gain abundant exercise experiences and cognitive benefits. Through the integration of diverse equipment and music, different sports scenarios are presented, immersing students in the activities and bringing them joy. The digital technology-enabled physical education during long breaks vividly interprets "what is happening" in the activity scenarios through body language, stimulating students' enthusiasm for participation. This extends physical exercise methods through active practice and effectively enhances the effectiveness of the long break activities.

According to educational neuroscience evidence, exercise monitoring can ensure appropriate exercise for students, leading to higher motivation and learning ability, and triggering positive changes in brain activation patterns related to executive function. The smart sports platform automatically generates exercise reports based on data collected from intelligent sports bracelets and exercise load monitoring systems. From grades to classes and individual students, and from horizontal to vertical perspectives, personalized and precise services are provided based on big data tracking, including content, format, methods, and platforms. Additionally, personalized attribution diagnoses are conducted for students within the class, categorizing them as "students with physical excellence, students with physical shortcomings, students with physical difficulties, students with body fat issues, and students with technical issues." Through the data terminal, a "personalized diagnosis and feedback class report framework" is created based on the class and an "exercise prescription" is formed based on individual students. Teachers use scientific and objective data to help classes and students overcome their weaknesses, enabling students to form proper exercise habits, improve their athletic performance, and solve the problems of "difficult quantification, difficult recording, difficult supervision, and difficult analysis" in traditional physical education.

14.4 Innovations

This case represents a pioneering integration of educational neuroscience and physical education during long breaks. Based on educational neuroscience theories, it provides a solid research foundation for designing physical education activities during these long breaks. This approach not only enriches the theoretical connotations of physical education but also offers a novel research perspective and explanatory mechanism. In terms of teaching content and activity design, innovative, diversified, and personalized teaching strategies are proposed based on the impact of different types of sports on brain regions. By optimizing activity content, creating diverse exercise scenarios, and accurately regulating exercise load, the developmental needs of different students' brains and motor skills are met, promoting their comprehensive cognitive and motor development.

Furthermore, this case introduces the innovative application of digital technology in evaluating and optimizing physical education. Through advanced technologies such as data analysis and virtual reality, students' learning states and athletic performances are monitored in real-time and accurately, providing a scientific basis for teaching adjustments. Public demonstrations and outcome verifications not only prove the effectiveness of digital technology in physical education but also provide a model and reference for future technological innovations in the field.

14.5 Promotion

The "Smart+" physical education design during long breaks, empowered by digital technology and based on the principles of educational neuroscience and the overall requirements of educational reform and development, is replicable, imitable, and scalable, possessing certain promotability.

By establishing safety rules and standards for exercise through digital technology, utilizing intelligent sports bracelets and exercise load monitoring systems to oversee students' entire activity process, and adjusting exercise duration, frequency, and rules based on visualized data, the system ensures reasonable regulation of exercise intensity and prevents the occurrence of sports injuries.

After exercise, it provides sport reports and personalized diagnostic prescriptions, achieving automated data interaction, routine physical sign recording, and digitized physical health data, thereby safely and scientifically regulating students' exercise behavior and independent exercise habits. The long-term practical process has fully demonstrated the safety and controllability of this case, making it easier to gain trust and support from parents and schools during promotion. Since 2020, the overall physical health status of all students in the school, including indicators such as physical fitness test scores, BMI index, lung capacity, and vision, has significantly improved and enhanced. The practical results have fully proven the effectiveness of the case, showcasing its strong adaptability and promotion potential.

14.6 Summary

Based on the principles of educational neuroscience, the "Smart+" physical education break design case addressed the issues of insufficient engagement, unscientific exercise load, and low participation in traditional activities by integrating neuroscientific theories with digital technology. Core findings showed that endurance running significantly promoted the development of prefrontal/parietal cortex in adolescents, while activities such as rope-jumping and martial arts exercises activated brain regions associated with cognition-related and executive functions. These outcomes provided empirical evidence for the role of physical exercise in improving brain development. The case introduced an innovative seasonal model—"3 km endurance running in winter/spring + themed activities (1 + 1 + 1 + X model) in summer/autumn." Smart bracelets were used to monitor and adjust students' heart rates (target range: 140–160 bpm, ensuring moderate-intensity exercise conducive to cognitive benefits. In addition, a digital resource library of micro-lesson videos and multimodal materials enhances motor imagery, activating both prefrontal and cerebellar regions to improve motor learning and skill acquisition. Over three years of implementation, the program significantly improved students' BMI indices, lung capacity, and physical fitness test scores. The significance of this work lay in pioneering integration of educational neuroscience with school physical education practice, using digital tools to address traditional challenges of difficult quantification and analysis, and establishing a "body-brain integrated" educational approach through personalized exercise prescriptions (e.g., stratified interventions for fitness levels). Future recommendations include developing affordable smart device, promoting the "themed activity" resource library, conducting longitudinally studies on the cognitive effects of structured exercise, and investigating brain responses across different physical profiles. Further research could focus on exploring how interactive digital environments can support dynamic and personalized exercise scenarios, which would advance school physical education from experience-based to evidence-based practices.

References

Brooks, S. J., Parks, S. M., & Stamoulis, C. (2021). Widespread positive direct and indirect effects of regular physical activity on the developing functional connectome in early adolescence. *Cerebral Cortex, 31*(10), 4840–4852.

Herting, M. M., Keenan, M. F., & Nagel, B. J. (2016). Aerobic fitness linked to cortical brain development in adolescent males: Preliminary findings suggest a possible role of BDNF genotype. *Frontiers in Human Neuroscience, 10*, 327.

John, A., & Schöllhorn, W. I. (2018). Acute effects of instructed and self-created variable rope skipping on EEG brain activity and heart rate variability. *Frontiers in Behavioral Neuroscience, 12*, 311.

Li, L., Zhang, X., Xu, Y., Zong, B., Zhao, W., Zhao, G., & Fan, M. (2023). Dynamic changes on brain function during early stage of Tai Chi training: A motor imagery-based fMRI study. *Acta Psychologica Sinica, 55*(8), 1243.

Chapter 15
Developing Health Education Resources for Tibetan Farmers' and Herdsmen's Children Using Digital Technology Applications

Xiao Jiang and Liangni Lai

Abstract This study endeavors to probe into the practice of developing health education resources by leveraging digital technology for Tibetan herdsmen and farmers' children in the realm of infectious disease prevention and control. Through in-depth analysis of the educational background and health needs of Tibetan farmers and herders' children, this study develops a comprehensive set of health education resources—"Goodbye, Little Monsters of Infectious Diseases" based on the practice of digital technology application. The research content mainly includes designing promotional illustrated text and creative products, producing audio illustrated books, carrying out online cloud health courses and offline fun courses, as well as developing a knowledge testing system for infectious diseases to help children of farmers and herders self-test their learning effects. Through the development and application of this set of resources, this study expects to effectively enhance the knowledge of Tibetan children of farmers and herders about infectious diseases and strengthen their self-protection ability, thus reducing the incidence of infectious diseases in the region and contributing to the healthy growth of Tibetan children of farmers and herders.

Keywords Health education · Digital technology · Tibet · Child health · Resource development

15.1 Background

The Central Committee of the Communist Party of China and the State Council have emphasized the importance of promoting students' healthy growth as an ultimate educational goal in the "National Medium- and Long-Term Education Reform and

X. Jiang · L. Lai (✉)
South China Normal University, Guangzhou, China
e-mail: lailiangni@126.com

D. Liu et al. (eds.), *Empowering Educational Innovation Through Design: Volume One*, Smart Computing and Intelligence, https://doi.org/10.1007/978-981-95-4621-3_15

Development Plan Outline (2010–2020)"[1] and the "Outline of the Healthy China 2030 Plan."[2] These documents propose enhancing health education in schools, integrating it into the national education system, and recognizing it as a vital component of quality education at all educational levels. In 2021, the Ministry of Education published the "Guidelines for Integrating Life Safety and Health Education into Curriculum and Teaching Materials for Primary and Secondary Schools,"[3] emphasizing the necessity of incorporating life safety and health education into a various academic discipline. These policies signify national-level recognition of the importance of health education, reflecting China's commitment and fundamental mission in cultivating students who are physically and mentally healthy and are comprehensively developed.

Nevertheless, in Tibet's agricultural and pastoral regions, geographical, climatic, and transportation constraints, combined with traditional mindsets, have resulted in a general lack of awareness and capacity for disease prevention among farmers, herders, and their children. This vulnerability makes them susceptible to common infectious diseases such as tuberculosis, hepatitis, and chickenpox. In these regions where infectious diseases are prevalent, the absence of health education and protective measures represents a significant challenge.

As a model junior high school in Tibet, the Second Junior High School of Karuo District in Changdu City, Tibet Autonomous Region, has leveraged the collective input of local and external teaching staff to lead the advancement of health education resources in the region. The school is actively engaged in the exploration of prospective pathways for the construction of comprehensive health education resource architecture. The school has accelerated the transition of health education and teaching into the digital age by developing and integrating health education resources such as the "Goodbye Infectious Disease Monsters" initiative. By harnessing smart education and modern technology, the health education system has been empowered, which provides a replicable model for in-depth exploration of the integration of health education with smart education. Therefore, utilizing contemporary digital technologies to facilitate targeted and diverse health education activities (França et al., 2022) is of paramount importance for enhancing health literacy and disease prevention capabilities amongst farmers, herdsmen and their children. In 2018, the first batch of members from the South China Normal University Graduate Student Volunteer Teaching Corps Changdu Branch arrived in Tibet. During their teaching assignments, they suffered from infectious diseases due to inadequate precautionary awareness.

---

[1] Office of the Working Group for the National Medium- and Long-Term Education Reform and Development Plan. (2010). National Medium- and Long-Term Education Reform and Development Plan (2010–2020) [EB/OL]. http://www.moe.gov.cn/srcsite/A01/s7048/201007/t20100729_171904.html.

[2] The Central Committee of the Communist Party of China, The State Council. (2016). "Healthy China 2030" Plan Outline [EB/OL]. https://www.gov.cn/zhengce/2016-10/25/content_5124174.html.

[3] Ministry of Education. (2021). Notice of the Ministry of Education on Issuing the Guide for Life Safety and Health Education in Primary and Secondary School Curriculums and Textbooks [EB/OL]. http://www.moe.gov.cn/srcsite/A26/s8001/202111/t20211115_579815.html.

In-depth research and field observations revealed weaknesses in the prevention and control of infectious diseases among the children of local farmers and herdsmen. To gain a deeper understanding of the local health situation, the team conducted systematic demand research. The findings revealed that disease prevention and control levels in rural and pastoral areas were low. Healthy hygiene habits and disease prevention awareness had not yet been established among the locals, especially among school-aged children of farmers and herdsmen, who showed extremely weak awareness of preventing common infectious diseases. Moreover, local schools have an urgent need for health education, with over 147,000 students estimated to require improved health knowledge.

15.2 Design Process

15.2.1 Basis of the Health Education Resources

To ensure the professionalism of health education resources, the team leveraged expert teams from institutions such as the Changdu Municipal Center for Disease Control and Prevention, the South China Normal University School Hospital, the Karuo District Center for Disease Control and Prevention in Changdu, and the South China Normal University Red Cross Society to establish an authoritative health knowledge resource library. The team recruited technical talents from South China Normal University to develop a health knowledge testing system application and build an online platform for cloud-based health courses. With the help of art majors at South China Normal University, they completed the illustration, coloring and layout of picture books. Additionally, the school assembled local Tibetan language teachers to create a reading system suitable for farmer and herdsman children. Furthermore, the university cultivated a group of specialized young volunteer speakers and invited authoritative experts to conduct multiple professional training sessions for the 316 volunteers involved in the project.

With the support of the local municipal government, the school has officially incorporated the "Goodbye Infectious Disease Monsters" course into its curriculum. This initiative includes well-equipped classrooms and a diverse teaching staff, primarily composed of volunteer teachers from the South China Normal University Graduate Volunteer Teaching Corps, supplemented by local Tibetan teachers and school nurses.

In implementing the health education project, the school fully leverages its online resources to present course content online and form a comprehensive learning resource library. It uses online platforms to carry out flexible and diverse online teaching activities, making learning more efficient. Additionally, the school actively collaborates with external institutions such as South China Normal University to introduce cutting-edge health education resources to promote the healthy growth of students. By integrating these resources, the school develops health education resources, forming an organic whole that provides students with comprehensive

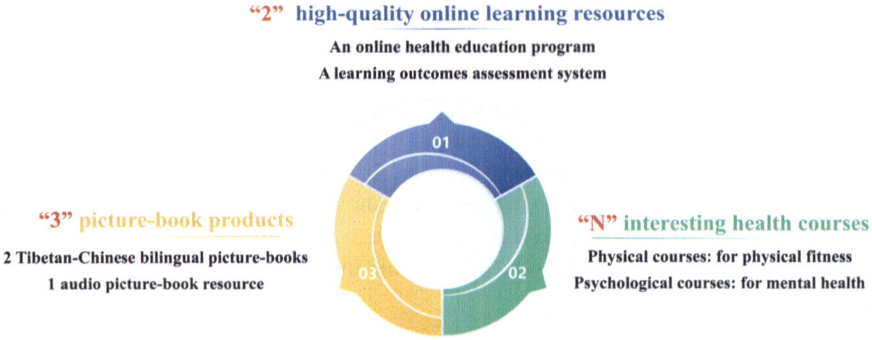

Fig. 15.1 "3 + 2 + N" curriculum system

and rich learning resources on health knowledge. Moreover, this initiative also enables resource optimization and sharing, improving educational effectiveness and sustainable development and utilization of resources.

15.2.2 "3 + 2 + N" Curriculum System

Within the framework of curriculum resources approved by the Ministry of Education, the school has effectively leveraged its unique advantages to meticulously construct the "Goodbye Infectious Disease Monsters" health education resource system (see Fig. 15.1.).

This system achieves comprehensive expansion and innovation through three core pathways. Firstly, it carefully selects three sets of high-quality picture books that vividly present infectious disease knowledge in a graphic and illustrated format, stimulating students' interest. Secondly, it introduces two complementary high-end online learning resource platforms, utilizing modern information technology to provide rich and diverse interactive learning experiences, enhancing the convenience and depth of learning. Thirdly, it innovatively develops multiple sets of curriculum plans that integrate health education with fun elements. These plans employ diversified teaching methods such as situational simulations and practical operations to deepen students' understanding and application abilities regarding infectious disease prevention and control. This series of initiatives not only demonstrates the school's student-centered educational philosophy but also deeply integrates modern educational ideas such as contextualized teaching, innovative integration, and practice-oriented learning. It effectively promotes the optimal allocation of educational resources and the implementation of personalized teaching. Through this resource development model, the school enriched the content and forms of health education, taken solid steps in cultivating students' health awareness, and enhanced their self-protection abilities, laying a solid foundation for building a healthy, safe, and harmonious campus environment.

15.2.2.1 "Three Sets" of Picture Book Products

Utilizing its resources and professional talent pools, the school has compiled Tibetan-Chinese bilingual picture books such as *Goodbye Infectious Disease Monsters* and *A Knowledge Handbook for Preventing COVID-19*, along with accompanying audio picture book resources (see Fig. 15.2). The development of these audio picture books complements the printed versions, enriches learning formats, and enhances students' autonomous learning abilities. This approach fully embodies the concept of personalized education and meets the diverse learning needs of students. For example, using images, videos, and other teaching scaffolds helps students understand more intuitively. The content and explanations in the picture books are also adjusted according to the students' age and cognitive level, making it easier for them to accept and comprehend. Currently, this series of picture books is being promoted and used in Changdu. The compilation of these three sets of picture books provides scientific reference and teaching content for local health education, as well as professional reference materials for teachers conducting health education in the local area.

15.2.2.2 "Two" Complementary Online Learning Resources

Based on the concept of shared distance education, the school has developed two complementary high-quality online learning resources: an online health education course and an evaluation system for learning outcomes. By utilizing internet platforms, the school aims to facilitate distance education and resource sharing, enabling the children of farmers and herdsmen to learn autonomously and self-assess their learning outcomes.

- **Online Health Education Course**

The online course of the "Goodbye Infectious Disease Monsters" initiative is ingeniously designed. Firstly, it leverages AI-powered writing technology to intelligently adjust the course content based on factors such as the students' age and cultural background, making it more relatable and enhancing knowledge acceptance. Secondly, by using Seewo animation technology, the course transforms technical infectious disease prevention knowledge into lively and engaging animations and games, including level-crossing challenges and competitions. By simulating scenarios of monster invasions, the course guides students to recognize and understand the transmission routes, prevention methods, and treatment principles of various infectious diseases. Additionally, the team uses audio synthesis technology to create contextual scenes and vivid character dialogues, providing students with an immersive learning experience. Utilizing PR video editing software, the team produces course

Fig. 15.2 *Goodbye Infectious Disease Monsters* and *A Knowledge Handbook for Preventing COVID-19*, etc.

videos with smooth visuals, compact pacing, and natural transitions, offering high-quality learning resources to students. Furthermore, the online platform has rich interactive features, allowing students to ask questions online and exchange ideas with teachers and classmates. The course also emphasizes parental involvement and interaction. Through online parent-teacher meetings and family health knowledge lectures, parents are informed of the course content and teaching progress, which encourages them to actively participate in their children's health education. As shown in Fig. 15.3, the online "Cloud Health" course combines interactive design and visual storytelling to create an engaging digital learning experience.

- **Evaluation System for Learning Outcomes**

To comprehensively evaluate the effectiveness of students' online learning, the school has developed an evaluation system on the health knowledge. As shown in Fig. 15.4, this system integrates student evaluation models, student recording systems, and learning log sheets, embodying the educational philosophy of diversified evaluation. Under the guidance of professional institutions such as disease control centers, hospitals, and the Red Cross, the school has established a scientific health knowledge question bank and developed a corresponding testing system application. Students can download and install the application to conduct self-tests anytime and anywhere, allowing them to promptly understand their mastery of health and hygiene knowledge.

The student evaluation model incorporates diversified, personalized, and comprehensive evaluation methods. Besides, the team has introduced online classroom discussions, group collaborations, and project presentations to assess students' mastery of infectious disease prevention knowledge, innovative thinking, and practical skills comprehensively.

The student recording system includes students' basic information, learning progress, assignment completion status, health test scores, etc. As shown in Fig. 15.5, it uses the Access 2000 database for data management and data analysis and reporting functions, making it convenient for teachers and students to understand their learning situations.

The learning log sheet includes information such as the student's study time, content, methods, and achievements. Students can fill it out and upload it based on their learning progress. Teachers can analyze the learning log sheets to understand students' study habits and progress, providing personalized learning suggestions accordingly. Through these four aspects of the evaluation system, we can comprehensively and objectively assess students' effectiveness in the "Goodbye Infectious Disease Monsters" course. Meanwhile, these evaluation results will also serve as a reference and basis for local teachers to conduct education on infectious disease health knowledge. The team has established multiple user feedback channels to encourage children of farmers and herdsmen, as well as their parents, to actively provide valuable opinions and suggestions.

- **"N Sets" of Fun Health Courses**

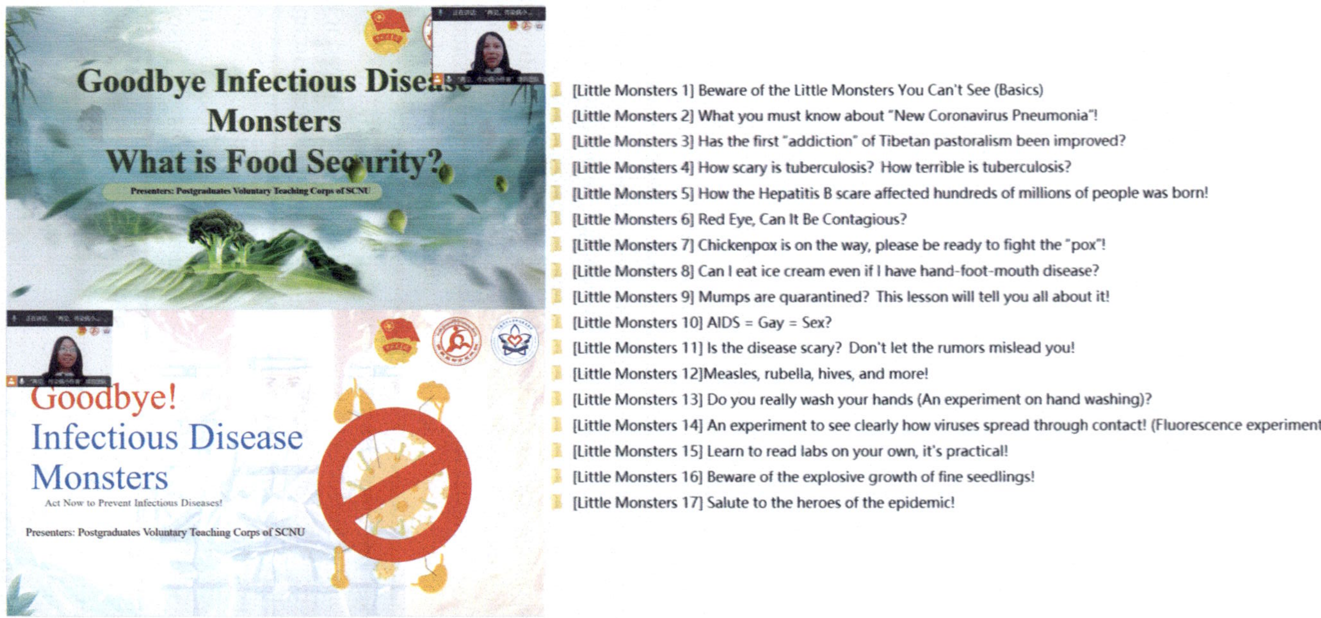

Fig. 15.3 Online "Cloud Health" course live screenshot and course design

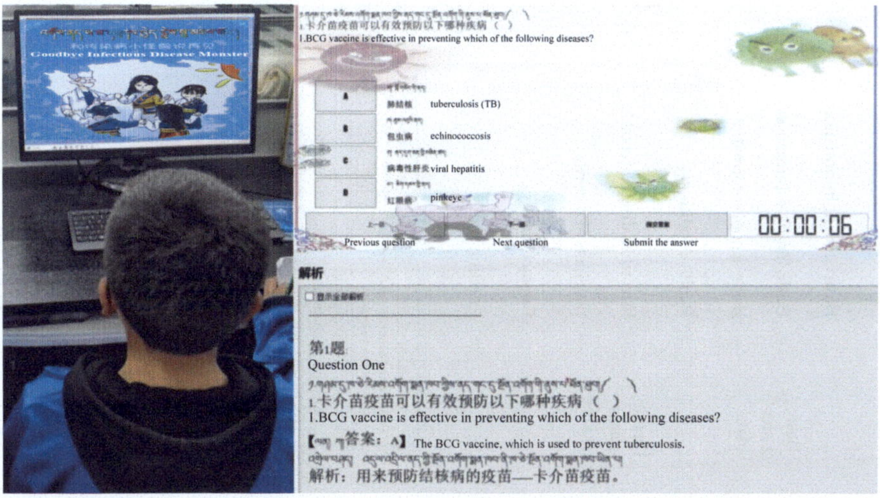

Fig. 15.4 "Goodbye Infectious Disease Monsters" health knowledge test system

Adhering to the concept of promoting the all-round development of students and taking "helping to build a healthy China" as its core, the school has formulated its mission around "teaching farmers' and herdsmen's children well and benefiting farmers' and herdsmen's family. Addressing the pain points of weak health awareness among children in farming and pastoral areas, the school has constructed "N sets" of fun health courses that prioritize health education and integrate multiple disciplines. Based on interdisciplinary projects, thematic activities, and cultural events, with club classes and class meetings as the main carriers, the school incorporates infectious disease prevention and control knowledge through physical education to strengthen physical fitness and psychology to adjust mental states in case of illness. Additionally, it integrates elements of art and collaborates with information technology. Considering the psychological characteristics of rural and pastoral teenagers, a series of fun health courses have been designed, reflecting the concepts of integration, innovation, and personalized education. As shown in Figs. 15.6 and 15.7, volunteer teachers lead engaging sessions, and local students actively participate in these health education activities.

To expand its influence of health education, the school actively develops and integrates various health education resources. It shares, disseminates, and exchanges the diverse digital resources of the "Goodbye Infectious Disease Monsters" curriculum system. It aims to achieve widespread coverage of high-quality health education resources through resource integration and optimization, thereby injecting new momentum into the development of health education in Tibet.

- **Digitization of Works**

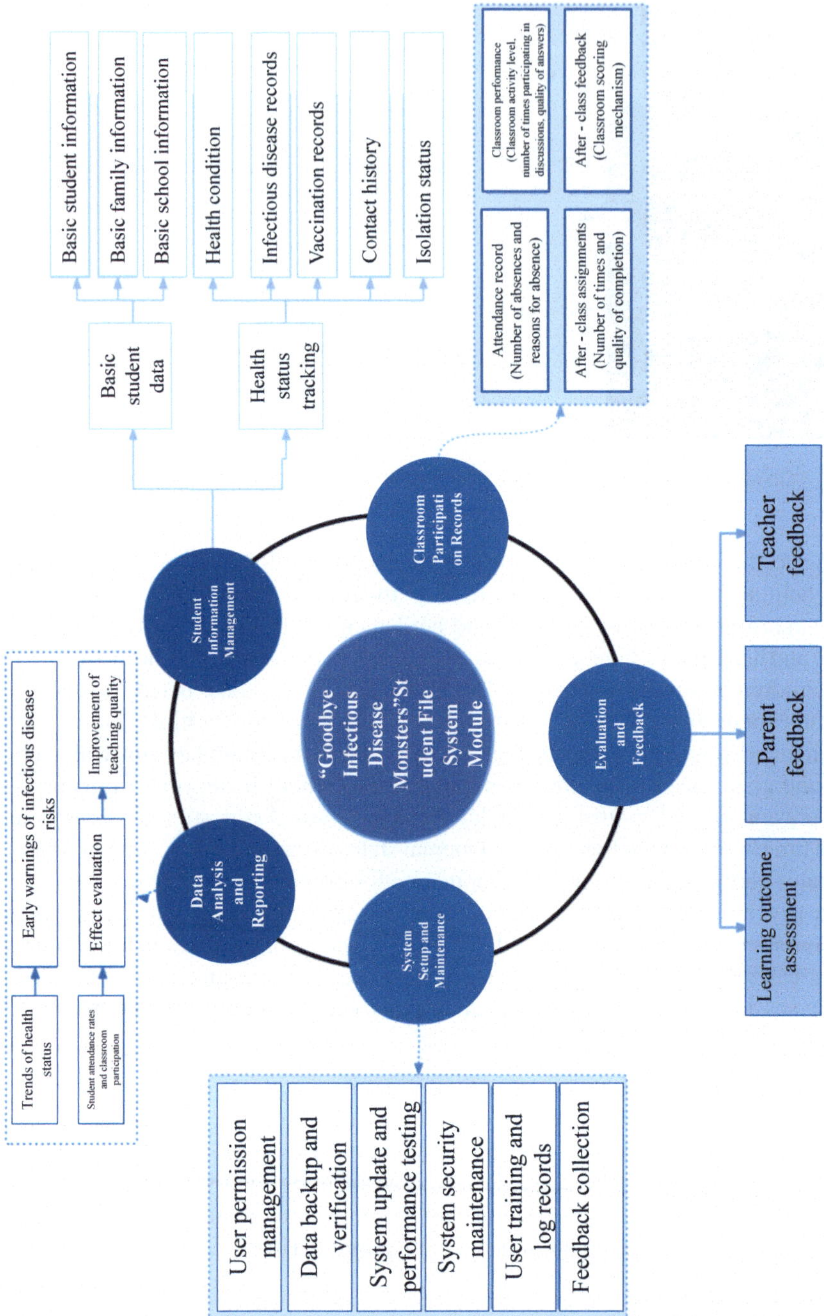

Fig. 15.5 "Goodbye Infectious Disease Monsters" student file system module

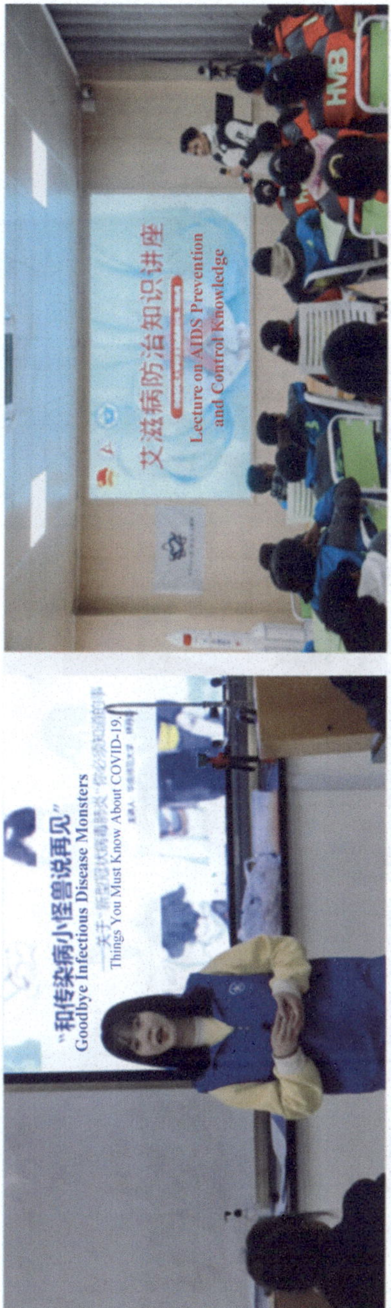

Fig. 15.6 Volunteer teachers conduct fun health courses

Fig. 15.7 Local students actively participate in health education classes

Apart from *Goodbye Infectious Disease Monsters* and *A Knowledge Handbook for Preventing COVID-19,* the school has further developed other audio picture books. After careful production, these audio picture books were uploaded to the Xizang Everest Qiyun Education Platform, enabling online sharing of the picture book resources (see Fig. 15.8).

- **Digitization of Activities**

To disseminate health knowledge, the school fully leverages online education platforms and integrates resources from external institutions and various parties to consolidate specialized knowledge on common infectious diseases in Tibetan areas. As shown in Fig. 15.9, the school has invited professionals in the field of health to conduct a series of lectures and presentations on specialized knowledge of common infectious diseases for students.

The design and implementation of the curriculum system above fully embody educational concepts such as student-centeredness, contextualized teaching, integration and innovation, practice orientation, resource integration, and personalized education, providing strong support for improving the health literacy and disease prevention capabilities of Tibetan children of farmers and herdsmen.

15.3 Implementation Outcomes

So far, over 3,150 farmers' and herdsmen's children have completed the course, mastering infectious disease prevention knowledge. It has enhanced students' awareness and, via the "educate one, influence one family" effect, improved rural families' public health awareness in Changdu. With support from local authorities and institutions, the "Goodbye Infectious Disease Monsters" resources are well-recognized. Through fundraising, 28,000 RMB has been raised for activity sustainability. The school has held 30 sessions with over 150 activities across 10 schools for such children. As shown in Fig. 15.10, the program's promotion events have been expanded. Using the Internet, it has extended resources like picture books and online courses to remote areas via relevant organizations, benefiting over 4,810 students. An outstanding student presentation group has been formed based on the "Eagle Soaring Volunteer Team," creating a virtuous cycle to boost the program's impact.

Furthermore, through the collaboration on this project, exchanges and interactions between the Han and Tibetan ethnic groups have been further strengthened, promoting ethnic integration and harmony. As illustrated in Fig. 15.11, this kind of cross-ethnic cooperation and exchange not only contributes to advancing public health in ethnic regions but also plays a significant role in fostering ethnic unity and building a harmonious society.

Fig. 15.8 The audio picture book *Goodbye Infectious Disease Monsters*

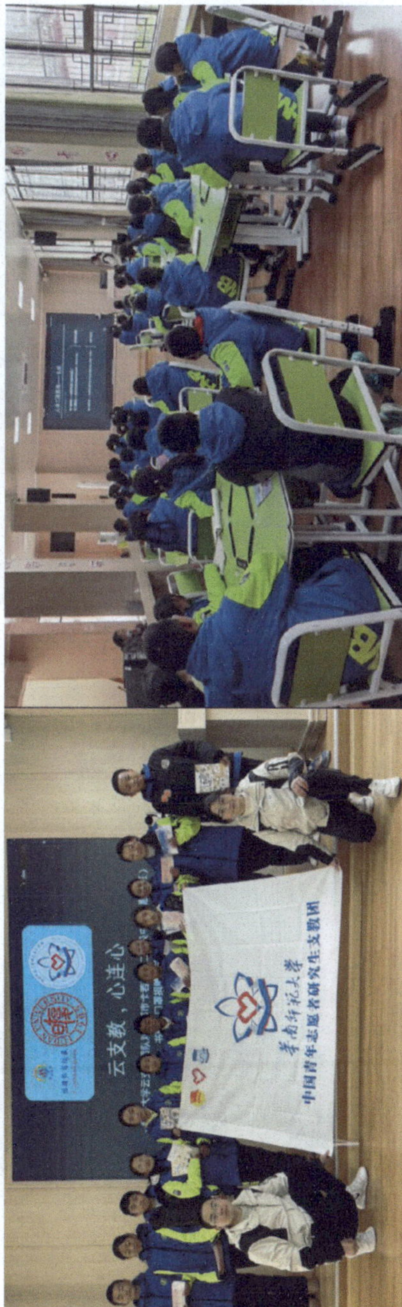

Fig. 15.9 Professional teachers from Fudan University were invited to give a cloud lecture

Fig. 15.10 “Goodbye Infectious Disease Monsters” promotion event

Fig. 15.11 Group photo of Han volunteers and Tibetan students

15.4 Innovations

With AI and big data, personalized learning paths are precisely customized. The system tracks the learning trajectory, slows down the animation rhythm and splits up knowledge points for slow learners, and expands cutting-edge research and complex cases for fast learners, so that the curriculum matches individual rhythms and learning efficiency jumps dramatically, and traditional "one size fits all" shortcomings are discarded.

The team adopted a "child-leads-parent" approach, providing health education to teenagers while also organizing special lectures for parents in agricultural and pastoral areas and primary and secondary school teachers. This home-schools collaboration model effectively strengthened the joint efforts of homes and schools, increasing parents' and teachers' emphasis on preventing common infectious diseases and injecting new impetus into health poverty alleviation work in Tibetan agricultural and pastoral areas.

While focusing on infectious disease prevention, the team also addressed teenagers' mental health. They simultaneously promoted the popularization of both infectious disease prevention and mental health knowledge, adding science courses on common mental illnesses. They also innovatively developed the "Goodbye Mental Illness Monsters" curriculum system, aiming to help children learn about mental health, safeguard their psychological growth, and build a strong mental health defense.

15.5 Promotion

The "Goodbye Infectious Disease Monsters" health education resource has achieved remarkable results in Tibet and demonstrates strong scalability. This resource targets children from agricultural and pastoral families, whose health status is directly related to Tibet's long-term development. Therefore, when similar projects are implemented in other regions, the experience and methods from this project can be utilized and appropriately modified and adjusted according to local conditions. Secondly, the resource adopts digital presentation and gamified learning approaches, making the content more engaging, interesting, and easier to understand and accept. This learning method aligns with the psychological characteristics and interests of teenagers and can be promoted and used in other regions and groups. Finally, the resource fully considers factors such as cultural integration and interactive design during the design process, allowing children from agricultural and pastoral families to learn about infectious disease prevention and control in a relaxed and enjoyable atmosphere. This design concept and method can also be applied and promoted in other health education resources.

15.6 Summary

This case aimed to explore the application of digital technologies to enhance the effectiveness of infectious disease prevention education for children in Tibet's agricultural and pastoral areas, which would strengthen their self-protection awareness and capabilities. Building upon distance education principles and digital technology applications, the project developed a digital-based "3 + 2 + N" health education resource framework titled "Goodbye Infectious Disease Monsters." This framework included three sets of illustrated books, two online learning resources, and multiple offline interactive courses. Its core value showed in three dimensions: first, significantly improving the target group's knowledge level and protective skills regarding plateau-region infectious diseases; second, promoting the overall enhancement of public health literacy among families in Changdu's agricultural and pastoral communities through a "student-family" linkage mechanism; finally, establishing a replicable model for integrating smart education with health education, providing regional implementation evidence for China's "Healthy China" strategy.

The case validated the adaptive innovation of situated learning theory and distance education models within specific geographical and cultural contexts. Its innovative nature was shown through three key breakthroughs: first, digital technologies overcome geographical constraints to build a targeted educational resource delivery system; second, it went beyond the unidirectional transmission of traditional health education to establish interactive learning loops; and third, it extended beyond single-course frameworks to develop multidimensional intervention strategies. These advances provided significant reference points for integrated education innovation supported by technology in underdeveloped regions.

Regarding practical application, the primary task involved establishing cross-regional promotion mechanisms in Tibet and similar agricultural-pastoral regions, while simultaneously enhancing localized teacher training systems and the development of Tibetan-Chinese bilingual resources to ensure sustainable implementation. It was recommended to progressively extend this validated model to other critical public health domains such as nutritional health and emergency response skills.

Future research should prioritize the development of AI-powered adaptive learning systems supported by intelligent technologies, employing longitudinal studies to evaluate long-term knowledge retention and changes in health behavior, and exploring the interactivity of blended learning through lightweight mobile applications.

Reference

França, C., Santos, F., Martins, F., Lopes, H., Gouveia, B., Gonçalves, F., Campos, P., Marques, A., Ihle, A., Gonçalves, T., & Gouveia, E. (2022). Digital health in schools: A systematic review. *Sustainability, 14*. https://doi.org/10.3390/su142113848

Chapter 16
Project-Based Course for the Automatic Treatment of the Waste Liquid in the Secondary School Chemistry Laboratory

Wen Zhao and Li Ma

Abstract Under the background of the K-12 "double new" reform (new curriculum, new textbook) issued by the Ministry of Education of the People's Republic of China (http://www.moe.gov.cn/fbh/live/2022/54639/sfcl/202207/t20220705_643405.html), combined with modern education equipment, information technology and innovations in secondary school chemistry curriculum development, it is essential to explore how project-based teaching and learning methods can promote students' core literacy and teachers' experimental literacy. Based on this premise, the topic of "the automatic treatment of waste liquid in the secondary school chemistry laboratory" was selected. Resources such as TI calculators, digital sensors, and the Arduino open-source electronic prototype platform were integrated to design and explore project-based course activities, as well as practical research. On the one hand, this case is intended to provide a general teaching strategy with high universality, strong operability, replication, and promotion for the design and implementation of inter-disciplinary project-based learning activities, on the other hand, it also intended to provide a reference for the global promotion of the deepening reform of primary and secondary school education and teaching oriented toward literacy.

Keywords Literacy-oriented · Science education · Waste liquid treatment · Environmental education · STEM

W. Zhao (✉)
Educational Technology Equipment Center Shanghai Education Committee, Shanghai, China
e-mail: zhaowengood@163.com

L. Ma
DaJing High School Affiliated to Shanghai, International Studies University, Shanghai, China

D. Liu et al. (eds.), *Empowering Educational Innovation Through Design: Volume One*, Smart Computing and Intelligence, https://doi.org/10.1007/978-981-95-4621-3_16

16.1 Background

The Chemistry Curriculum Standards for Senior High School (2017 edition, 2020 revision) emphasize the importance of literacy-based teaching in high school chemistry education. These standards advocate for creating real problem situations through chemical experiments and various investigations to develop core competencies in chemical disciplines. This approach aims to stimulate students' interest in chemistry, transform their learning methods, and cultivate their innovative spirit and practical abilities. By leveraging students' existing experiences and the social issues they will encounter, the curriculum guides students to focus on social problems related to chemistry, fostering a sense of social responsibility, participation, and decision-making skills.[1] The Opinions on Strengthening and Improving Experimental Teaching in Primary and Secondary Schools issued by the Ministry of Education in 2019 emphasizes the following in experimental teaching:

- "We should follow the characteristics of the subject.
- Actively encourage students to carry out research-based, task-based, project-based, problem-based, and cooperative learning.
- Expand innovation, by constantly integrating cutting-edge knowledge and the latest technological achievements into experimental teaching, enriching content, and improving methods.
- Prioritize practical effects by enhancing students' practical operation, situational experience, knowledge-seeking, personal perception, and innovation and creation.
- Focus on improving students' observation ability, practical ability, creative thinking ability, and teamwork.
- Cultivate students' interests, innovative spirit, scientific literacy, and quality of intention."[2]

16.2 Challenges and Problems

Project-based teaching and learning is an experiential learning mode where students construct and apply knowledge by engaging in experiential activities. This approach allows students to cross knowledge barriers between different disciplines, utilize interrelated knowledge, and solve problems from multiple perspectives, fostering interdisciplinary learning. In developed countries project-based teaching has made significant progress in K-12 education, accumulating many mature practice cases [1–3]. However, it remains a new teaching and learning method in China. At present,

[1] Ministry of Education. General High School Chemistry Curriculum Standards (2017 edition, 2020 Revised) [S]. Beijing: People's Education Press, 2020.

[2] Ministry of Education Opinions on Strengthening and Improving Experimental Teaching in Primary and Secondary Schools [EB/OL]. http://www.moe.gov.cn/srcsite/A06/s3321/201911/t20191128_409958.html.

it remains a challenge for secondary school teachers to cultivate students' innovative spirit and practical abilities, which are the key goals of the K-12 education reform. Addressing the need to shift chemistry teachers' one-sided understanding of experimental teaching and improve their project-based teaching skills is urgent [4]. As curriculum reform and classroom transformation in K-12 education continue to deepen, introducing project-based teaching and learning in chemistry experiments is becoming a developing trend. This project-based course, focusing on "the automatic treatment of the waste liquid in the secondary school chemistry laboratory," aims to conduct a preliminary exploration. The goal is to provide references for project-based experimental teaching and learning, K-12 educational technology application, and related research in secondary school chemistry education.

16.3 Aim of the Project-Based Course Activity

The new curriculum standards emphasize the importance of literacy-based teaching, emphasizing that teachers should "fully understand the unique value of chemical experiments and carefully design practical course activities." Chemistry, being an experiment-based subject, uses experiments as a unique and irreplaceable tool to develop students' core chemistry literacy. Project-based experimental teaching and learning serve as a vital medium for students to acquire and apply chemical knowledge, an essential method to cultivate chemical thinking and scientific literacy, and an effective approach to nurture innovative talents [5]. The new curriculum standards take scientific inquiry and innovation consciousness as core qualities of science, which highlight the value and status of chemistry experiment teaching and learning.

This project-based course activity focuses on the theme of "the automatic treatment of waste liquid in the secondary school chemistry laboratory," integrating TI technical resources. Emphasizing students' practice and experience, it addresses real-life problems and develops project-based experimental teaching and learning activities suitable for secondary school chemistry classrooms. The process follows: "course activity theme determination → course activity target design → course activity implementation strategy design → course activity content design → course activity evaluation key points formulation → course activity content implementation and results → course activity optimization reflection." This approach has allowed for preliminary exploration and practical research into project-based experimental teaching and learning in secondary school chemistry courses.

16.4 Exploration and Implementation of the Project-Based Course Activity

16.4.1 Determination of Project-Based Course Activity Theme

This course activity focuses on the theme of "the automatic treatment of waste liquid in the secondary school chemistry laboratory," addressing real problems encountered in daily chemistry experiments at secondary schools. The recent emphasis on chemistry experiments has inevitably led to a significant increase in the volume and complexity of waste liquid in these laboratories. Improper disposal of this waste liquid poses significant risks to the health of teachers, students, and the environment. Faced with this real problem, the authors combined programming education, maker education, artificial intelligence education, social practice to design an exploratory experimental course activity on the treatment of chemistry laboratory waste liquid with students. In this course activity, we explored solutions together with students, integrating various knowledge such as inspection, separation and purification of substances, preparation of molarity solutions, ion reaction and balance in aqueous solutions, colloid and electro flotation water purification, chemical fertilizer, solution acid–base, and tail gas treatment. A set of automatic waste liquid treatment systems based on Vernier digital sensors, the TI system, and Arduino open-source electronic prototype platform was designed. Through the innovative integration of technology and chemistry knowledge, this approach helps students use technology as an effective tool to acquire knowledge, process information, and solve real problems. By designing various experiment-related course activities, it aims to change students' learning styles, guide them to apply modern science and technology to solve real-life problems, improve their independent inquiry abilities, and develop their core chemistry literacy.

16.4.2 Design of Course Activity Objectives and Implementation Strategies

The Course activity objectives and implementation strategies are shown in Tables 16.1 and 16.2.

Table 16.1 Design of course activity items

Project-based course activity objectives	Reflect the core quality of the discipline	Attribute	Attribute description
	– Macroscopic identification and microscopic analysis – Change ideas and balance concepts – Evidence reasoning and model cognition – Scientific inquiry and innovation consciousness – Scientific attitude and social responsibility	Knowledge and Skill	To understand ion reactions, their conditions, and to learn the micro perspective of substance existence and reactions in aqueous solution To learn to test common anions and cations and design experiments for the separation and purification of various substances
		Process and Method	Understand acid–base neutralization reactions and their environmental impacts Initially master the use of digital experimental instruments (pH meter, ion electrode sensor, Arduino UNO, TI graphics calculator, TI innovator system), obtain and analyze experimental data, and appreciate the convenience of digital tools in chemical experiment exploration
		Emotional attitude and values	Improve environmental awareness, deepen understanding and attention to environmental issues, cultivate a sense of social responsibility, and develop a mindset focused on sustainable development

16.4.3 The Designation of the Activity Content and Evaluation Key Points

16.4.3.1 The Designation of Course Activity Content

Before the experimental activity, students should understand the type and quality of waste liquid produced by the chemistry laboratory through material research, reading, and analysis. They then design experimental schemes and carry out collaborative group experimental activities, as outlined below: Example of Course Activity Content Design:

- **Course Activity Theme**: Project-based course activity for the automatic treatment of waste liquid in the secondary school chemistry laboratory
- **Course Activity Tasks**: Statistics and analysis of the type and quantity of waste liquid produced by chemistry laboratory, based on relevant material research, reading, and analysis. Combining existing chemical knowledge and data collected by the group (such as wastewater discharge standards), we aim to explore

Table 16.2 Implementation of strategy design of the course activity

Activity theme	Project-based course activity for the automatic treatment of waste liquid in secondary school chemistry laboratories		
Activity time and space	Extension classes, interest group activities, research learning activities (1 ~ 2 h)		
Activity level	Why to do it, how to do it, and how to do it		
Activity tools	Resource	Experiment	Equipment, place
		Medium	Task paper, projection
	Circumstances	Source	Real life
		Function	Stimulate interest, experimental inquiry
Implementation strategy	Implementation of the layout	Activity list, projection	
	Implementation dimension	Scheme design, experimental operation, activity record, exchange and cooperation	
Evaluation scheme	Valuator	Self-evaluation, mutual evaluation, teacher evaluation	
	Evaluation method	Oral evaluation, evaluation table evaluation	
	Evaluative dimension	Scheme design, experimental operation, activity record, exchange and cooperation	

waste liquid treatment methods and solutions for achieving automatic wastewater treatment in secondary school chemistry laboratories.

- **Design Each Group's Experimental Activity Scheme**:

  (1) Regulate Acid–Base of Waste Liquid: Explore schemes for regulating the acid-based waste liquid, including automatic treatment methods for heavy metal ions (e.g., chemical precipitation method, electrolysis method, metal replacement method).
  (2) Recycling Waste Liquid: Investigate waste liquid recycling schemes.
  (3) Absorption of Exhaust Gas: Develop schemes for the absorption of exhaust gas produced during waste liquid treatment.
  (4) Automatic Control: Explore methods for achieving automatic control. Design and set up automatic treatment devices for waste liquid in secondary school chemistry laboratories using TI graphics calculator, the TI innovator system, the Arduino open-source electronic prototype platform, and various Vernier digital tools. Employ variable control and classification methods to analyze, process, reason, and induce experimental data.

Group Discussion and Communication: optimize and improve each group's experimental activity design schemes. Develop different treatment methods for waste liquid in secondary school chemistry laboratories and summarize the common and special treatment methods for laboratory waste liquid. To be specific:

Exploring the best scheme to regulate the acid and alkalinity of waste liquid, which is the most common type of waste in the chemistry laboratory. If it is not recycled, direct discharge can cause serious corrosion to pipelines and harm the self-purification function of the water bodies, destroying natural ecology. We design experiments to explore the effectiveness of hydrochloric acid, sulfuric acid, acetic acid, and other acids in neutralizing alkaline waste liquid, and sodium hydroxide, calcium hydroxide, potassium hydroxide, ammonia, and other alkalis in neutralizing the acid waste liquid. This exploration combines data access, experimental data analysis, and considerations of economy and environmental protection.

Exploring the best treatment methods for heavy metal ions in laboratory waste liquid. Direct changes of such waste into the environment often lead to biological enrichment, making it difficult to eliminate and posing significant health risks, potentially causing serious harm or even death. This study investigates the effectiveness of various treatment methods, including alkali precipitation, sulfide precipitation, electrolysis, and metal replacement, aiming to identify the best solution from both economic and environmental perspectives.

Exploring the absorption scheme for exhaust gas produced during waste liquid treatment. This course activity design investigates the possible types of waste gas, develops treatment and collection schemes, and explores methods to reduce waste gas production by controlling waste liquid pH or other conditions.

Exploring the best solutions for recycling waste liquid, which often contains recyclable ions. The threshold range can be examined from economic and environmental protection perspectives. For example, investigating how waste liquid with certain pH ranges can be directly collected and used as acid–base treatment liquid, and how waste liquid containing certain concentrations of nitrate ions, potassium ions, and high dissolved oxygen can be used as plant nutrient liquid.

Exploring how to automate control using pH sensors, nitrate ion electrodes, the Arduino open-source electronic prototype platform, TI graphics calculator, TI data collector, and other digital equipment. Additionally, investigating the use of TI graphics calculator to write related programs for forming automatic control systems, and so on. Groups of students independently design automatic waste liquid treatment devices for secondary school chemistry laboratories using TI and Arduino UNO for programming. They apply the concept of variable control and classification in experimental exploration to analyze, process, reason, and summarize the experimental data. Group members discuss and communicate with each other during experimental activity to optimize and improve their design schemes, develop different waste liquid treatment methods, and summarize both common and special treatment methods for chemistry laboratory waste liquid.

16.4.3.2 Formulate the Key Points of Course Activity Evaluation

Based on the evaluation key points formulated in Table 16.3, teachers can evaluate the project-based course activities from four dimensions: that is, program design, experimental operation, experimental record, communication, and cooperation. Teachers should also pay attention to guiding students to conduct objective and scientific self-evaluation and mutual evaluation.

16.4.3.3 Teaching Practice and Results of the Course Activity

In specific implementation, according to the inquiry content, the program file is run on the TI graphics calculator and the Arduino UNO to collect data from the sensor system (e.g., pH, nitrate ion concentration, potassium ion concentration, dissolved oxygen, liquid level, turbidity). This information is then fed back to the Arduino UNO and the TI graphics calculator (see Fig. 16.1). Through students' practice and continuous optimization, the complete automatic treatment system consists of four units: heavy metal ion identification and treatment unit of waste liquid, pH automatic adjustment unit of waste liquid, recovery and treatment unit of waste liquid, and hydroponic plant growth environment automatic adjustment unit. To be specific:

- Heavy metal ion identification and treatment unit of waste liquid

As shown in Fig. 16.2, the working principle of the waste liquid heavy metal ion identification and treatment unit is as follows:

- After the laboratory waste liquid is filtered into the heavy metal identification tank, the No. 1 water pump automatically injects 5 mL of sodium sulfide solution.
- When the turbidity sensor value is less than 50, the No.2 water pump starts to transfer the waste liquid to the No.1 pretreatment tank for subsequent treatment.
- When the turbidity sensor value reaches 50, pump 1 continues injecting the sodium sulfide solution. When the pH level exceeds 7, the No. 3 pump starts to transfer the waste liquid out, discharging it after multilayer filtration.

As shown in Fig. 16.3, the pH automatic regulation unit of waste liquid is mainly divided into two intelligent processing and selection steps:

Step 1: When the pH value of waste liquid is 12 in the No.1 pretreatment tank, the No.1 water pump performs the corresponding program command issued from the intelligent program system to automatically transfer the waste liquid to the alkali liquid collection tank, which has a pH value of 2. Then, the No.2 water pump automatically moves it to the acid collection tank.

When the No.1 pretreatment tank has $2 < pH < 12$, the No.3 water pump performs the corresponding program command issued by the intelligent program system to automatically transfer the waste liquid to the treatment tank.

Table 16.3 Key points of course activity evaluation

Dimension	Specification requirements	Evaluation points
Conceptual design	① To design the experimental protocol ② To explore the best solution for adjusting waste acid and alkaline solutions (such as energy efficiency, speed, and effectiveness) To investigate the best treatment methods for heavy metal ions (such as chemical precipitation, electrolysis, and metal replacement) ③ To explore the best solutions for recycling waste liquid (such as collecting waste liquid within a certain pH range for use as an acid–base treatment solution and using waste liquid containing certain nitrate ions as plant nutrient solutions) ④ To explore the absorption of exhaust gas produced during waste liquid treatment (such as methods to reduce exhaust gas production by controlling waste liquid pH) ⑤ To explore how to achieve automated control ⑥ To ensure the experiment process is rigorous ⑦ To pay attention to variable control during the experiment	Observe students 'design experiment, control variable ideas Observe students' experiment design ability and problem-solving ability
Operations in experiment	① Ensure the experimental manipulation is completed within 120 min ② Students have initially mastered the use of digital experimental instruments, such as pH meters, ion electrode sensors the Arduino open-source platform, TI graphics calculator, and the TI innovator system	Observe whether the students can correctly use the digital instruments and standardize the operation in the experimental activities

(continued)

Table 16.3 (continued)

Dimension	Specification requirements	Evaluation points
Experimental activity record	Ensure that the experimental data are acquired, recorded, classified, analyzed, and processed	Observe whether the students' activity recording and recording is accurate, whether they can record the activity process from multiple angles, whether there is the consciousness of classification and induction, and the ability, whether they have the reflection and improvement of the experiment activity and scheme
Communication and cooperation	Facilitate group discussions Ensure a clear division of tasks among group members Optimize the experimental scheme and make improvements based on the implementation results	Observe whether the students can actively express their opinions, whether the division of labor is clear, and whether there is a sense of team cooperation

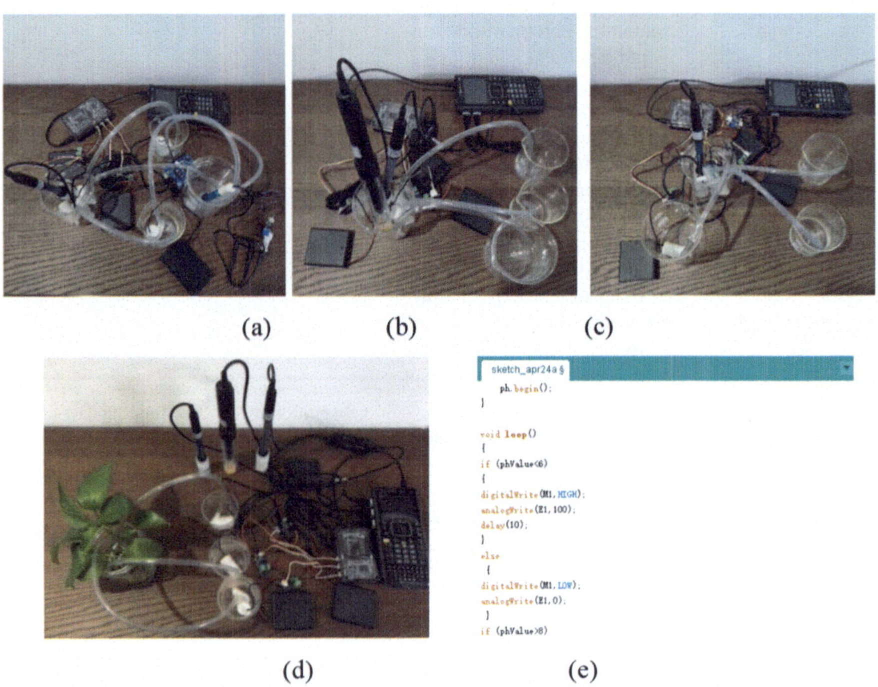

Fig. 16.1 Examples of experimental device (**a**–**d**) and program code (**e**)

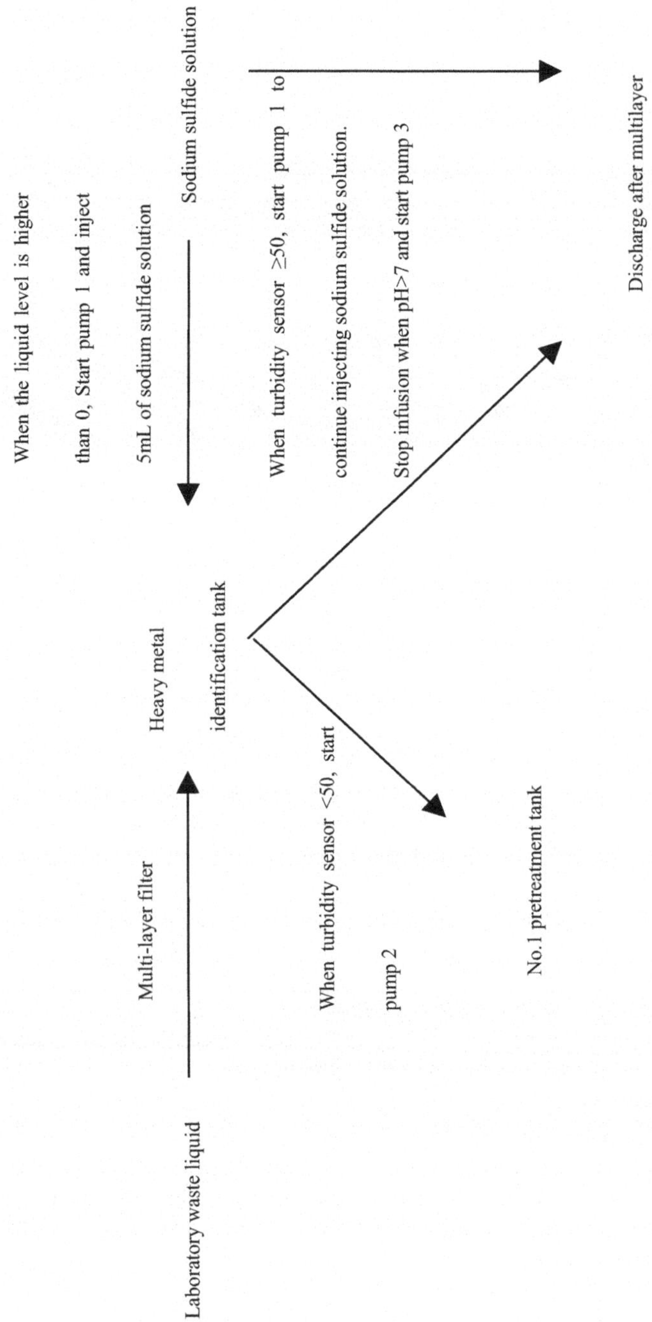

Fig. 16.2 Heavy metal ion identification and treatment unit of waste liquid

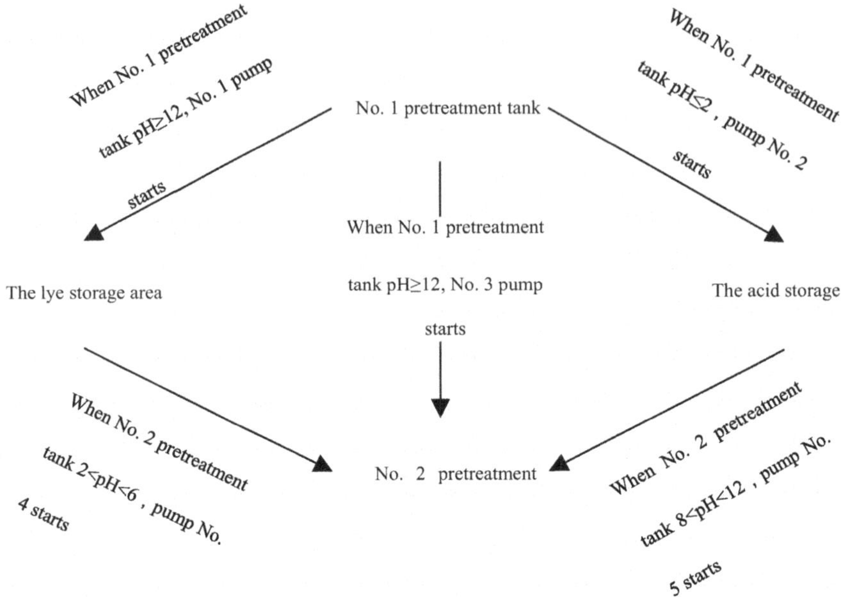

Fig. 16.3 pH automatic regulating unit of waste liquid

Step 2: When the waste liquid in No.2 tank has $2 < pH < 6$, pump No.4 performs the corresponding program command issued by the intelligent program system, automatically injecting the liquid from the alkali fluid collection tank into No. 2 tank and adjusting its pH to 6.

When the waste liquid has $8 < pH < 12$, pump No.5 performs the corresponding program command issued from the intelligent program system, automatically injecting the liquid from the acid collection tank into No. 2 tank and adjusting its pH to 8.

As shown in Fig. 16.4, the intelligent treatment of the waste liquid recovery unit is as follows:

- Recovery and treatment unit of waste liquid.
- When the pH value of the waste liquid in the No. 2 treatment tank is between 6 and 8, the liquid is transferred to the No.3 treatment tank.
- In the no. 3 treatment tank, if the concentration of nitrate ion (NO3-) is more than 500 mg/L or the concentration of dissolved oxygen (DO) is more than 100 mg/L, both pumps 6 and 7 execute the corresponding program commands issued by the intelligent program system to automatically collect the waste liquid as hydroponic plant nutrient solutions. If these conditions are not met, the waste liquid is discharged directly.

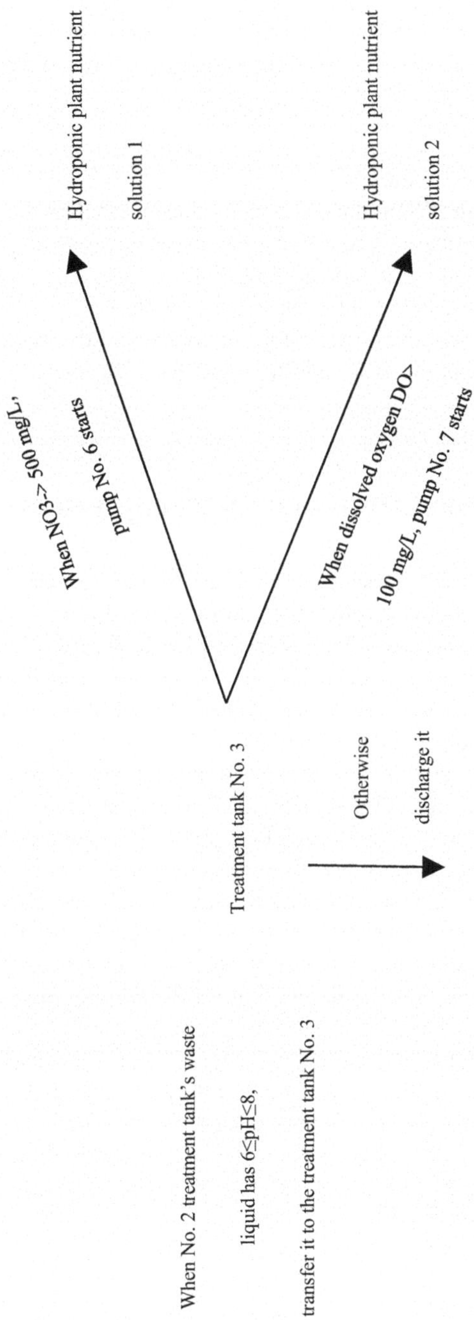

Fig. 16.4 Recovery and treatment unit of waste liquid

16.4.3.4 Automatic Regulation Unit of Hydroponic Plant Growth Environment

As shown in Figs. 16.5 and 16.6, the operating principle of this automatic regulation unit is

- **Data Collection**: Digital sensors collect the system's pH, dissolved oxygen, and nitrate ion concentration data.
- **Automatic Regulation**: When the measured value is less than the preset parameters, the pumps automatically execute the corresponding program command issued by the TI system and add regulated solutions to the hydroponic plant system, until the parameters reach the target set values (see Fig. 16.6).
- **System Shutdown**: Once the target values are achieved, the system switches off automatically, and the pump stops adding regulated solutions.

16.5 Implementation Effects and Spreading Values

Overall, this project-based course activity uses digital sensors, Arduino open resources, and the TI system to explore the automatic treatment of the waste liquid in the secondary school chemistry laboratory. Through program instructions, data collection, graphics processing, and analysis, the activity can improve students' scientific literacy, promote understanding of the principles and essence of chemistry phenomena, and enhance the ability to analyze and solve real-life problems. Of course, this project-based course activity can be further expanded to explore additional influencing factors, enriching students' learning experiences to the fullest. By leveraging digital experiments, we can quickly capture information during chemical reactions, making the process more intuitive, especially in observation analysis. This helps uncover the underlying experimental principles behind the datasets. Moreover, this approach not only promotes effective learning for K-12 students but also transforms their learning methods and teaching strategies. It enhances students' ability to independent explore and lays the foundation for developing their core literacy in chemistry.

At present, this project-based course activity has been successfully piloted in many secondary schools in Shanghai. The results proved that it can be generalized and easily spread. Moreover, this project-based course activity is available for all secondary students in Shanghai through the digital curriculum on the science and innovation curriculum platform of the Teaching and Research Department of Shanghai Education Commission.

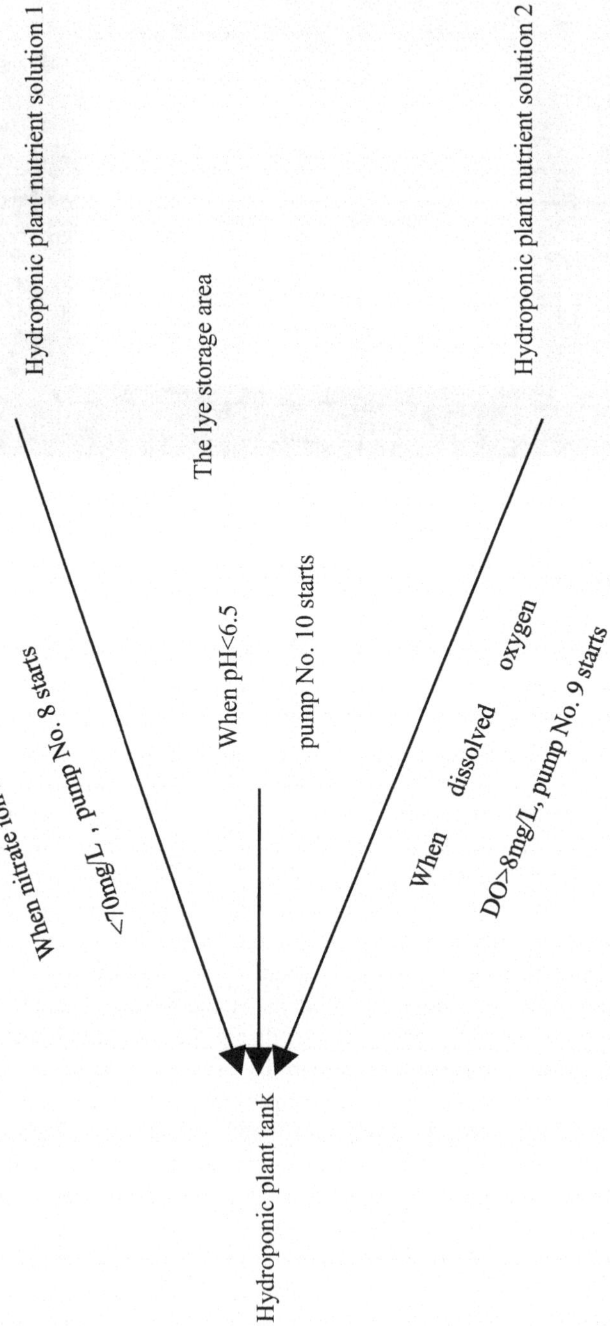

Fig. 16.5 Automatic regulatory unit of hydroponic plant growth environment

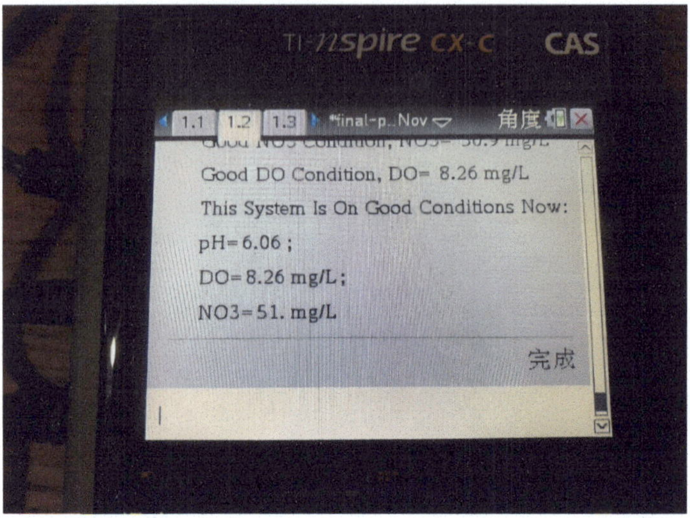

Fig. 16.6 Screen display of the TI graphics calculator after the program runs

16.6 Conclusion

The project-based experimental course activity, focusing on the "automatic treatment of waste liquid in the secondary school chemistry laboratory," successfully launched a real-world scenario and established a structured inquiry-based teaching and learning process, integrated with information technology. Throughout various course activities, teachers employed a "task-driven" approach to guide students in engaging their brains, hands, and verbal cooperative learning. Despite differences among students in course activities, all the students were able to leverage their existing knowledge during problem-solving experiences. This allowed them to consolidate old knowledge, acquire new knowledge, deepening their understanding, and master relevant chemistry principles, ultimately developing students' core literacy in chemistry.

Secondly, in implementing the project-based course activity, the integration of various disciplines constantly improves students' experiences. This effective integration of real problem-solving across various disciplines not only boosts students' environmental awareness and social responsibility but also contributes significantly to secondary school chemistry education. This project-based course guides students to focus on environmental issues, equipping them with the knowledge, methods, and abilities people needed for harmonious coexistence with the environment. It strengthens students' environmental consciousness and encourages them to adopt environmentally friendly learning habits and lifestyles. Furthermore, it underscores the importance of continued exploration in this research area by education workers and scholars. Thirdly, a comprehensive analysis of the designation of this project-based course activity reveals that the effective implementation of project-based experimental course activities largely depends on teachers' own experimental literacy and teaching ability. To achieve this, chemistry teachers should search for or read academic research literature about chemistry experiments, actively participate in national and provincial high quality experimental teaching and learning innovation seminars, actively explore modern educational equipment applied to secondary school experiments and constantly improve their experimental literacy and experimental teaching ability. For provincial and municipal education departments, developing teachers' experimental teaching ability through training workshops, teaching ability improvement programs, experimental operation skills sessions, and experimental teaching lectures, as well as providing online training courses, is crucial. This approach promotes the development of chemistry teachers' experimental teaching ability and literacy. Of course, addressing the current lack of project-based experimental course activities and curriculum should begin with cultivating normal university students, ensuring future teachers are well-equipped with the necessary skills and knowledge.

16.6.1 Research Purpose

In the context of China's K-12 "double new" reform (new curriculum and new textbook), this study aimed to explore how project-based teaching and learning can enhance students' core literacy and teachers' experimental literacy in secondary school chemistry. By focusing on the "automatic treatment of waste liquid in the secondary school chemistry laboratory," it integrated modern education resources such as TI calculators, digital sensors, and the Arduino platform to design and implement project-based course activities.

16.6.2 Core Values

This study offered a highly universal, workable, replicable, and promotable teaching strategy for interdisciplinary project-based learning. It not only contributed to the specific field of secondary school chemistry education but also provided a valuable reference for the global promotion of literacy-oriented reforms in primary and secondary education.

16.6.3 Significance for Educational Theory and Innovation

Theoretically, it improved the application of project-based learning theories in the chemistry curriculum, demonstrating how modern technology can be effectively integrated into traditional teaching methods. Practically, it innovated the way chemistry experiments were conducted, making them more engaging and relevant to real-world issues, thus promoting students' practical skills and critical thinking.

16.6.4 Suggestions

For practical application, schools should invest in modern education equipment and provide teacher training to effectively implement such project-based learning. In future research, more interdisciplinary projects could be explored, and long-term studies on students' literacy development could be conducted to further validate the effectiveness of this approach.

References

[1] Diawati, C., Liliasari, L., Setiabudi, A., et al. (2018). Using project-based learning to design, build, and test student-made photometer by measuring the unknown concentration of colored substances. *Journal of Chemical Education, 3*, 468–475.
[2] Juhl, L., Yearsley, K., & Silva, A. J. (1997). Interdisciplinary project-based learning through an environmental water quality study. *Journal of Chemical Education, 12*, 1431–1433.
[3] Nagarajan, S., & Overton, T. (2019). Promoting systems thinking using project-and problem-based learning. *Journal of Chemical Education, 12*, 2901–2909.
[4] Zhao, W. (2024). The exploration and practical research of Maker-based experimental teaching in primary and secondary schools. *Journal of Shanghai Educational Research, 8*, 69–74.
[5] Zhao, W., & Ma, L. (2025). Exploration of educational reform in experimental teaching aimed at cultivating innovative creativity—fabrication and application of electronic balance based on open source hardware. *Educaiton in Chemistry, 1*, 65–69.

Part III
Showcase of Educational Service Designs

Chapter 17
Intelligent After-School Extended Care Programme for Primary School Students

Guobing Qian, Ao Dong, and Fang Chen

Abstract The end time of after-school care services often conflicts with parents' off-duty hours, leading to challenges in coordinating the pick-up of children. To address this issue, the "SafePick" intelligent system has been developed, which integrates a range of functionalities, including automatic notifications, online reservations, information exchange, location sharing, and duty clock-in. The primary goals of the system are to facilitate convenient transportation for parents, ensure the safety of both teachers and students, provide data to inform management decisions, streamline transportation logistics, and enable precise child pick-up. The system has received positive feedback from parents, teachers, and students, with favourable responses from the broader community.

Keywords After-school services · Intelligent hosting · Digital governance · School safety management · SafePick

17.1 Background

In China, a common issue faced by working parents is the misalignment between the end time of after-school services and their work hours, which complicates the process of child pick-up (Gao & Xu, 2019). Furthermore, during peak after-school hours, school gate areas tend to become crowded with a complex mix of people, which raises safety concerns. This problem is particularly pronounced in rural schools, where many caregivers are elderly, often grandparents (Zhang, 2022). To address these challenges in after-school care services, Chang Xi Primary School, a member of the Ai Shan Primary School Education Group in Huzhou City, has adopted intelligent technology to create the "SafePick" after-school childcare system.

The "SafePick" intelligent system at the school operates on the DingTalk YiDA platform, facilitating the development of a comprehensive transportation service

G. Qian (✉) · A. Dong · F. Chen
Hu Zhou Ai Shan Primary School Education Group Chang Xi Primary School, Wu Xing District Education Bureau, Huzhou, China
e-mail: asqgb@163.com

D. Liu et al. (eds.), *Empowering Educational Innovation Through Design: Volume One*, Smart Computing and Intelligence, https://doi.org/10.1007/978-981-95-4621-3_17

system. Key features of this system include automatic notifications, online reservations, information sharing, location tracking, and duty clock-in. Following its implementation, the school established a designated waiting area and installed facial recognition gate devices, all of which are connected to the "SafePick" after-school service platform. When students enter the school gate, the system automatically recognizes them and sends a notification to parents. Parents can then use their mobile devices to select the time they plan to arrive at school. This system not only enhances student safety but also reduces the time parents spend waiting at the school gate, thus streamlining the pick-up process. The result is a more efficient and timely pick-up experience, thus achieving the goal of "stop, pick up, and go" (Xiong & Wu, 2015).

17.2 System Overview

This intelligent hosting service system, built on the DingTalk YiDA low-code development platform, is compatible with PC, mobile, and web platforms. It facilitates real-time information synchronization between devices and incorporates multiple features such as automatic notifications, online reservations, information exchange, location sharing, and duty clock-in.

17.2.1 Overall System Architecture

The system includes a large electronic screen located in the waiting room, which displays real-time pick-up and drop-off information for students and parents. Additionally, a voice broadcast system is in place to provide reminders to students about their pick-up details (see Fig. 17.1).

17.2.2 SafePick Parent Mobile Application

The parent mobile application offers two user-friendly features: the ability to send a voice message and a "Parent has arrived at the school gate" notification to the waiting room (see Fig. 17.2).

17.2.3 SafePick Parent Mobile Application

The teacher mobile application allows educators to view the specific status of students in the waiting room for the day. It also enables the setting of start and end times for the waiting room's operation. The "Teacher on Duty Record" feature allows teachers

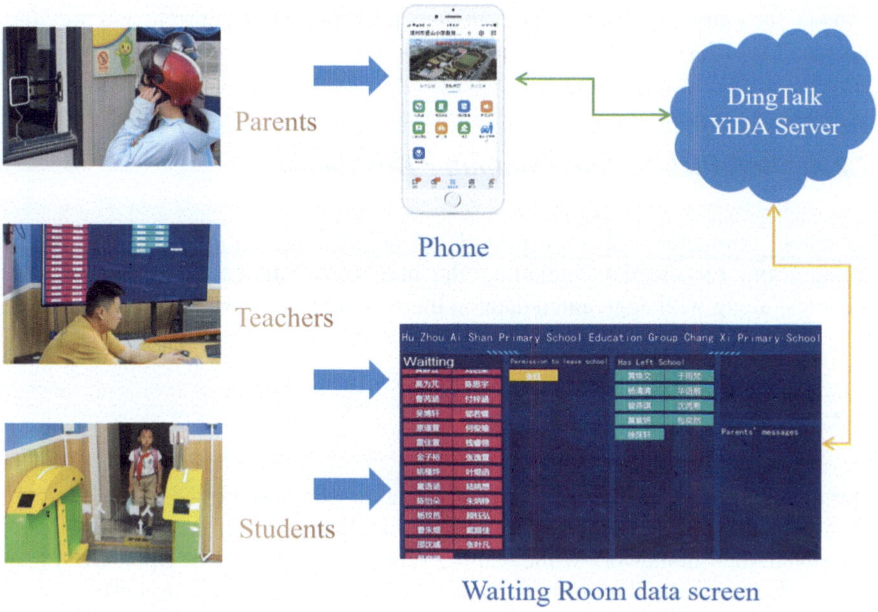

Fig. 17.1 Overall system architecture

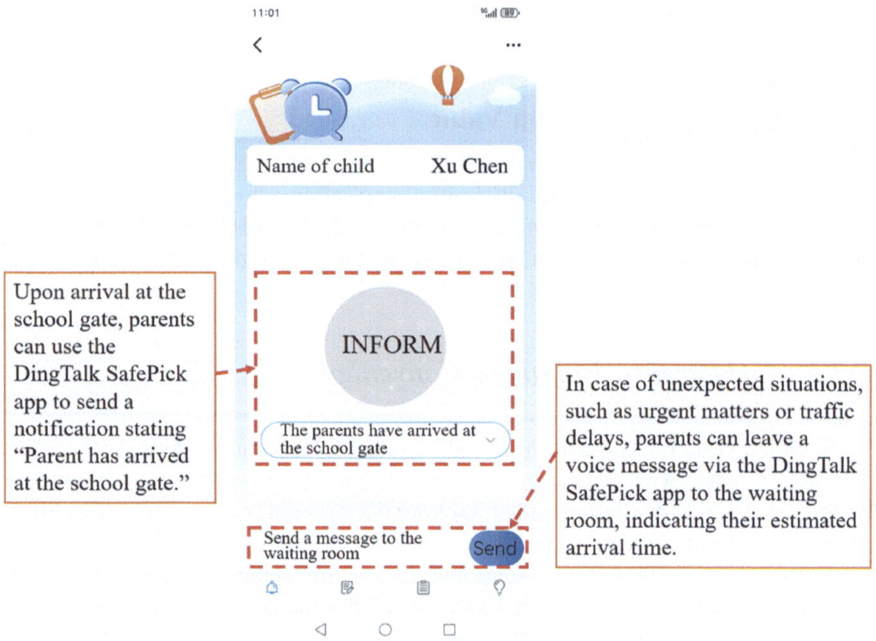

Fig. 17.2 Parent mobile application

to review the entry and exit times of duty teachers for both the current and previous days (see Fig. 17.3).

17.2.4 SafePick School Computer Dashboard

The school computer dashboard provides detailed information on students in the waiting room for the day, including the number of students and teacher duty schedules, along with aggregated data on the waiting room's usage (see Fig. 17.4).

17.2.5 Implementation Process

After the conclusion of after-school care, students who have not been picked up on time enter the waiting room via facial recognition. In the waiting room, they can read or complete their homework while waiting safely for their parents. Upon arriving at the school, parents send a notification via the DingTalk SafePick app. The system then transmits a voice message to the waiting room to inform the students, who will use facial recognition to exit the room, thereby completing a secure, closed-loop process (see Fig. 17.5).

17.3 Practical Application Value

Since its implementation, the SafePick intelligent after-school care service system has gained widespread recognition from parents, students, and teachers for effectively addressing the common challenge of delayed student pick-ups.

17.3.1 For Parents: Enhanced Convenience

The SafePick system has greatly reduced incidents of anxious or disruptive parents at the school gate. With the system in place, parents are now able to pick up their children in a calm and orderly manner, without the stress of rushing to arrive on time. Additionally, by leveraging backend big data, the system identifies students who frequently require extended after-school care. This information data is shared with class teachers, allowing for targeted analysis of the underlying causes, improving communication with parents, and enhancing overall satisfaction with the school (Gong, 2021).

Fig. 17.3 Teacher mobile application

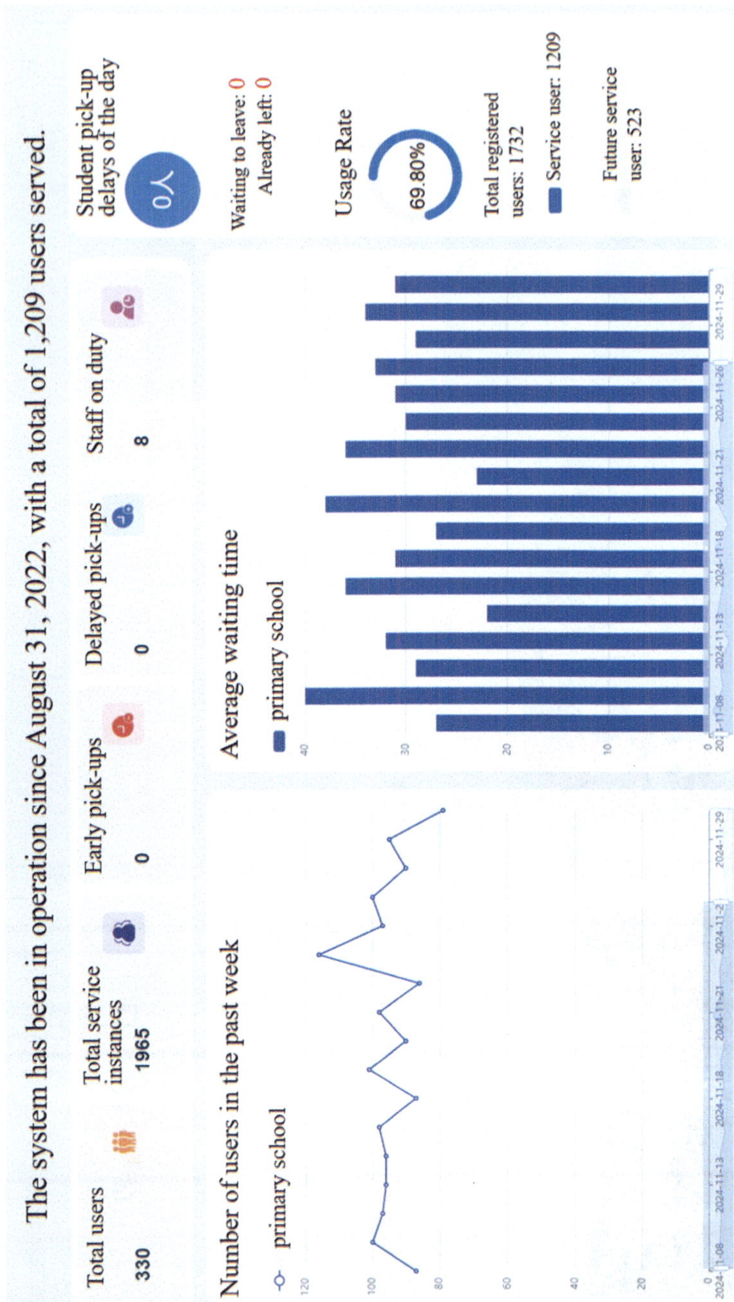

Fig. 17.4 School computer dashboard

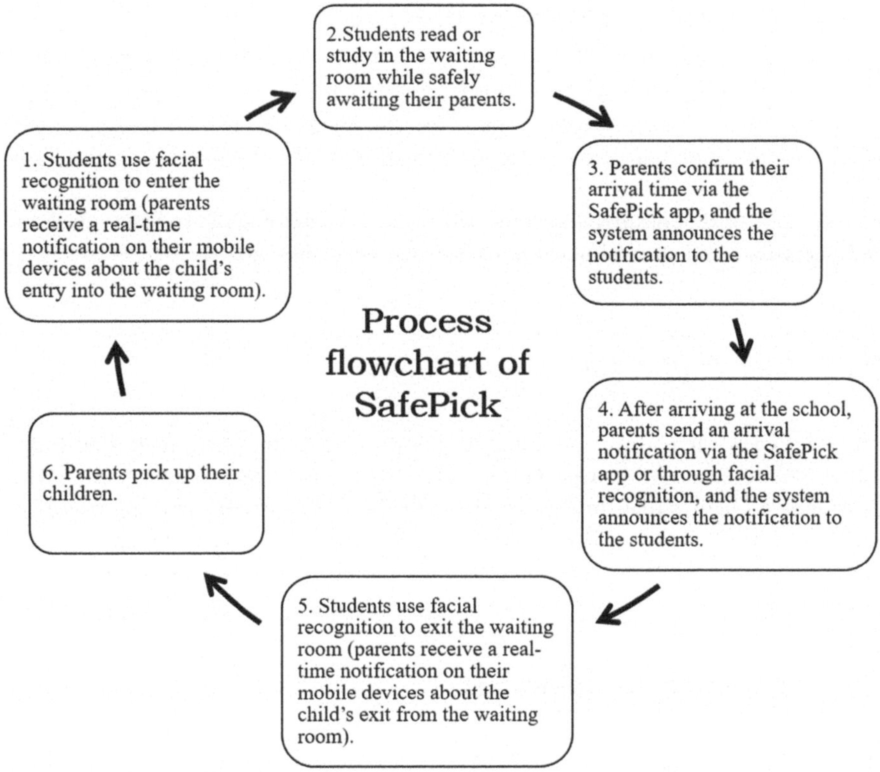

Fig. 17.5 Process flowchart of SafePick

17.3.2 For Students: Improved Safety

The SafePick system addresses the inefficiencies of students waiting outside the school gates. By streamlining the waiting process, the system ensures that students' time is used productively and, most importantly, enhances their personal safety while waiting for their parents.

17.3.3 For the School: Smarter Management

The system collects real-time data, enabling the school to monitor campus safety promptly and accurately. By comparing this data with historical trends, the system aids informed decision-making regarding student management. This data-driven approach strengthens communication between the school and parents, improving the overall effectiveness of school operations.

17.3.4 For Teachers: Reduced Workload

The SafePick system reduces the workload of teachers by shifting away from the traditional method where parents wait at specific locations, such as the school gate or classrooms, to pick up their children. Instead, the system introduces a more efficient, centralized waiting room management approach. This shift from decentralized management to a centralized system, managed by school staff, reduces teachers' responsibilities and streamlines school operations (Zhang, 2018).

17.3.5 For Traffic: Coordinated Efforts

The SafePick system fosters collaboration with local authorities, including traffic police, public security, transportation, and urban management departments. This coordination enables real-time adjustments to staff deployment, ensuring resources are allocated efficiently and effectively. As a result, the system enhances traffic management around the school and improves overall school-related activities.

17.4 Promotional Significance

The SafePick intelligent after-school childcare system developed by our school has effectively addressed the challenges faced by rural parents in picking up their children, ensuring student safety after school. The innovative approach has received wide recognition from regional schools for its effectiveness and practicality.

The system was first featured on March 1st, 2022, in the official account of the local News Media Center, where it quickly garnered over 80,000 views and was shared across multiple platforms. Following its debut, major outlets such as *China Education News*, *Zhejiang Daily*, and *Zhejiang Education News* reported on the system, generating significant attention and sparking widespread interest.[1]

The system is cost-effective, relying on hardware like large screens and facial recognition devices, with software developed on the DingTalk YiDA platform. By leveraging existing data infrastructure and communication channels within the DingTalk home-school framework, the system remains both accessible and familiar to parents.

In addition to resolving the challenges of after-school care coordination, the SafePick system offers a new model for digital governance in schools, setting a precedent for future innovations in educational administration.

[1] http://www.jyb.cn/rmtzcg/xwy/wzxw/202203/t20220304_682308.html

17.5 Summary

This case explored innovative applications of intelligent management approaches in after-school extended care services at the primary school level. It addressed important issues such as the difficulties parents faced in picking up their children, possible safety risks, and management problems, which improved the collaborative governance capacity among schools, families, and society.

The core contribution of this case was the use of "SafePick" intelligent pickup system, which allowed for a more intelligent and data-informed approach to student management. By protecting students during after-school care, the system significantly improved pickup efficiency for parents and enhanced schools' overall administrative capacity. It showed the significant potential of educational informatization to improve both the quality and delivery of educational services.

This case showed the effective integration of educational technology with management theory, bringing information technology into school administrative practices. It introduced innovative approaches to traditional models of home–school communication and student supervision, offering both practical insights and theoretical support for the development of digital governance in education. Furthermore, the initiative created a new model of school governance, defined by smart management and targeted, responsive services.

Looking ahead, further efforts should focus on combining advanced technologies, such as big data and artificial intelligence, to improve management scenarios and expand service offerings, allowing more personalized and precise student oversight. Equal attention should be given to improving user experience and strengthening information security to ensure data privacy. Future research should also consider the diversity of school types and regional contexts when assessing the system's ability to adapt and scale. Ultimately, the goal is to develop a universally applicable digital after-school care model that supports ongoing innovation in educational digital governance.

References

Gao, H., & Xu, F. (2019). Study on after-school care for primary school students. *Elementary Education Studies, 3*, 3.

Gong J. (2021). Analysis of problems and optimization strategies of after-school extended hours service in primary schools. *Journal of Qiqihar Teachers College,* (02), 3–5. https://doi.org/10.16322/j.cnki.23-1534/z.2021.02.002.

Xiong G. & Wu K. (2015). Design and implementation on kindergarten shuttle management system combining RFID with face recognition. *Journal of Huaiyin Teachers College: Natural Science Edition, 14*(1), 4. https://doi.org/10.3969/j.issn.1671-6876.2015.01.010.

Zhang, G. (2018). From “Out of Control” to “In Control” of after-school care for primary school students: An analysis of after-school care services for primary school students. *Teaching Management and Education Research, 3*(21), 113–114.

Zhang, B. (2022). Problems and countermeasures of after-school extended hours service in rural primary and secondary schools. *Gansu Education Research, 7*, 29–31.

Chapter 18
Innovative Practices in Cultivating a Sense of Social Responsibility Among Rural Students: Taking *a Touch of Blue* as an Example

Xuequn Xu, Wei Tong, Kewei Wang, Caiwei Zheng, Hongxia Yu, and Zongyao Lu

Abstract Recent years have seen significant challenges in cultivating a sense of social responsibility among rural students, partially due to the lack of family education and the influence of Internet culture. These students often have limited understanding of local cultures and lack the labor awareness necessary for dedication. Ningbo Zhenhai Vocational Education Centre (ZVEC) has adopted the theory of "Service Learning." Service-learning is an educational method that integrates academic coursework with community service, offering students practical experiences to deepen their comprehension of the material while promoting civic involvement and social responsibility (Felten and Clayton, New Directions for Teaching and Learning 2011:75–84, 2011). This approach aims to strengthen co-operation among schools, village communities, enterprises, and social organizations, creating a symbiotic educational environment. In response to social needs, the school also guides students to improve their professional skills, problem-solving abilities and cultivate a sense of social responsibility.

Keywords Service learning · Social responsibility · Rural education · Cultural heritage · Vocational education

18.1 Background

The Chinese government has long been promoting East–West pairing-off cooperation for poverty alleviation, ultimately aiming to achieve the goal of common prosperity. Ningbo, located in the economically developed eastern region of China's Zhejiang Province, and Qianxinan Buyi and Miao Autonomous Prefecture in the

X. Xu (✉) · W. Tong · K. Wang · C. Zheng · H. Yu · Z. Lu
Ningbo Zhenhai Vocational Education Centre, Ningbo, Zhejiang, China
e-mail: 123694514@qq.com

D. Liu et al. (eds.), *Empowering Educational Innovation Through Design: Volume One*, Smart Computing and Intelligence, https://doi.org/10.1007/978-981-95-4621-3_18

economically underdeveloped southwestern region of China's Guizhou Province, have engaged in a 25-year-long partnership for targeted assistance. This collaboration encompasses various aspects including industry, education, society, culture, and ecological environmental protection, achieving remarkable accomplishments in promoting people's escape from poverty, as well as driving rapid economic, cultural, and social development that have attracted worldwide attention.[1]

Among various poverty alleviation measures, education poverty alleviation stands as a fundamental strategy. Ningbo focuses on the principle of supporting both "aspiration" and "intelligence," not only by assisting in the construction of schools, donating educational equipment, and continuously providing high-quality teachers and courses, but also by attracting students from impoverished areas to study in Ningbo and learn advanced skills. For instance, Zhenhai Vocational Education Center (ZVEC) in Ningbo has provided free vocational education to students from impoverished mountainous regions such as Pu'an County and Jinyang County in western China for ten consecutive years, enabling over 120 students to enter universities, secure employment, or return to their hometowns to start businesses.[2] ZVEC is particularly taking their educational advantages to offer extensive innovation and entrepreneurship education, benefiting these students for lifelong development.

18.2 Design Process

The innovative project combines fashionable embroidery techniques from the eastern region with traditional dyeing art from the western region. It aims to revive traditional textile products and make them fashionable and popular. The exhibition hall and workshop of A Touch of Blue at ZVEC visually represent this integration of regional crafts (see Fig. 18.1).

It initially involved collaboration between ethnic craft masters from the eastern and western regions. Together with students, they sourced traditional plants suitable for indigo dyeing from the western mountainous areas and incorporated modern dyeing techniques to create unique color palettes tailored for young urbanites. Subsequently, teachers and students collectively designed patterns that blended fashion with artistry, utilizing hand-embroidery techniques such as embroidery, to further develop Chinese tea art decorations, fashionable apparel, business gifts, and more. Finally, by participating in exhibitions and setting up high-end shopping malls for contextual marketing, the project significantly enhanced the added value of the products, directly benefiting the artisans economically and stimulating their further

[1] https://kjj.ningbo.gov.cn/art/2018/9/28/art_1229596276_58956479.html.

[2] https://www.baidu.com/link?url=AnGaFbCSFQDzWb9uzI4iYl_iq8OBa_23Il3gypat0G1wJ-9f6H0kKp1QGNya6pVjaHangHFwbZLhR9sfq4Y8GUKAJqGQ5T2Y-Z-yICOlVvi&wd=&eqid=fe056f70001f303c0000000667641795.

Fig. 18.1 Exhibition hall and workshop of *A Touch of Blue* at ZVEC

creativity. The series of activities was also turned into skill-based courses, prioritizing the training of vocational school students, who then went on to start their own businesses and promote employment among women in the mountainous regions.

Over five years, the project has undergone significant development. In 2018, ZVEC and vocational schools in Guizhou Province reciprocally dispatched skilled masters for study exchanges, marking the initial exploration of curriculum directions. In February 2019, a team named *A Touch of Blue* was formed under the guidance of the teachers. Subsequently, a fabric course based on these techniques was introduced and has since become a selective course available to all students and teachers.

The course follows a three-step development. Firstly, the course objectives have been redefined to integrate handicraft training with the nurturing of social responsibility, enabling students to apply their acquired skills in diverse volunteer activities.

Furthermore, this endeavor is dedicated to assisting locally unemployed middle-aged individuals in initiating their own enterprises and enhancing their quality of life through the effective utilization of these handicraft competencies.

Secondly, the course emphasizes the preservation and promotion of traditional crafts while cultivating a sense of social responsibility. In addition to teaching dyeing and embroidery techniques, the curriculum provides an overview of the history, current status, and trends of Chinese traditional culture. Students are guided to deeply understand the origins of various ethnic traditional textile arts and to contemplate strategies for preserving non-material cultural heritage. Furthermore, students are encouraged to develop innovative products that align with modern aesthetics by incorporating fashion elements. This approach not only introduces traditional culture to young people but also inspires more individuals to participate in its protection.

Thirdly, an innovative teaching methodology has been adopted, blending short-term training with long-term development. Apart from offline training sessions and volunteer activities, the online "Hand Maker" app has been integrated to enhance and spread the handicraft instruction courses. By offering free online courses, remote education has been conducted in local evening schools, women's skill training centers, and skill teaching sites in western mountainous regions. Online guidance services are also provided, enabling a wider audience to acquire skills, independently market their products, and generate income.

The curriculum team has collaborated extensively with social organizations to integrate traditional Intangible Cultural Heritage elements into the *A Touch of Blue* project and developed a series of school-based teaching materials. These materials include cultural popularization, plant dip-dyeing, hand-rolled flowers, and hand embroidery. Students are guided to understand the rural environment, contribute to rural revitalization, and practice their sense of social responsibility. In terms of content development, the course development team has designed a "layered training" program and set "three stages" objectives by taking into account the basic conditions of students.

In the FIRST stage, students will master the basic skills of embroidery and simple pattern embroidery on blue cloth. In the SECOND stage, students will learn to dye the cloth independently and complete the entire process of dyeing and embroidery. In the THIRD stage, students will focus on cultivating their innovation abilities, design patterns independently, attract a certain group of fans, and establish their personal style and brand influence. Since 2021, the course development team has added an *A Touch of Blue* handbook to the existing teaching materials. This handbook is distributed to unemployed women to help them quickly learn the basic skills. Additionally, teachers and students have collaborated to launch a series of experiential courses on online platforms to expand the reach of the course.

18.3 Implementation Outcomes

The teaching activities of the course are divided into two aspects: intensive teaching and practical experience. Students join the course on a voluntary basis according to their interests. Intensive teaching sessions are held twice a week for skills training. A comprehensive assessment is conducted once a month to evaluate learning. Students who pass the assessment remain in the course, while those with good or excellent grades are awarded the title of "(Gold/Silver) A Touch of Blue Tutor." These tutors carry out external volunteering activities and teach handicraft skills to rural women. One example is Zheng Xiao, a student from Qianxinan Prefecture, who is learning embroidery techniques (see Fig. 18.2).

The practical experience aspect is shared between teachers and some students, with team members coming from various professional backgrounds, including tourism, arts and crafts, e-commerce and digital media. As shown in Fig. 18.3, He Xingling, a student from Qianxinan Prefecture, introduces handmade tea mats to customers in a mall in Ningbo. Before the volunteering activities commence, contact is strengthened with the community to investigate and map out the basic situation of potential recipients. After the initial list is determined, the village committee issues an application notice and individual targets are contacted on a peer-to-peer basis to ensure precise mapping, targeting, and assistance.

Additionally, online courses are published on web-based platforms to aid in promotion and publicity. Micro-courses are available for free learning through the "Hand Maker" app (see Fig. 18.4). The service activities are divided into three parts: Pre-course team study, During the course, and After-class follow-up counseling. During the Pre-course team study, we enhanced the process, designed the course, recorded micro-courses and promote them online to raise awareness of the *A Touch of Blue* series. To ensure the authenticity and quality of the teachings, professional craftsmen or cultural heritage inheritors are hired to guide the teaching of

Fig. 18.2 Zheng Xiao, a student from Qianxinan Prefecture, Guizhou Province, is learning embroidery techniques

Fig. 18.3 He Xingling, a student from Qianxinan Prefecture, Guizhou Province, is introducing handmade tea mats to customers in a mall in Ningbo

the curriculum and the development of teaching materials. The school also collaborates with village communities to accurately select eligible recipients and connect with the market to bring traditional cloth products closer to consumers and increase the income of recipients. The school leverages the widely acclaimed "Hand Maker" app, which is popular among cloth art enthusiasts. It helps promote the product and assists recipients in placing their products in hotels, shopping mall display cases, large exhibitions, and craft shops.

During the second part, the team focuses on the dissemination of information. Focusing on practical teaching skills, we establish beginner, intermediate and advanced classes for unemployed rural women (service recipients). An increasing number of rural women are participating in the course during their free time (see Fig. 18.5). We fully cooperate with grassroots units to help the service recipients achieve the "three things" goals, i.e., improve skills, adjust mentality, and improve their quality of life.

The third part involves long-term guidance to resolve any confusion and assist with orders, sourcing goods, managing brands, trademarks, and other conventional difficulties. We also support mature service recipients in transitioning to become part of the volunteer team forming a virtuous cycle and further enhancing the program's impact.

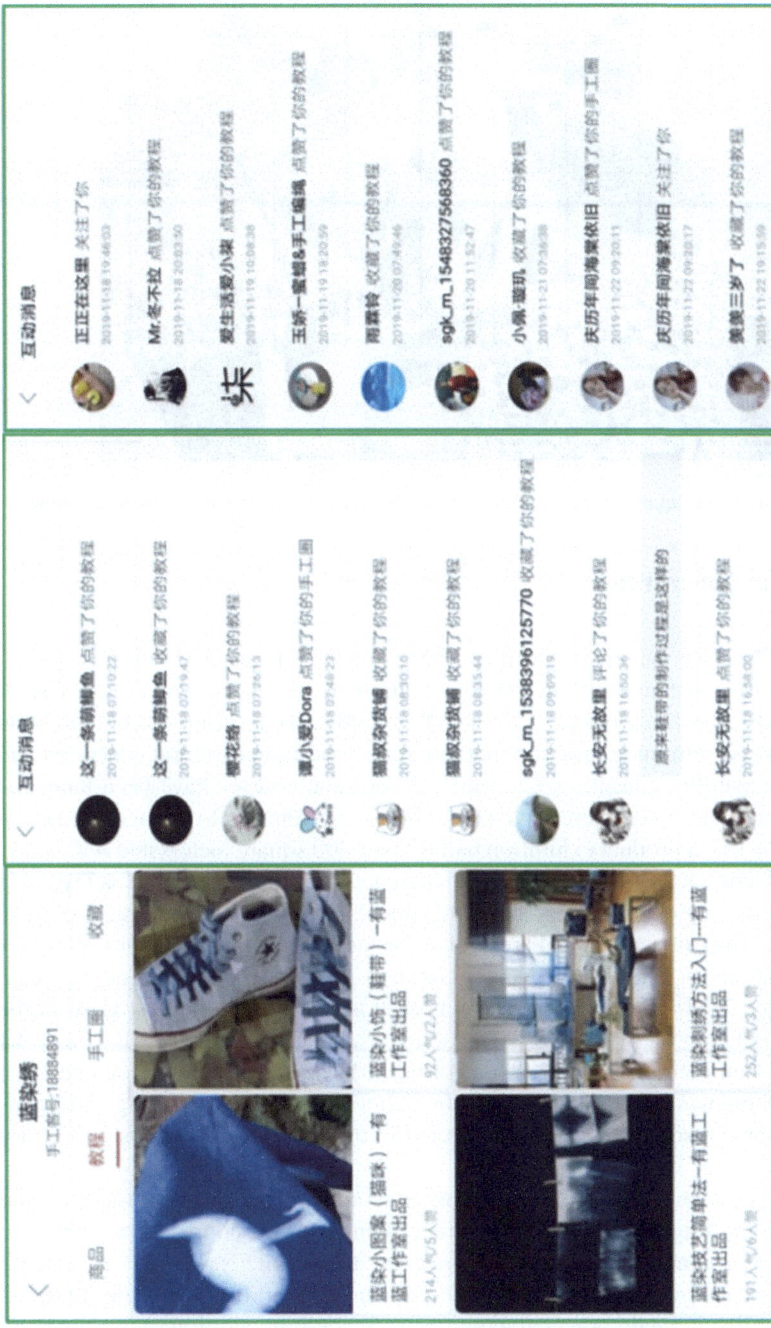

Fig. 18.4 Micro-courses are available on the "Hand Maker" app for free learning by interested citizens

Fig. 18.5 An increasing number of rural women are learning the course during their free time

18.4 The Resources

A Touch of Blue has established a cultivation chain for artisans that follows a "block linkage—strip assistance—point supplementation" model. It streamlines and visualizes the craftsmanship process, introducing an innovative level-up teaching method and a progressive training program that integrates traditional cultural education and social responsibility education. Currently, 26 original courses have been launched online, along with e-commerce live streaming sales channels to expand marketing reach. At ZVEC, a product exhibition hall of over 200 square meters and *A Touch of Blue* R&D center have been established, where new products such as "Tea Dyeing", "Song Dynasty-style Embroidery", and "Mythical Creature Lion-like Creature" have been successfully developed and are soon to be launched. The project has set up six training stations in Zhejiang province, Guizhou province, Sichuan province, etc., training over 2,000 local women. It has been honored as the 4th Best Case in the Global Solicitation on Best Poverty Reduction Practices (see Fig. 18.6), won in the 5th Global Competition on Design for Further Education and been recognized as a National Demonstration Project for Middle School Student Volunteering. Additionally, the course resources have been selected in the UNESCO Teachers' E-Library platform (see Fig. 18.7).

The course team plans to focus on marketing and promotion, team training, and branding in the future. First, we will initiate public service cooperation with more handicraft shops and hotels to enhance coverage and publicity. We will also record public service announcements, so that students can have more opportunities to showcase their skills and gain practical experience. Second, in addition to teaching and volunteering, the course will be promoted to the public through experience days,

Fig. 18.6 The project has been honored as the 4th best case in the global solicitation on best poverty reduction practice

Fig. 18.7 The project won in the 5th global competition on design for further education

themed lectures, and photo exhibitions on non-heritage science and technology. These initiatives aim to attract talents with mature professional skills and rich social experience to join the course team.

18.5 Conclusion

The project has successfully accomplished three primary objectives. Firstly, through the integration of the theory of service learning into traditional craft technology curriculum, it has significantly enhanced students' understanding of the material, while simultaneously cultivating a sense of civic engagement and social responsibility. Secondly, the project has established vocational and technical education as a crucial platform for the preservation and promotion of ethnic cultural heritage, specifically by safeguarding and popularizing the blue dyeing techniques of ethnic minorities among the broader populace. Thirdly, within the context of China's East–West Paired Assistance Initiative, the project has presented a replicable model for public welfare entrepreneurship, directly contributing to the economic empowerment of thousands of students and rural women. The curriculum has been promoted across multiple regions in Zhejiang, Guizhou, and Sichuan provinces.

This research examined Zhenhai Vocational Education Center's "A Touch of Blue" project to demonstrate how vocational education can integrate cultural heritage preservation, poverty alleviation, and civic development within China's East–West paired assistance framework. Its core value lay in creating a replicable model that empowered marginalized communities, particularly rural women, while revitalizing intangible cultural heritage, especially traditional dyeing and embroidery techniques from ethnic minority groups.

At the theoretical level, the case was innovative by embedding service-learning pedagogy into vocational training, moving beyond technical skill development to cultivate social responsibility through real-world applications. Practically, it validated vocational education as a dynamic platform for ICH safeguarding, demonstrating how modernizing traditional crafts for contemporary markets could generate sustainable livelihoods and directly link cultural preservation to economic resilience.

The case used a training system called "block linkage-strip assistance-point supplementation" that showed how skills build on each other through both online and in-person teaching methods. This provided a scalable blueprint for equitable regional development. It was recommended that service-learning be institutionalized in vocational curricula nationwide, with an emphasis on integrating social impact metrics to encourage ethical entrepreneurship. Secondly, it improved technology use through the "Hand Maker" platform by adding better online selling and certification tools, and build partnerships with businesses for ongoing market access. Thirdly, it systematized long-term exchanges between Eastern educators and Western cultural bearers to maintain pedagogical authenticity. Finally, it adapted this collaborative, market-responsive model to other regions, by using local cultural resources. The case repositioned vocational education as a transformative engine for cultural sustainability, social equity, and poverty reduction, offering globally relevant insights into education-driven development.

Reference

Felten, P., & Clayton, P. H. (2011). Service-learning. *New Directions for Teaching and Learning, 2011*(128), 75–84.

Chapter 19
Indigenized Implementation of Service-Learning in Secondary Schools

Shuya Xiao, Wanhua Wu, Mei Zheng, Jialin Luo, Wanzhen Xu, and Guanyu Wang

Abstract The case is based on the concept and method of service-learning, aiming to improve the mental health of students transitioning from elementary to middle school, particularly addressing the growth challenges faced by immigrant children in Shenzhen. Guided by the principle of "helping people to help themselves", this initiative targets ten 8th-grade students at the Shenzhen Luohu Education and Scientific Research Institute Affiliated School. The case centers around the theme of "Adolescent Mental Health" to organize a youth book exhibition. The aim is to assist students in enhancing their psychological adjustment abilities during the transition from elementary to middle school, while also raising public awareness about adolescent mental health. This activity leverages information technology as a medium to conduct extracurricular moral education activities, enriching students' learning experience, stimulating their learning interest, and promoting their academic development. Students will enjoy extracurricular moral education activities through the use of information technology and will have the opportunity to learn professional skills as aspiring curators.

Keywords Service-learning · Growth challenges · Adolescent mental health · Book exhibition · Mental health awareness

19.1 Background

According to Jacoby (1996), "service-learning is an educational practice that centers on community service and enhances learning outcomes through reflection." This educational approach not only contributes to students' academic achievements but also enhances their social awareness and leadership skills (Eyler & Giles, 1999). The

S. Xiao (✉) · W. Wu · M. Zheng · J. Luo · W. Xu · G. Wang
Guangzhou Xinhua University, Dongguan, Guangdong, China
e-mail: xiaoshuya3308@xhsysu.edu.cn

W. Xu
e-mail: xuwanzhen2020@sina.com

D. Liu et al. (eds.), *Empowering Educational Innovation Through Design: Volume One*, Smart Computing and Intelligence, https://doi.org/10.1007/978-981-95-4621-3_19

concept of service-learning dates back to the 1960s. In recent years, service-learning has been widely adopted and promoted globally (Billig & Eyler, 2002). In the United States, most colleges and universities have established specialized service-learning institutions, such as Campus Compact, the National Service-Learning and Civic Engagement Conference, and the Service-Learning Resource Center. As globalization progresses, an increasing number of countries are recognizing the application of service-learning in local communities and global society (Kiely, 2005). Despite variations in practice across different countries, the positive impact of service-learning is universally acknowledged (Kendall & Associates., 1990).

Shenzhen is a rapidly growing city predominantly populated by migrants, characterized by high population mobility, which makes it difficult for newcomers to find a sense of personal belonging in the city. Most parents of migrant children work long hours in manual labor, sacrificing time with their children and missing out on experiencing their children's achievements and challenges as they grow. These children have no outlet for their emotions and feelings in life, and "initial emotional neglect" gradually evolves into "emotional neglect after growing up," further affecting children's realization of self-worth within the family. Due to the highly restrictive community environment, migrant children often live with their parents in rented accommodations in urban villages where migrant workers gather. Our school is located in one such an urban village in Shenzhen, and the majority of our students are migrant children. These students often face unstable living environments and family situations, which can lead to additional pressures and challenges to their lives.

Based on the mental health test results of our school's new students, this service-learning case addresses the growth challenges faced by migrant children in Shenzhen. This case leverages students' participation in organizing a book exhibition at Shenzhen's largest public library as an opportunity to strengthen the emotional connection between students and the community, fostering a sense of "being a young master of the city" among them. Guided by the philosophy of "helping people to help themselves," this initiative aims to provide students a sense of belonging in the city, enhance their interest in learning, help them gain a sense of self-worth, and advocate for societal awareness of adolescents' mental health issues.

19.2 Design Process

This case focuses on the theme of adolescent mental health, utilizing the concepts and methods of service learning, and combining practical needs to carry out course design and practice in various forms. The specific design concept is as follows.

19.2.1 Case Objectives

This activity focuses on adolescent mental health and aims to help students better understand mental health issues, improve their mental well-being, and enhance their sense of social responsibility and teamwork skills. Through the curation practice, students will deeply experience the concept of service learning, thereby strengthening their awareness of social issues and self-growth.

19.2.2 Interdisciplinary Literacy Goals

This activity covers content from multiple disciplines, including Chinese language, moral education, mental health, information technology, and visual arts. In the Chinese language component, students will be required to write book recommendations and exhibition descriptions to express their understanding and views on mental health. The use of language not only helps students enhance their logical reasoning and expression skills but also stimulates their interest in reading, improving both reading comprehension and critical thinking abilities.

First, moral education runs throughout the entire activity. The book exhibition focused on adolescent mental health serves as a platform to guide students to pay attention to their mental well-being, enhance their self-regulation abilities, and strengthen their sense of social responsibility and moral character. Through real-life social practice scenarios, the activity allows students to deeply experience and reflect, thereby internalizing moral awareness and promoting personal development.

Secondly, in the Chinese language component, this activity deeply integrates the development of language expression skills by requiring students to write book recommendations and exhibition descriptions. In this process, students need to express their understanding of mental health through precise and engaging language, while using the art of language to attract readers' attention to the exhibition content. Through organizing reading lists and writing exhibition materials, students enhance both their reading and writing abilities, while also developing their skills in crafting compelling narratives and presenting ideas effectively.

Information technology plays a crucial role in this activity. As digital natives, students use tools such as search engines and digital libraries to gather resources related to adolescent mental health. At the same time, by utilizing technologies like PowerPoint for presentations, video design, and poster creation, students incorporate their own creativity to design more personalized and contemporary exhibition content. This not only enhances the visual impact of the book exhibition but also strengthens its promotional effectiveness, ensuring that the content resonates with a modern audience and effectively conveys the theme of mental health.

Visual arts play a key role in the curation process. Students are required to apply their knowledge of art to design the layout, color schemes, and display forms of the exhibits. With mental health as the central theme, students use artistic expressions to enhance the exhibition's appeal, providing the audience with a more profound visual experience. Throughout this process, students' aesthetic taste and ability to appreciate art are effectively developed and refined.

Finally, centered around the theme of mental health, the activity allows students to gain in-depth knowledge of mental health, develop healthy psychological habits, improve their psychological resilience, and explore ways to prevent and address mental health issues. The overall activity, through interdisciplinary collaboration, forms an educational model for the all-around development of students—covering moral, intellectual, physical, aesthetic, and labor education—providing important support for their growth. Through the authentic curatorial experience, students gain a profound understanding of the importance of mental health and enhance their moral character while engaging in practical service to society.

19.2.3 Implementation Path

The implementation of this case is divided into several stages, including planning, data collection, exhibition design, exhibition setup, and summary. In the planning stage, the core goals of the activity are defined, with a focus on mental health as the entry point, addressing the growth needs of adolescents. During the data collection and exhibition design stages, students work in groups to complete tasks such as organizing reading lists, designing creative ideas for themed exhibits, and producing the exhibition content. The setup phase involves the physical arrangement of the exhibits and ensuring everything is ready for the final display. Finally, the summary and reflection phase allows students to evaluate their work, identify areas for improvement, and reflect on their learning experiences throughout the project.

19.3 Implementation Outcomes

The case spans a total of 19 class hours over a period of two months (see Table 19.1).

Table 19.1 Phases and distribution of class hours

Phase		Class hour
Planning phase		2
Preparing for the book exhibition	Information technology learning	4
	Information technology application	8
Staging the book exhibition		3
Summary and reflection phase (Information Technology Application)		2

19.3.1 Planning Phase

Firstly, the task must be clarified: to write a planning proposal for a junior high school mental health-themed book exhibition. The content should include the activity background, significance, slogan, theme, purpose, and preliminary preparations. Additionally, students should utilize information technology to search for relevant literature on book exhibition planning as references. Lastly, in the grouping stage, students will be divided into three groups based on their personality traits and individual strengths. The first group will be responsible for the activity background and significance, the second group for the preliminary preparations, and the third group for the activity slogan, theme, and purpose.

19.3.2 Preparing for the Book Exhibition

19.3.2.1 Technical Learning

Staff from the Shenzhen Library conducted a group training session for students and introduced them pathways and methods for information technology retrieval. They provided a detailed explanation of the preliminary work involved in exhibition layout, including methods for searching, organizing, and summarizing booklists, requirements for poster design, and key points for news release content. Furthermore, the Shenzhen Library staff took the students on a tour of existing book exhibitions within the library, explaining the layout of each exhibition in detail. This provided students with excellent examples and inspired their design ideas (see Figs. 19.1, 19.2 and 19.3).

19.3.2.2 Technological Application

The students responsible for book information collation initially searched for books suitable for junior high school students to read on websites such as Douban, Baidu Wenku, and Dangdang. They then used the Shenzhen Library's official website to verify the availability of these books, removing any that were unavailable. Finally, they used a table to summarize the book information, compiling a list of 210 books related to mental health and other themes suitable for teenagers. The book information included the book title, author, publication year, ISBN number, call number, and location within the library. During this process, the ten students were divided into three groups: the first group of three completed information for 63 books; the second group of three completed information for 84 books; and the third group of three completed information for the remaining 63 books.

The students involved in poster design were divided into three groups, with each group providing three poster titles and at least one paragraph of poster text, each

Fig. 19.1 Students having group training in the library

containing no fewer than 300 words. The students responsible for poster design began by searching for past poster texts on the Shenzhen Library's official We Chat account. They examined previous examples and found essays related to adolescent mental health and book exhibitions on Baidu Wenku. They then used Word documents to edit the poster titles and body text.

The students were divided into three groups, with each group providing three news releases of no less than 300 words each, as well as book recommendations and reading impressions of no less than 200 words. The students responsible for editing news releases searched for news releases related to book exhibitions on Baidu Wenku and studied exemplary past news releases on the Shenzhen Library's official WeChat account. They then used Word documents to edit their own news releases. After multiple revisions, the students ultimately shared their reading experiences of excellent books with their peers, which became the main content of the news releases (see Figs. 19.4 and 19.5).

19.3.3 Staging the Book Exhibition

Firstly, the students selected heartwarming ornaments that fit the theme from the Shenzhen Library's warehouse and completed the initial arrangement. Next, they

Fig. 19.2 Students having group training in the library

found their recommended books and placed them in prominent positions within the exhibition. They then organized and arranged the remaining 200 + books and affixed the posters. Additionally, the students had the opportunity to write their sincere thoughts on sticky notes and attach them to the message board, offering encouragement to their peers. Finally, the students actively wrote news release copy for the Shenzhen Library to publish, inviting all middle school students to come and read books that nourish the "soul" (see Figs. 19.6 and 19.7).

19.4 Summary and Reflection Phase

The students used Word documents to record their mood changes, psychological feelings, and gains during the preparation and setup of the book exhibition. After completing the tasks, they summarized their psychological feelings and gains and reflected on their performance (see Figs. 19.8 and 19.9).

Fig. 19.3 Staff explaining the layout of the book fair to the students

19.4.1 Learning Outcomes

- Creation of "Mind 'Book' Hole—Adolescent Mental Health Themed Book Exhibition"
- Compilation of 210 books suitable for adolescents, covering themes such as mental health, science, literature, and more, through information retrieval.
- Execution of the news titled "These Middle School Students Became 'Curators'—Join Them in Reading a 'Mind' Book!"
- "Mind 'Book' Hole—Adolescent Mental Health Themed Book Exhibition" (see Fig. 19.10)

19.4.2 Program Outcomes

Members of the project-based learning group on "Adolescent Mental Health" actively participated in the "Youth Action—You Are the 'Curator of the Themed Book Exhibition" event organized by Shenzhen Library. Their book exhibition plan, titled "Mind 'Book' Hole—Adolescent Mental Health Themed Book Exhibition," was selected as an excellent proposal and was displayed in the borrowing area on the second floor of Shenzhen Library from November 26, 2022 to December 15, 2022. A total of

Fig. 19.4 Students using information technology to complete book list summary, poster design and press release

465 books were exhibited during the event, attracting 13,069 visitors. According to official feedback from Shenzhen Library, during the exhibition period, the recommended books were borrowed (including renewals) 1,097 times. Notably, books such as *Explosive Experiments*, *Twelve-Night Journey*, *Mulan* and *The Great Blue Yonder* were borrowed (including renewed) more than 25 times each. This event leveraged the book exhibition as a platform and reading as a bridge to focus on adolescent mental health, fully harnessing the creativity and enthusiasm of young readers. The book exhibition was well-received by readers, particularly among adolescents.

19.5 Innovations

The application of information technology in middle school students' book curation activities is an important topic. It not only enhances students' interest in reading and their engagement but also helps them more efficiently understand and complete curation-related tasks. Students can use search engines, electronic whiteboards, social media, and other technologies for information retrieval, event management, and publicity. However, in this activity, we found that some students face challenges at home, lacking access to computers or tablets. They usually use their parents'

Fig. 19.5 Students using information technology to complete book list summary, poster design and press release

phones for online classes. Tasks such as text formatting and digital design require a computer to perform. While public service spaces like community libraries and party service centers do have computers, students are unaware of these available resources. The issue of educational equity in the era of information technology and digitization often focuses on the urban–rural gap, with concerns about the disparity in access to educational resources and digital tools between different regions. This highlights the need for greater awareness and accessibility of public resources to bridge this gap.

The educational background of parents, economic income, and the size of family living space can also contribute to disparities in education between urban areas. Even in the eyes of adults, families with the lowest income in cities can afford smartphones, and children can easily use online social tools, play online games, create short videos, and more. However, many applications of information technology require computer systems to function. For example, tasks like literature research, data analysis, digital content creation, and multimedia design often need the use of a computer or specialized software. These limitations can create further gaps in access to the full range of educational opportunities provided by technology, especially when students lack the resources to use the necessary tools for these activities.

At the same time, service learning provides a new possible pathway. By integrating classroom teaching with the community and even social issues, students can enhance themselves through community service while also becoming more aware of society

Fig. 19.6 Students arranging the book exhibition A

and engaging in positive interactions with it. China's "Double Reduction" policy not only aims to reduce students' academic burdens and homework load but also signifies that future education should not be limited to the teaching of subject knowledge. It should focus on the comprehensive development of students and emphasize practical learning experiences.

Service learning, as a learning and teaching strategy, is gradually being adopted worldwide. How should we implement service learning? First, we can establish a systematic service-learning program, which includes key elements such as designing learning objectives, collaborating with the community, and evaluating students' learning and service outcomes. Secondly, schools can collaborate with community organizations to jointly launch service-learning projects. Schools can design locally relevant projects based on the specific needs of the community, integrating local characteristics into the learning experience.

19.6 Promotion

The theme of this activity is adolescent mental health, where students have greater voice and participation, guiding all sectors of society to pay more attention to adolescent mental health issues. As young curators, students are not only organizers and

Fig. 19.7 Students arranging the book exhibition

participants but also important disseminators and advocates. This not only enhances their sense of responsibility and accomplishment but also provides society with more avenues and approaches to understand and focus on adolescent mental health issues.

From the perspective of Shenzhen Library's needs, this activity can help the library expand its promotion and outreach, attract more readers, and increase reader borrowing rates and reading interest. It also assists the library in establishing closer ties with schools, promoting the in-depth development of library services in education.

From the students' perspective, this activity helps them better comprehend adolescent mental health issues, improve their mental health, and gain positive psychological influences and insights. Additionally, students can cultivate their reading interests and abilities, enhance their overall quality, and strengthen their sense of social responsibility. This effectively achieves the transfer and replication of service-learning from outside the school to within its walls.

Service learning, as a new educational model, inevitably presents some challenges for educators during the implementation process. For example, determining service projects that are both educationally valuable and meet community needs, ensuring student engagement and service quality, guaranteeing student safety during service activities, ensuring the quality and effectiveness of service learning, and evaluating students' learning outcomes effectively. However, these challenges, rather than being

Fig. 19.8 Student using information technology to complete the project diary

seen as problems, should be viewed as opportunities for educational reform. Implementing service learning requires the joint efforts of schools and teachers. Teachers need to possess a solid understanding of the principles and methods of service learning, while schools must provide the necessary resources and policy support. Only through continuous reflection, evaluation, and improvement by both schools and teachers in the practice of service learning can this approach truly be realized.

19.7 Summary

This case examined the implementation of service-learning in secondary schools with the specific aim of improving adolescent mental health among immigrant children in Shenzhen as they transitioned from elementary to middle school. The project involved ten 8th-grade students who organized a book exhibition on mental health, with the dual goals of enhancing their psychological adjustment skills while raising public awareness. The core contribution of this initiative lay in its innovative integration of service-learning with mental health education, empowering students through hands-on curation. This approach fostered a sense of belonging, self-worth, and social responsibility, effectively addressing the unique challenges faced by migrant children in urban settings.

Fig. 19.9 Student using information technology to complete the project diary

Fig. 19.10 Book fair scene photos

The significance of this study lay in its application of service-learning principles, which combined community engagement with reflective learning. It adapted this framework to the local context of Shenzhen, addressing the specific needs of the city's immigrant population. The study also integrated interdisciplinary elements to promote students' holistic development and align with contemporary educational objectives. In addition, the use of digital tools for research and design enhanced the project's relevance in technology-rich learning environments.

For practical application, schools can replicate this model by partnering with community institutions such as libraries to create service-learning projects that connect classroom learning to real-world issues. The case highlighted the importantce of equitable access to technology, as some students faced barriers due to limited resources at home. Schools were encouraged to ensure all students had the necessary tools to participate fully.

For future research, the study recommended exploring the long-term effects of service-learning on students' mental health and academic success. It also suggested investigating the model's scalability across diverse educational settings and cultural contexts to broaden its impact. Additionally, studying the role of technology in service-learning could address issues of digital equity and also provide insights into overcoming resource disparities.

References

Billig, S. H., & Eyler, J. S. (2002). Introduction: Emerging visions of service in higher education. In J. S. Eyler & S. H. Billig (Eds.), *Evaluating Service-Learning Activities and Programs: Concepts and Practices* (pp. 1–14). IAP.

Eyler, J., & Giles, D. E., Jr. (1999). *Where's the learning in service-learning?* Jossey-Bass.

Jacoby, B. (1996). Service-learning in today's higher education. In B. Jacoby, & Associates (Eds.), *Service-Learning in Higher Education: Concepts and Practices* (pp. 3–25). Jossey-Bass

Kendall, J. C., & Associates. (1990). *Combining Service and Learning: A Resource Book for Community and Public Service* (Vol. 1). National Society for Experiential Education.

Kiely, R. (2005). A transformative learning model for service-learning: A longitudinal case study. *Michigan Journal of Community Service Learning, 12*(1), 5–22.

Part IV
Insight into Educational Mechanism Designs

Chapter 20
Design and Implementation of Multi-course Blended Teaching in Smart Education

Xianlian Zou, Renjiao Xing, Xiao Huang, Zhiyou Chen, Yuqin Tan, Jiaping Lan, Mengjing Tang, Xueping Xiang, and Lu Tan

Abstract With the advancement of technology, education faces significant challenges. The classroom, as the primary battleground of education, should undergo transformation. The traditional singular approach to classroom types is insufficient to meet new learning demands. Even though some schools have explored various classroom types sporadically, without a systematic framework, they cannot achieve optimal alignment between educational content and classroom formats. Against this backdrop, schools are leveraging smart teaching platforms to deconstruct and reconstruct classroom elements. By expanding learning spaces, making classroom durations flexible, redefining teacher-student roles, and restructuring class sizes, five new instructional models have emerged: single-class innovation, mixed-ability grouping,

X. Zou (✉)
Jinzhou Primary School, Chongqing, China
e-mail: 379216839@qq.com

R. Xing · X. Huang · Z. Chen · Y. Tan · J. Lan · M. Tang · X. Xiang · L. Tan
Xingyuan Primary School, Chongqing, China
e-mail: 342833304@qq.com

X. Huang
e-mail: 940572872@qq.com

Z. Chen
e-mail: 454712103@qq.com

Y. Tan
e-mail: 619367893@qq.com

J. Lan
e-mail: 364323027@qq.com

M. Tang
e-mail: 835834223@qq.com

X. Xiang
e-mail: 674262886@qq.com

L. Tan
e-mail: 932551877@qq.com

D. Liu et al. (eds.), *Empowering Educational Innovation Through Design: Volume One*, Smart Computing and Intelligence, https://doi.org/10.1007/978-981-95-4621-3_20

outdoor inquiry, dual-mode dual-teacher, and online live broadcasting. These changes effectively transform teaching and learning methods, establishing new paradigms for the classroom.

Keywords Multi-course mixing · Classroom transformation · Systematic model

20.1 Brief Introduction

20.1.1 Core Concept

The core method adopted by Xingyuan Primary School for classroom structural reform, aimed at effectively changing teaching and learning methods and upgrading overall teaching, is known as "Multi-Course Mixing." This involves choosing and combining different classroom organization methods according to different learning conditions, subject natures, curriculum contents, etc. The method is also a new exploration to conduct school-based implementation of the national curriculum and digital transformation of the classroom.

20.1.2 Analysis of Core Issue

Traditional education can no longer meet the demands of social development and talent training in the modern era due to the rapid advancement of science and technology, the continuous expansion and broadening of teaching boundaries, and the continuous iteration and renewal of learning ecologies. The 2022 edition of the new curriculum standard clearly points out "giving full play to the advantages of new technologies, exploring the in-depth integration of online and offline formats, and orienting toward personalized learning."[1]

However, up to now, most classrooms still adopt a one-size-for-all organization mode (Liang, 2007) of "one classroom + one teacher and one cohort of students". There have not been fundamental breakthroughs or structural reforms in the learning space, curriculum content, curriculum resources, and teaching and learning methods.

Based on this, Xingyuan Primary School in Liangjiang New Area of Chongqing has continuously made innovations and explorations over the past three years. Utilizing the smart teaching platform, the classroom has been deconstructed and reconstructed. The learning space has been expanded; class hours have become flexible; the roles of teachers and students have been reshaped; and the class size has been redefined. Eventually, five types of class were established, including Creative-Learning Single Classes, Joint Classes, Outdoor Exploring Classes, Online/Offline

[1] http://www.moe.gov.cn/srcsite/A26/s8001/202204/t20220420_619921.html

Co-Teaching Classes, and Livestreaming Classes (see Fig. 20.1). Classrooms can adopt multiple organizational modes based on the subject, learning foundation, and teaching materials, fostering a Multi-Course Mixing mode.

20.2 Design Process

20.2.1 Design of Multi-Course Mixing

How is the Multi-Course Mixing designed and constructed? This case deconstructs and reconstructs elements of a classroom. Multi-Course Mixing will be built and analyzed from four critical perspectives: the expansion of learning spaces, the roles of teachers and students, the integration of technologies and platforms, and the innovation and creation of teaching contents based on a wider range of resources.

20.2.1.1 Learning Expansion in Time and Space

Extending learning time and space is the premise of exploring the Multi-Course Mixing model. We have broken the traditional offline teaching mode by upgrading the classroom from offline to online, developing "virtual and physical" learning spaces, and integrating online/offline teaching. Different courses are set up based on various hardware facilities and online teaching platforms. In terms of study time, considering students' physical and mental growth and subject characteristics, the "one-size-fits-all" schedule has been abandoned (Ye, 2003). Class hours have been divided into short courses (15 or 20 min), medium courses (30 or 35 min), and longer courses (50 or 60 min). Based on this flexible schedule of "long + medium + short class hours," different courses and class types are forming a synergy (see Fig. 20.2).

20.2.1.2 Roles of Students and Teachers

Transforming the roles of teachers and students is a fundamental requirement for Multi-Course Mixing. In this model, the roles of teachers and students can be flexibly switched as needed. Through the deep integration of online and offline teaching activities, the purpose of diversified and flexible roles between teachers and students is to enable individuals grow into:

- "Learners" who will adapt to society.
- "Digital citizens" who will adapt to the future (Zhang, 2011).
- Finally, "Lifelong learners orienting toward the future" (see Fig. 20.3).

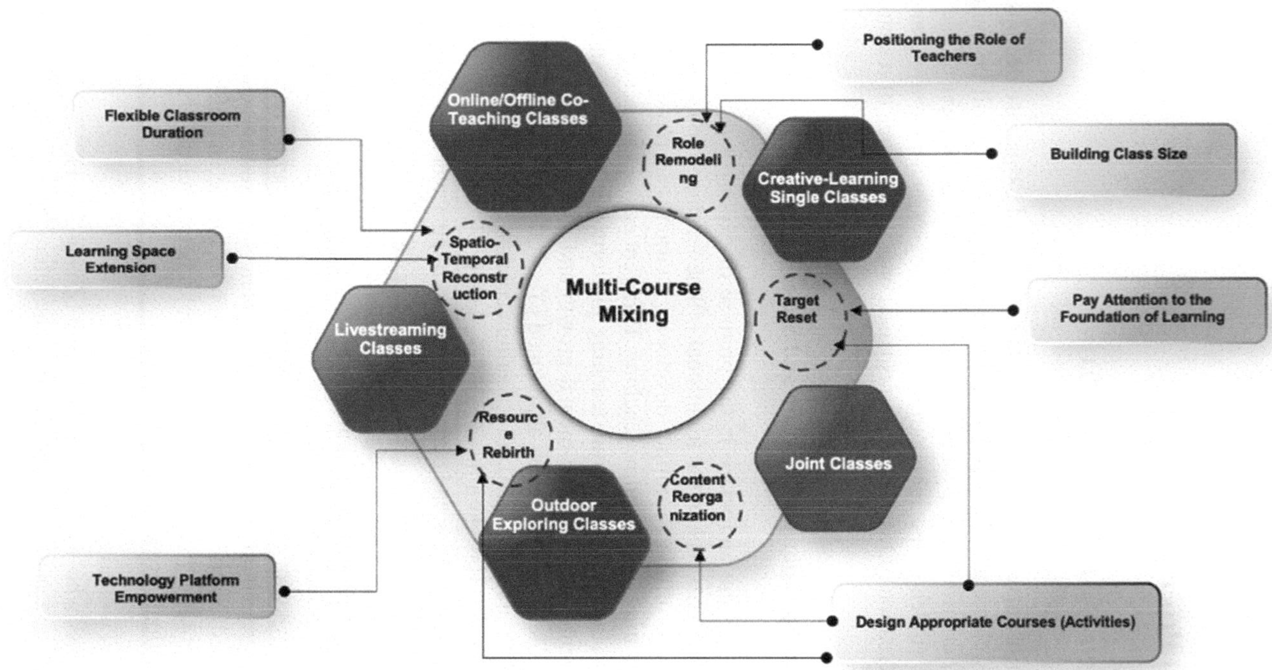

Fig. 20.1 Reconstruction of elements for five types of classrooms

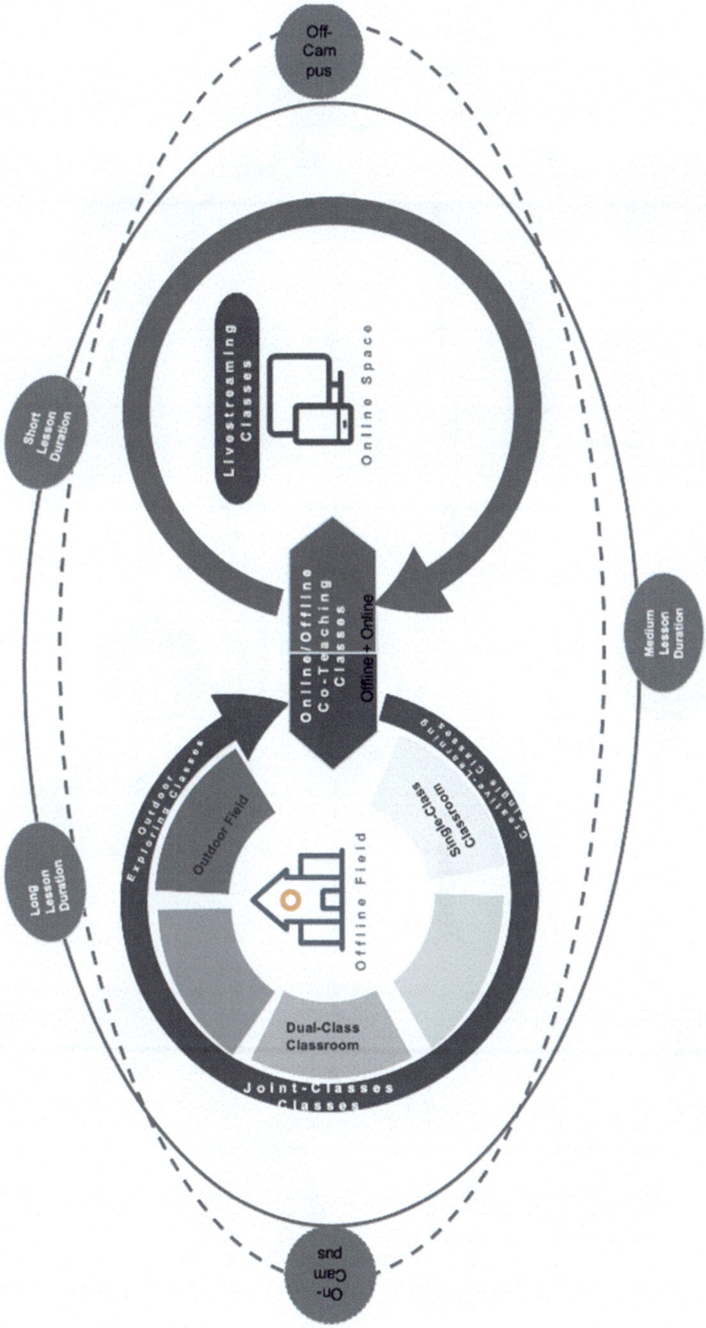

Fig. 20.2 Time–space expansion model of the teaching in five types of classes

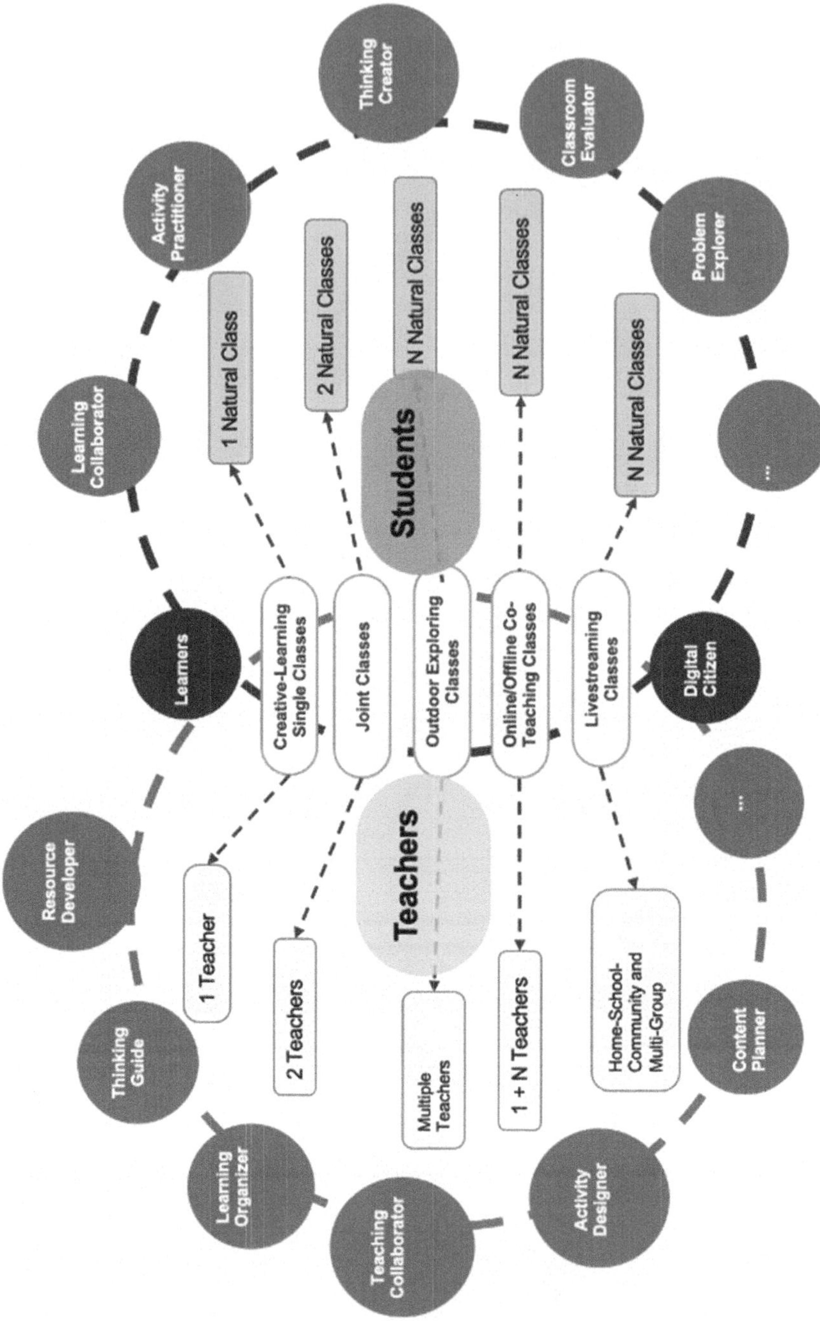

Fig. 20.3 Roles of teachers and students in the five class types

20.2.1.3 Smart Integration of Technologies and Platforms

Mastering new technologies and platforms has become an effective pathway for Multi-Course Mixing. More focus should be placed on applications and digital equipment. By addressing Infrastructure foundations, platforms and applications, and end-user terminals, we can reconstruct the technological pathways for classroom reform. This will fully tap into students' wisdom and enhance digital literacy (see Fig. 20.4).

20.2.1.4 Creation and Reorganization of Resource Content

The creation and reorganization of resources is a vital strategy for exploring Multi-Course Mixing (Luo & Guo, 2013). Resources inside and outside the school have been integrated to explore those factors that are necessary to meet educational needs. These Resources have been given new applications, and contents have been reconstructed using four strategies: a) Giving full play to people's talents and joining efforts from various parties, b) Improving efficiency with multiple types of classes c) Maximizing the potential of available spaces, and d) Exploring the basics of subjects and fostering interdisciplinary integration (see Fig. 20.5).

20.2.2 Practical Logic of Multi-Course Mixing

How to practice Multi-Course Mixing? Case analysis brings education back to its very nature, explaining the fundamental concepts of various courses and their practical logic. It involves choosing and applying various courses according to different classroom content, leveraging the strengths of multiple courses, and modifying teaching and learning mode (see Fig. 20.6).

20.2.2.1 Practical Logic of Creative-Learning Single Course

Creative-Learning Single Class is based on the class teaching system and is systematically taught or tutored by one teacher + one class of students. Learning is made more open and flexible through module teaching by creating a high-quality teaching resource repository, an all-scenario learning field, an intelligent interactive circle, expanding learning channels, enriching learning content, and broadening students' horizons. These efforts also foster the open creation of teachers' and students' thinking, improve classroom teaching effectiveness, and enhance students' core literacy (see Fig. 20.7).

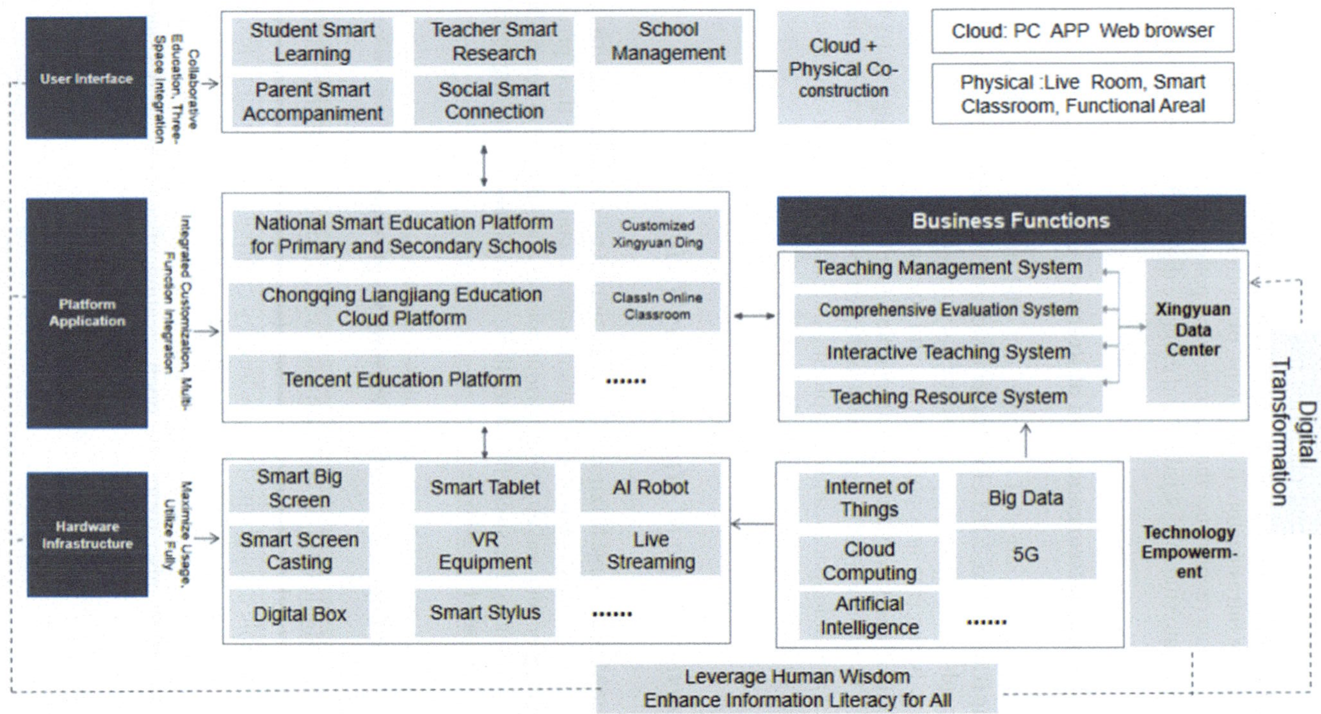

Fig. 20.4 Integration model of five curriculum technology platforms

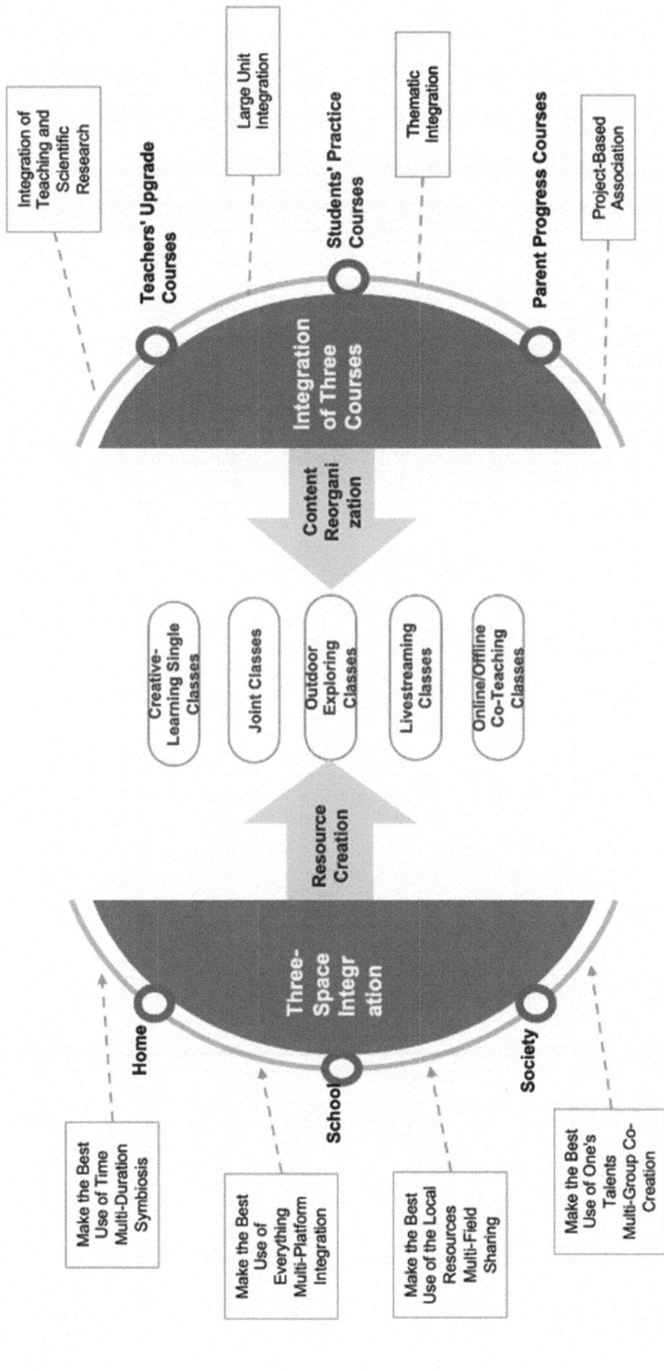

Fig. 20.5 Content creation and innovation model for five types of classes

Course Type / Elements	Creative-Learning Single Classes	Joint Classes	Outdoor Exploring Classes	Online/Offline Co-Teaching Classes	Livestreaming Classes
Class Size	1 natural class.	2 natural classes, combined with adjacent classes.	N natural classes, with a dynamic learning community as needed.	N natural classes, with grade, year, whole-school or cross-regional classes as needed.	N natural classes, with N learning ends as needed.
Teacher' s Role	1 teacher, focusing on the development of resources and the guidance of students' thinking.	Two teachers (one leading and one cooperating), with complementary advantages and cooperative division of labor.	Many teachers raising funds, focusing on the design and organization of activities.	1 online tutor + N offline companions, Dual-teachers realize seamless cooperation under technical empowerment.	Teachers come from many groups in home, school and community, focusing on live teaching and planning and organization of online activities.
Learning Space	Ordinary Classroom	Joint-class Classroom (Open the Partition Walls of Two Adjacent Classes)	Outdoor Space	Live Room + Ordinary Classroom	Live Room + N Learning Spaces
Technical Resources	According to the course content, multimedia technology is adopted to screen high-quality resources and available terminal equipment.	According to the content, multimedia technology is selected, resources are screened, and teaching platform and terminal equipment are available.	Multimedia technology is selected before class, and high-quality resources are screened for application. Terminal equipment can be used for practical exploration in class.	With terminal equipment, multimedia technology, and high-quality resource applications, online teaching is conducted on a teaching platform.	Multimedia technology is selected to screen resources, and teaching platform and terminal equipment are needed.
Class Duration	Medium Lesson Duration: 35 or 40 minutes according to the year	Long Lesson Duration: 50 or 60 minutes according to content requirements	Long Lesson Duration: 50 or 60 minutes according to content requirements	Duration set according to the course content Medium Lesson Duration: 35-40 minutes Long Lesson Duration: 50-60 minutes	Long Lesson Duration: 50 or 60 minutes according to content requirements
Appropriate Courses (Activities)	Unlimited, depending on the subject content.	Homework guidance, class competition teaching activities (class meeting, debate competition, etc.).	Labor courses, science courses, research courses, etc.	Subject courses and moral education courses can be selected according to needs.	School opening/closing ceremony, parents' meeting and other home-school community activities.
Feature Description	Selection and Processing of Resources & Open Creation of Thinking	Dual-class competition and hierarchical counseling make learning value-added.	Space Extension, and the World is Our Classroom	Online guide refinement, offline companion for study, and building a learning field of mutual learning and symbiosis.	Home-school-community cooperation and multi-group linkage.

Fig. 20.6 List of core elements of five classes

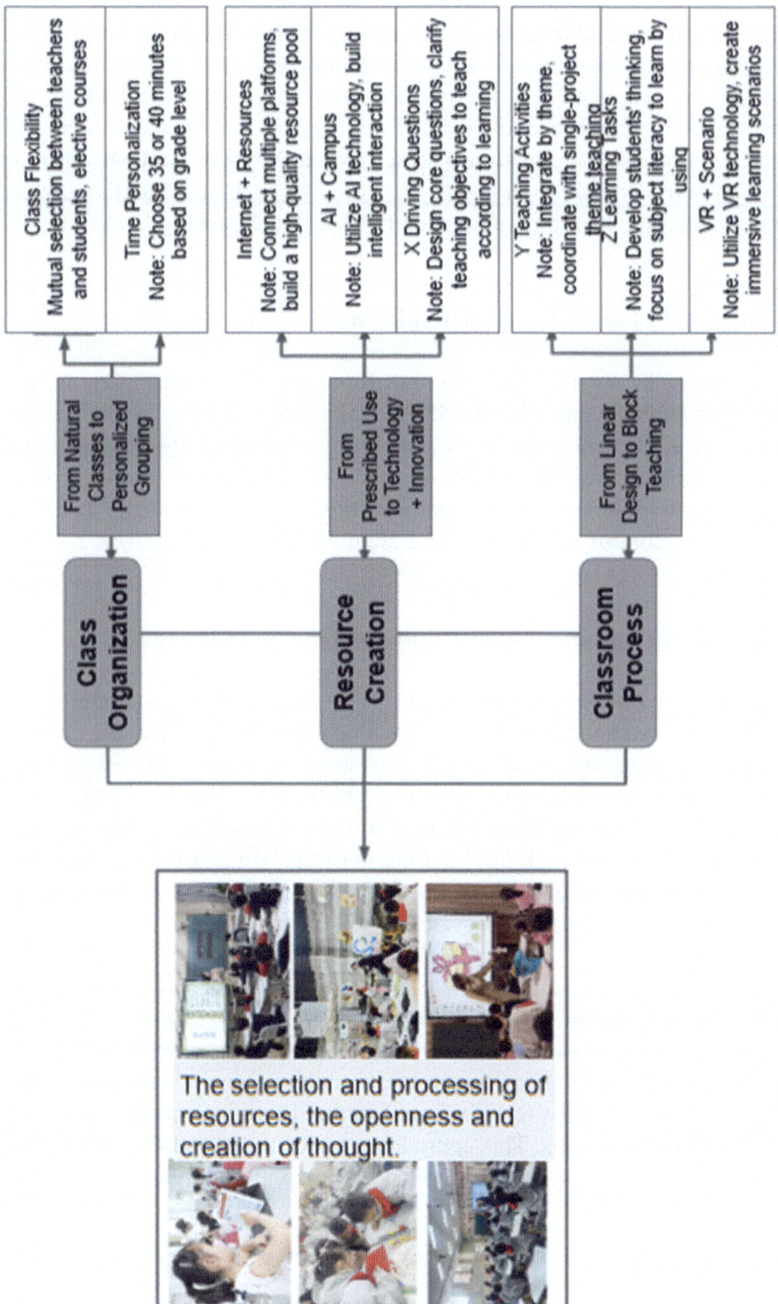

Fig. 20.7 Practical logic diagram of creative-learning single classes

20.2.2.2 Practical Logic of Joint Classes

Joint Classes break through the boundaries of traditional class structures, reorganize the learning space, and select multimedia technology and terminal equipment according to the teaching content and activity theme. This approach involves one lead teacher and one cooperating teacher. Teacher mixing and dual-teacher cooperation help achieve the optimal combination of teachers and improve students' learning enthusiasm and pertinence through mentoring based on each individual's progress. This method realizes the learning value-added of $1 + 1 > 2$ (see Fig. 20.8).

20.2.2.3 Practical Logic of Outdoor Exploring Classes

Outdoor Exploring Classes fully use both on-campus and off-campus learning environments, so that students can go beyond the classroom and enjoy a more expansive learning space. By engaging in and carrying out exploration activities such as discipline expansion, life practice, and research experience, the learning content can become closer to nature and real scenarios, thus cultivating students' practical, cooperative, and exploratory abilities (Ma & Nan, 2007) (see Fig. 20.9).

20.2.2.4 Practical Logic of Online/Offline Co-Teaching Classes

Through online and offline integration, the "1 + N" learning community, and teacher cooperation, the online/offline Co-Teaching Class builds a learning environment of mutual learning and mutualism, and enables interactive and collaborative learning among teachers, teachers and students, and students. The status of teachers' teaching and students' learning in the Online/Offline Co-Teaching Class is fed back to Xingyuan Data Center through the teaching platform. Comparing and analyzing the data benefits teachers' smart discussions, thereby developing their teaching and research abilities (see Fig. 20.10).

20.2.2.5 Practical Logic of Livestreaming Classes

Livestreaming Class is a class type through live broadcast room + N learning spaces with the help of live broadcast teaching platform and terminal, multimedia technology, and screening resources. This approach expands the boundary of education, and realizes the linkage and synergy of home, school, and society at any time (see Fig. 20.11).

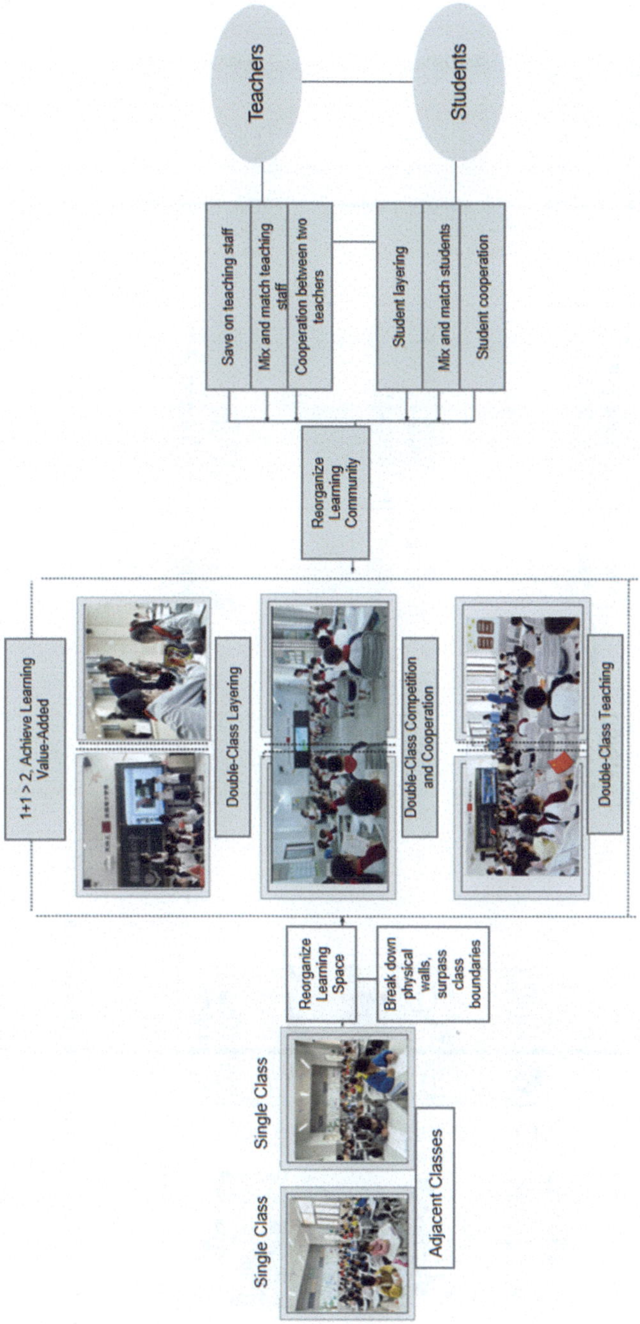

Fig. 20.8 Practical logic diagram of joint classes

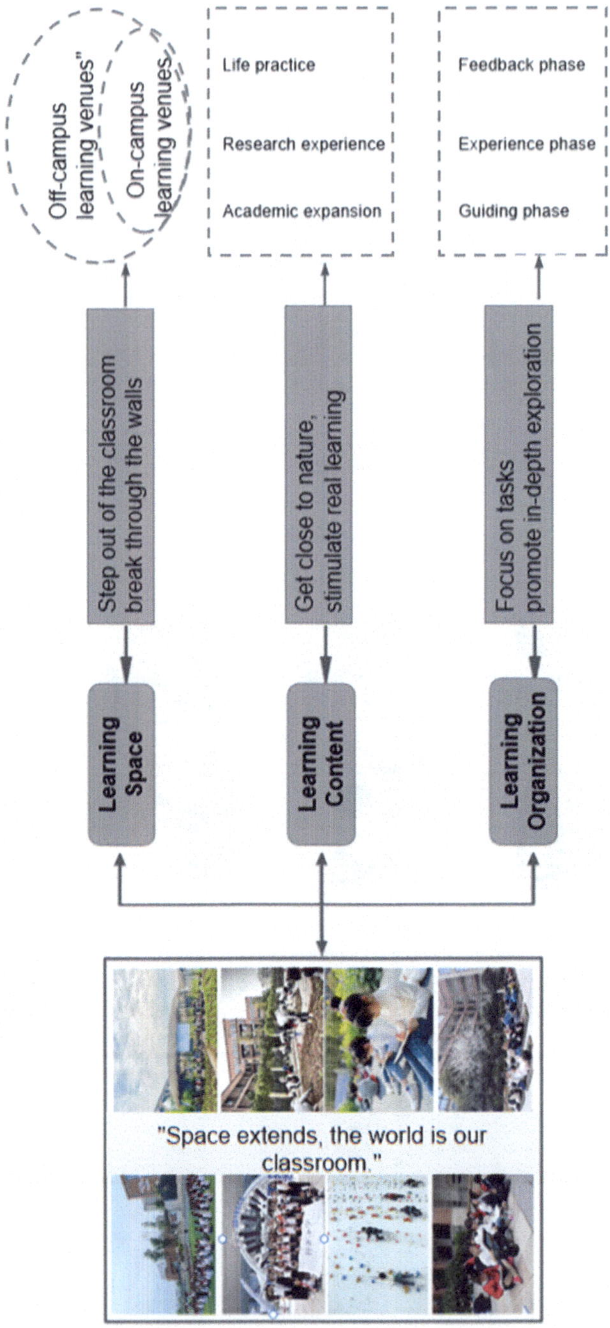

Fig. 20.9 Practical logic diagram of outdoor exploring classes

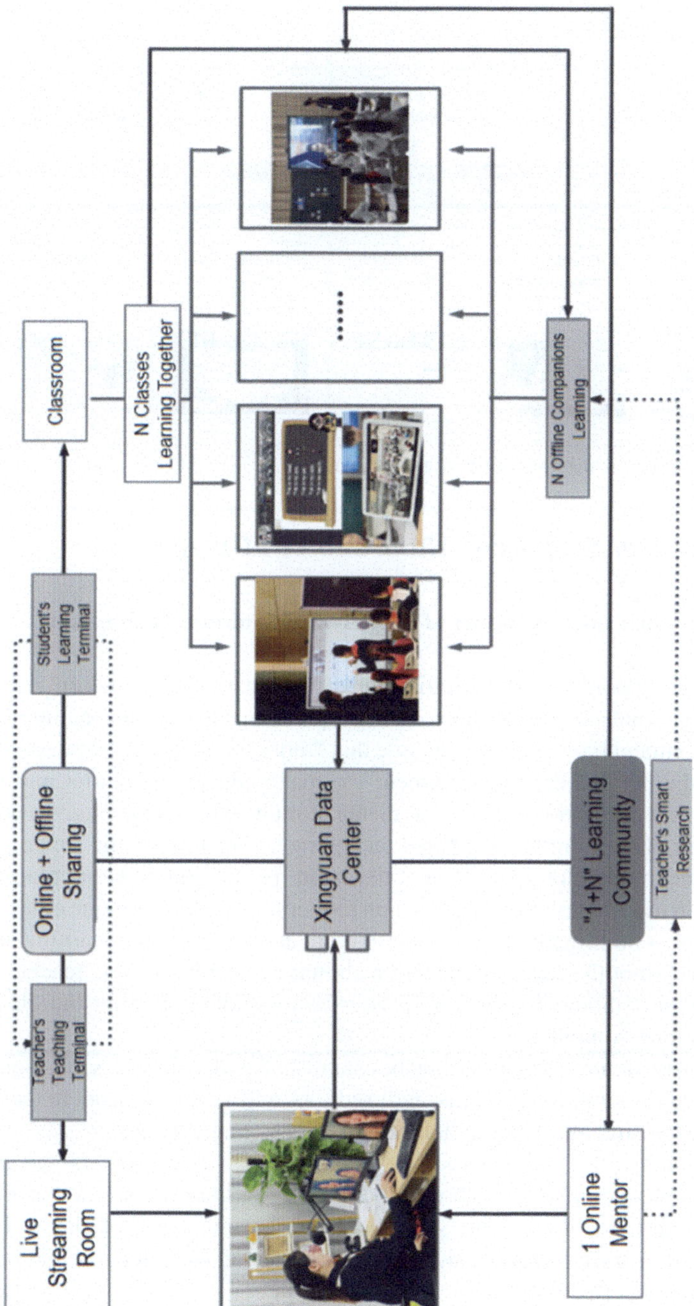

Fig. 20.10 Practical logic diagram of online/offline co-teaching classes

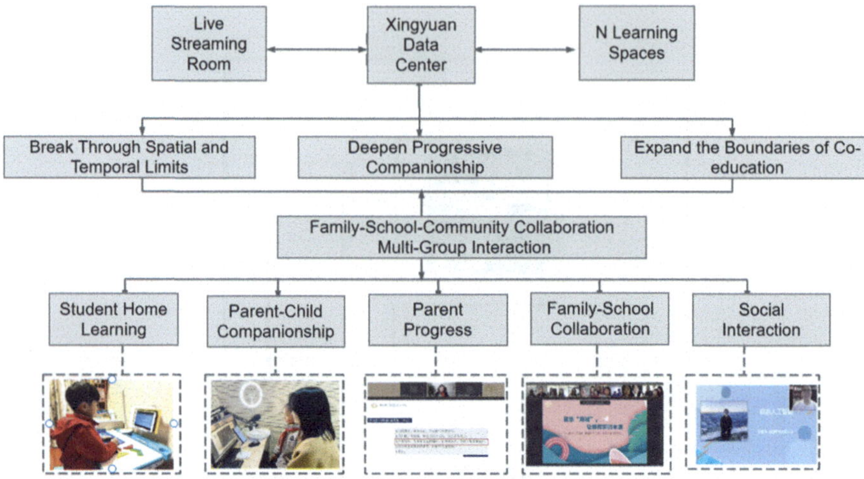

Fig. 20.11 Practical logic diagram of livestreaming class

20.2.3 Practical Support of Multi-Course Mixing

20.2.3.1 Environmental Support of Modern Information Technology

The "hardware support" for the Multi-Course Mixing mode is modern information technology. The school integrates multiple platforms and creates an intelligent learning environment based on the principle that "more focus should be put on applications with the help of digital equipment" to make education smarter and more effective. To meet the operational requirements of multi-course teaching, we adhere to the principle of "extremely simplified technology." We rely on the super-power digital base integrating cloud and Ding Talk, build five intelligent cloud platforms, connect class in teaching platform, and set up a smooth network, a unique course live broadcast room, a multimedia network classroom, and an "informal learning area."

A stable and smooth internet connection should be ensured so that teachers can deliver stable and smooth livestreaming courses to students joining the class on campus with tablet computers.

A physical livestreaming room should be established with computers, high-quality microphones, high-definition cameras, audio editors, and a cable network to facilitate orderly online livestreaming and online/offline co-teaching classes.

Multimedia network classrooms should be equipped with computers, electronic whiteboards, tablet computers, high-definition cameras, high-definition converters, radio microphones, and more. This setup facilitates various types of classes based on communication and discussion, significantly increasing student participation and improving teaching effectiveness.

20.2.3.2 Setting Flexible Teaching Arrangements

- **Flexibility in Study Time and Place**

Nowadays, the internet permeates in every part of our lives, and sole offline learning no longer fits the time. Consequently, the teaching paradigm needs an upgrade. Learning can take place offline with face-to-face, multi-dimensional conversations or in virtual spaces online, maintaining multi-dimensional conversations over screens. Additionally, learning can take place in both ways, breaking the traditional model by jointly developing "virtual and physical" learning spaces and integrating online/offline teaching.

In terms of the schedule, we have optimized the traditional "one timetable fits all" approach and adopted a more flexible timetable that considers the patterns of students' mental and physical development and the nature of subjects. One class hour is further divided into different small segments: short classes of 15 to 20 min, such as 8:30–8:45 am for morning reading; medium classes of 30–35 min, for subjects like Chinese and Math; and long classes of 50–60 min, usually scheduled in the afternoon for homework tutoring and club activities, etc. This new arrangement also includes morning exercises, a midday nap, and family time at evening.

The timetable defines the study and life of children. Since the establishment of the school, we have moved away from old arrangements and continually enriched students' school life by granting them more time for self-learning. These flexible arrangements of classes with different durations make school life more student-centered and give students a greater voice.

- **Upgraded Curriculum Arrangements**

The overall class arrangement will be renewed on a weekly basis, meaning the arrangements will differ each week. We have adopted this measure to explore the "tailored curriculum based on each school's realities."

Instead of using the same arrangements for all students, we hold a weekly class preparation meeting involving the teachers of all subjects, convened by the directors of each subject. The meeting discusses topics, contents, goals, class types, and teachers. An overall class arrangement is then formulated based on the results of the meeting.

20.2.3.3 Teaming up Teachers with Different Orientations

- Teachers sharing one discipline can do class preparation together, each contributing their own ideas.
- Interdisciplinary exchanges and discussions can integrate ideas from different class designs.
- Involvement in social groups and links with multiple resources can fully utilize their respective advantages and enhance class performance.

Multi-Course Mixing gives support to teaching and research with the collective intelligence of all members involved. Research on class types requires more sophisticated collaborative discussions, a professional and scientific division of labor, and more efficient dynamic adjustments. In the case of the online/offline co-teaching class, for example, each teaching and research group consists of one master teacher and several teaching assistants, working in four steps.

- brainstorming with the group,
- designing the first draft with the master teacher,
- providing support during class design challenges, and
- revising and optimizing the details for the final draft.

20.3 Implementation Outcomes

Multi-Course Mixing is a new path for classroom transformation and a new strategy to explore "tailored curriculum based on each school's realities." This approach effectively transforms teaching and learning methods, driving students to gain knowledge while learning and playing, improving teachers' teaching and research, and involving parents to accompany students and collaborate with the school for mutual progress.

20.3.1 Students Gaining Knowledge While Learning and Playing

Multi-Course Mixing offers students various class types, open learning spaces, independent learning styles, rich learning contents, and multi-dimensional evaluation systems. These elements stimulate students' interest in learning, enhance their learning motivation, and help foster their cognitive development.

20.3.2 Teachers Improving Teaching and Research

With the setup and application of each class type, teachers are changing and optimizing their educational view, upgrading their thinking, and accumulating professional knowledge and enriching practices. All teachers are involved in exploring Multi-Course Mixing Classes and are capable of teaching via independent livestreaming. Presentations of Multi-Course Mixing have been conducted with over 50 teachers participating. Additionally, more than 10 papers, such as "Practical Exploration of Online/Offline Co-teaching Integration Class in Post-Epidemic Era" and "Multi-Course Mixing: a Real Change to Teaching Paradigm," as well as a book,

"Multi-Course Mixing: A New Path to Digital Transformation of the Classroom," have been published.

20.3.3 Parents Accompanying Students for Common Progress

Multi-Course Mixing, along with multiple channels that link family and school resources, enriches the content students are learning and the methods used to teach them. The implementation of livestreaming classes, in particular, is a breakthrough in terms of time and space, deepening parent–child companionship, expanding the home-school co-education, and thus building a new model of home-school community collaborative education.

The achievements of the Multi-Course Mixing adoption have been reported by China Education News twice, becoming the first case in the new section of the China Institute of Education Innovation at Beijing Normal University. It has been awarded as the excellent case of 2022 Smart Education, the innovation project of online education application of the Central Education and Teaching Center, the typical case of Ability Enhancement Project 2.0, and the first prize in the Global Future Education Design Competition. The president of Xingyuan Primary School was also invited to give a keynote speech at the Global Smart Education Conference 2021. The school received more than 100 reports from 29 mainstream media such as People's Daily and Xinhua News Agency and became a model school for smart campuses in Chongqing.

It also became part of the online education application innovation project of the Central Education and Teaching Center, a director of Southwest University's Smart Teaching Experimental School Alliance, an experimental school of "AI + OMO" classroom of East China Normal University, and a national model school for the education research institutes of several technology companies. With profound exploration and solid implementation, the school successfully declared three national projects. Jiang Jianguo, Vice Minister of the Publicity Department of the CPC Central Committee, visited the school and said, "While Xinyuan Primary School aligns with national education policies and follows the leadership of the Party, it has made innovations and created something of its own, which deserves high recognition."

20.4 Wider Application

20.4.1 Systematic Model of Case Construction and Practice Logic

This case systematically analyzed Multi-Course Mixing in terms of the roles of teachers and students, technology resources, and learning spaces. It identified five class types: Creative-Learning Single Classes, Joint Classes, Outdoor Exploring Classes, Online/Offline Co-Teaching Classes, and Livestreaming Classes. These types extend learning time and space, redefine the roles of teachers and students, integrate technology and platforms, and innovate content creation based on wider access to resources. The logic behind teaching, resource application, teacher-student interaction, and technology integration for each class type is analyzed. The in-depth and systematic research and analysis clarifies the construction and practical application of Multi-Course Mixing, which can provide a more scientific and clearer framework for other schools to follow in their research and application.

20.4.2 Series of Replicable Cases and Resources

This case has been studied for more than three years. We have explored and analyzed how to mix and match the five types of classes and how to conduct each type of lesson. As a result, we have developed a series of practical cases and resources, including:

- analysis of the application of each subject type,
- teaching cases for each type (including teaching designs and teaching videos),
- sharing of teaching practices by front-line teachers, and
- a list of management organizations for multiple types of classes.

These resources can be directly applied to the promotion and implementation in various schools.

20.4.3 Wider Application of Case Achievements

The exploration of Multi-Course Mixing meets the needs of deepening classroom reform in the new era. By leveraging new technology and integrating home, school, and community resources, we can carry out online/offline co-teaching and implement the concept of tailoring education to teaching materials and students' talents, thus effectively improving teaching and learning methods. This approach represents a new way of thinking and a new path for curriculum reform, with experiences that can be widely applied.

In the future, we will explore the deep integration of technology and education, and realize the multi-dimensional interaction of physical, virtual, and interpersonal spaces through Multi-Course Mixing. We will leverage extensive high-quality resources and conduct timely, interactive cooperative learning to facilitate the collection and analysis of learning data. This will enable the digital transformation of the classroom and promote the consistency in the performance of teaching, learning, and evaluation.

20.5 Summary

The Multi-Course Mixing Teaching initiative at Xingyuan Primary School in Chongqing addressed long-standing issues in traditional education, including rigid classroom constraints, fixed teaching roles, and limited access to resources. By leveraging technology, it promoted school-based implementation of the national curriculum. Its core innovation lay in systematically redefining the teaching paradigm through flexible scheduling, with class durations ranging from 15 to 60 min, and the integration of physical, virtual, and social learning spaces. Teachers evolved into mentors and collaborators while students become active explorers. To support this shift, the model streamlines technology by using unified platforms such as DingTalk and ClassIn within the Five Intelligence Cloud framework. It also employed resource innovation strategies to reorganize and connect learning assets across home, school, and community.

This hybrid approach combined five distinct class formats: Creative-Learning Single Classes, Joint Classes, Outdoor Exploring Classes, Online-Offline Co-Teaching Classes, and Livestreaming Classes. Key contributions demonstrate:

- Theoretical advancement was achieved by expanding constructivism and connectivism principles, establishing an operational framework for digital-era learning ecosystems.
- Educational equity was enhanced through Joint Classes that enabled tiered instruction and Livestreaming Classes that bridged resource gaps.
- Collaboration was transformed where parents shifted from observers to engaged learning partners, gaining direct classroom access for community resources.

Implementation recommendations include:

- Developing intelligent class-matching systems that recommended optimal formats based on learning analytics.
- Creating regional cloud platforms for dynamic resource mapping, including outdoor exploration sites.
- Designing parent empowerment kits that featured co-learning guides.

Future directions should:

- Validate designs through learning science methodologies, including eye-tracking optimization.
- Explore Hybrid Classes combined with Cross-School Alliances to enable urban–rural teacher sharing.
- Establish three-dimensional evaluation metrics measuring learning efficacy, technology integration, and social connectivity supported by AI assessment tools.

These findings demonstrated that digital transformation in education requires fundamental restructuring rather than superficial technological additions. Xingyuan's hybrid teaching model catalyzed systemic change by promoting synergy among teaching, learning, and evaluation, and it offered a replicable foundational framework for smart education development.

References

Liang, J. (2007). Reflections on the reform of classroom teaching organization forms. *Educational Theory and Practice: Discipline Edition, 8*, 11–13.

Luo, Y., & Guo, L. (2013). The implementation path and effective mechanism of "Establishing Virtue and Cultivating People." *Ideological Education Research, 7*, 45–49.

Ma, L., & Nan, J. (2007). Exploring experiential teaching. *Contemporary Education Forum, 12*, 70–71. https://doi.org/10.3969/j.issn.1671-8305.2007.12.034.

Ye, L. (2003). Reconstructing the value system of classroom teaching. *Biology Teaching, 28*(2), 4–6. https://doi.org/10.3969/j.issn.1004-7549.2003.02.005.

Zhang, L. (2011). The educational appeal and response to "Digital Survival.". *Educational Development Research, 23*, 70–73.

Chapter 21
Exploring the Collaborative Model of School-Family-Community for Cultivating Creative Talents

Yakun Zhang, Xingli Zhang, Lina Huang, Yufei Zong, Jiannong Shi, and Wei Kang

Abstract The enhancement of children's everyday creativity holds implications for their mental well-being, overall achievement, and overall satisfaction. This case examines the collaborate development of school-family-community within the broader context of the future educational ecosystem. It proposes a curriculum system for cultivating creativity that links the nurturing of scientific and technological talents with collaboration among schools, families, and communities. This case introduces the Butterfly Model for Fostering Creativity to guide the design of creativity courses. Furthermore, to ensure the sustainable development of the creativity cultivation curriculum system, it suggests an institutional framework that encourages curriculum innovation and the mutual growth of teaching and learning. This framework lays a solid foundation for better collaboration among schools, families, and communities, and provides support for inclusive education.

Y. Zhang
Research Center for Supernormal Children Center, Institute of Psychology, Faculty of Education, Chinese Academy of Sciences, The University of Hong Kong, Beijing, China
e-mail: zhangyakun@connect.hku.hk

X. Zhang · L. Huang · Y. Zong · J. Shi (✉)
Research Center for Supernormal Children Center, Institute of Psychology, Chinese Academy of Sciences, Beijing, China
e-mail: shijn@psych.ac.cn

X. Zhang
e-mail: zhangxl@psych.ac.cn

L. Huang
e-mail: 13552043140@163.com

Y. Zong
e-mail: zongy11@uci.edu

Y. Zong
Department of Cognitive Sciences, University of California, Irvine, CA, USA

W. Kang
China Science and Technology Museum, Beijing, China

D. Liu et al. (eds.), *Empowering Educational Innovation Through Design: Volume One*, Smart Computing and Intelligence, https://doi.org/10.1007/978-981-95-4621-3_21

Keywords Creativity cultivation · School-family-community · Inclusive education · Course design · Creative motivation

21.1 Background

A healthy educational ecosystem calls for the joint participation of schools, families, and communities. Although the common goal of cultivating students' moral character and abilities serves as the link for tripartite collaboration, it is not easy to connect the three parties. Without sufficiently high-quality courses and theoretical resources, there will be no motivation for cooperation. At this point, communities, as the innovative and dynamic force with the advantage of acquiring and creating new knowledge, should actively step in to fill the gap, take on its corresponding responsibilities, propose forward-looking theories for innovation-driven education, and design high-quality courses. These courses should be sufficient to attract parents and schools to participate in the collaborative process, so that everyone benefits, gains inspiration, and grows.

As early as when Professor Shi Jiannong proposed the System Model of Creativity (Shi Jiannong, 1995) based on a synthesis of previous research and the Iceberg Theory of Creativity Cultivation (Shi, Qu, & Liu, 2007), practical schemes for cultivating creativity were already being contemplated. Initially, the research team, consisting of teachers, graduate students, and doctoral students at that time, pooled their wisdom to design and implement some courses based on their understanding of creativity. However, over time, several issues with the early developed courses became increasingly apparent: (1) The issue of creative motivation. (2) The constraint of methodologies. (3) The issue of fragmentation and lack of focus. In the past decade of exploration, we have focused on improving the stimulation of psychological motivation for creativity, by proposing the Butterfly Theory of Fostering Creativity, and thus advancing creativity cultivation to its 2.0 version.

21.2 Design Process

21.2.1 Creativity Cultivation

The Iceberg Theory of Creativity Cultivation (Shi et al., 2007), begins with the definition of creativity and deduces all the underlying conditions necessary for an individual to exert their creativity (as shown in Fig. 21.1). Initially, the research team brainstormed and jointly designed and implemented some courses based on their understanding of creativity at that time. However, overtime, several issues in the early developed courses became increasingly apparent:

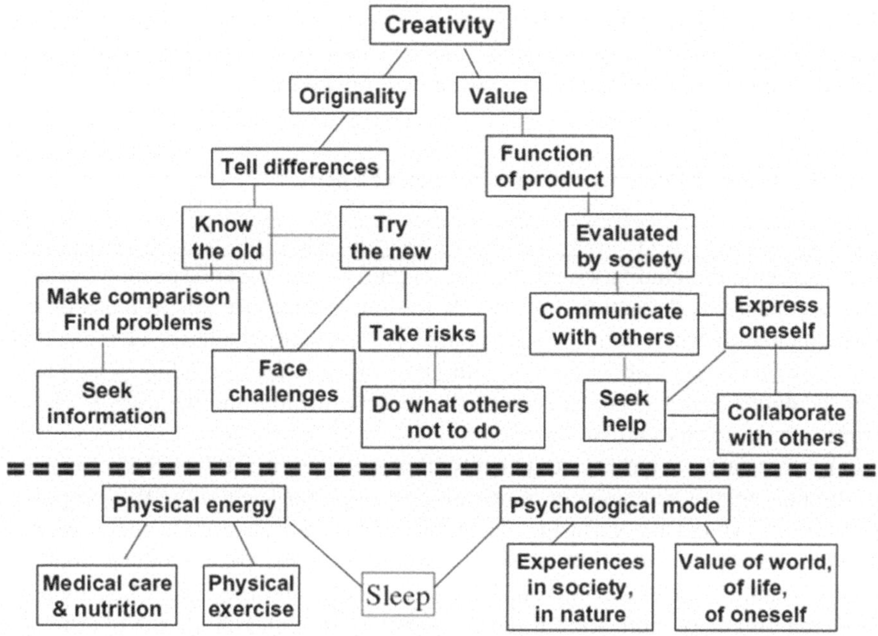

Fig. 21.1 Iceberg model of fostering creativity (Shi et al., 2007)

21.2.1.1 Issue of Creative Motivation

Some courses borrowed ideas from creativity assessments, such as presenting children with incomplete line drawings for them to complete with their imagination, or unfinished stories for them to fill in the blanks. While there is a rationale behind such approaches to creativity cultivation, some children, who entered the creativity class with expectations of gaining something new and experiencing novelty, may find themselves disappointed. After entering the course, they realize that they have gained little and have to expend considerable mental effort to complete divergent thinking tasks. With minimal gains and significant effort, some children may develop a sense of boredom. If fostering creativity fails to stimulate children's creative psychological dynamics and instead evokes boredom, has it truly achieved the goal of igniting children's creativity? If nurturing creativity does not effectively activate children's creative psychological dynamics, all efforts will be in vain.

21.2.1.2 Constraint of Methodologies

There are numerous methods for fostering creativity, such as brainstorming, divergent thinking, lateral thinking, reverse thinking, problem splitting, analogy, etc. (De Bono, 1970; Horng et al., 2005; Ma, 2006; Scott et al., 2004). In the process of teaching creativity, teachers tend to instruct students on specific methods. However, once methods become fixed, many students' thinking becomes constrained within these methods, ultimately failing to truly activate their creativity.

21.2.1.3 Issue of Fragmentation and Lack of Focus

In the 1.0 version of creativity cultivation, various elements related to creativity were aggregated, including cognitive, non-cognitive, and social aspects. However, when applying this theoretical framework, designers often struggle to focus on the key elements. Teachers tend to start course design with whichever element they encounter first in the iceberg model, without a core concept or guiding principle to steer the entire design process. As a result, they frequently find themselves lost in the maze of creativity cultivation.

In recent years, we have systematically improved theories and curricula focusing on stimulating the dynamic or motivating system for creativity. We proposed the Butterfly Model of Nurturing Creativity (Zhang et. al, 2021), which directs all elements of fostering creativity toward enhancing psychological motivational dynamics or other driving forces for creativity (as shown in Fig. 21.2), thereby advancing creativity cultivation to its 2.0 version. To ignite children's psychological motivational dynamics, the Butterfly Model offers a specific solution: continuously incorporating elements that are "unexpected yet reasonable" into the classroom, replacing boredom with surprises. By constantly introducing "unexpected plot twists" in class, children's interest can be sparked and sustained, thus addressing the motivational challenges in creativity cultivation. Focusing on psychological motivational dynamics, we aim to capture the core elements of creativity cultivation, addressing the fragmentation and lack of focus identified in the previous version. The Butterfly Model of Fostering Creativity tackles the issues arising from rigid methods. It does not endorse any specific, fixed routines but emphasizes that creativity cultivation has no set rules, and methods should not be applied dogmatically. In the process of cultivating creativity, the focus is not on the effectiveness of any particular method, but on how to activate children's psychological motivational dynamics, regardless of the approach. The goal is not to teach specific methods but to enable the enjoyment and effectiveness of creative activities.

Taking *The Art of Invisibility–Can We Truly Trust Our Eyes?* as an example, this lesson sparks children's interest with a "magic trick of invisibility." The trick involves drawing a pattern, placing it in a resealable bag, tracing part of it with a marker on the outside of the bag, and then immersing the bag vertically in water, making the untraced part appear invisible (as shown in Fig. 21.3). The disappearance of an object is highly intriguing to children who have developed the concept of object permanence, as the appeal of magic often lies in its unexpected effects. In the creative design phase, children are encouraged to use their imagination to design novel and unique (distinctive) yet practical (meaningful) magic props, leading them to break free from fixed patterns and inspiring new ideas. When children find the activity intriguing and try it themselves, they immediately generate a series of questions, hypotheses, and

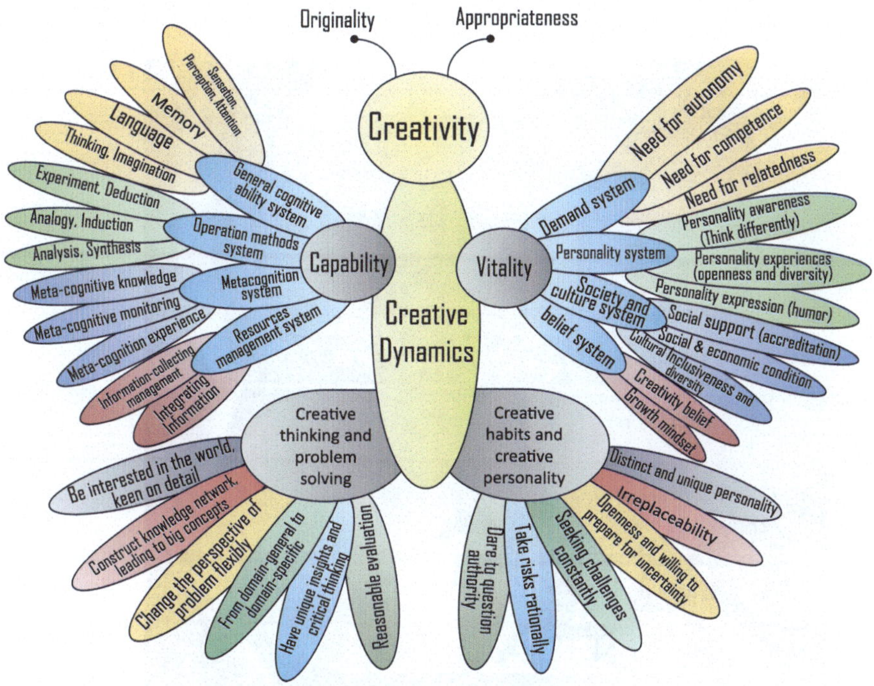

Fig. 21.2 Butterfly model of fostering creativity (Zhang et al., 2021)

causal inferences. This is a great opportunity to guide them in exploring causality and the underlying principles behind the phenomena, encouraging them to propose their own conjectures and then guiding them to test their guesses. After a series of explorations and attempts, students identify all the experimentally verified factors: the bag, water, the air inside the bag, the viewing angle, and light. However, they still need to delve deeper into why these factors are relevant. Finally, children are comprehensively guided to sketch the key parts for analysis, ultimately uncovering the principles behind the phenomena.

21.2.2 The Design of Creative Teacher Training System

The overall institutional arrangement begins with serving as a teaching assistant, progressing from imitation to improvement, and ultimately fostering a cultural atmosphere that encourages creativity (as shown in Fig. 21.4).

The teacher training system generally includes the following steps:

- **Step 1: Interview Stage**—Upon successful completion of the interview, candidates can proceed to the assistant teaching internship stage.

Fig. 21.3 Course: *The Art of Invisibility* in the water stealth or change

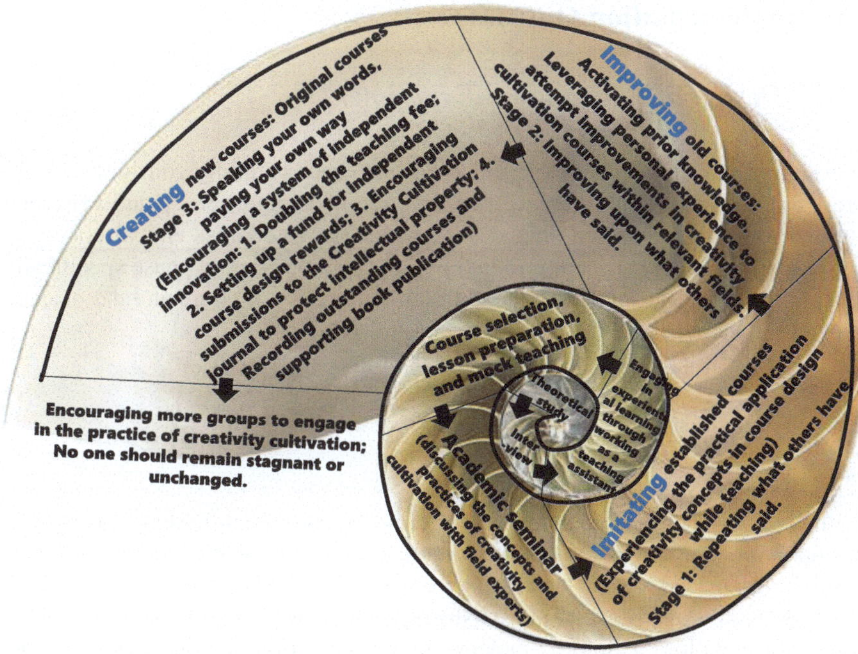

Fig. 21.4 Nautilus-inspired model for creative teacher training

- **Step 2: Internship Stage**—During the assistant teaching process, candidates engage in immersive, experiential learning through observation and participation. Afterward, they review relevant courses that interest them. Regularly, experts in the fields of creativity are invited to give lectures and engage in exchanges during the internship.
- **Step 3: Teaching Stage**—Once a teacher meets the standards, they can arrange a trial lecture, starting with imitating an established course. Then, they can gradually learn to design new courses while teaching, transitioning from improving and refining existing courses to independent innovation. It is necessary to provide new teachers with avenues for innovation and opportunities for trial and error, as well as personal growth.
- **Step 4: Development Stage**—Throughout the teaching process, everyone grows, refines their skills, and elevates themselves.

21.3 Implementation Outcomes

21.3.1 Design of Educational Experiment

The design of the educational experiment has also been carefully considered. 1. To address the issue of causal inference, we have set up a control group and a placebo group, as well as randomly assigned all subjects. 2. To address ethical concerns, we have divided the experiment into two phases, with the second phase specifically providing the same curriculum to the children in the control group and the placebo group (as shown in Fig. 21.5).

21.3.2 Test Results and Discussion

In both creativity cultivation experimental projects, a common finding was significant improvement in the divergent thinking test, indicating that encouraging divergent thinking can lead to short-term improvements. This is shown in Fig. 21.6, which presents the results of an educational experiment conducted at the Dream Space of the Institute of Psychology, Chinese Academy of Sciences; (Fig. 21.7 displays a comparison of the processes, measurements, and results of the two educational experiments). The latest research findings have been published in Educational Psychology (Zhang et al., 2025).

How do we interpret the other measurement results shown in Fig. 21.7? We can introduce some explanatory frameworks. Luthans et al. (2007) classified common variables in psychological research into several categories: very stable and difficult-to-change trait variables (such as intelligence and heritable personality characteristics); followed by trait-like variables (such as the Big Five personality traits and virtues); then the more easily changed and developed state-like variables (such as psychological capital and happiness); and finally, the very unstable and easily changed state variables (such as positive emotions). We can use this classification to explain the results of this study (Wen, 2017).

In the results of this study, spatial reasoning, which falls under the category of intelligence, is close to a trait variable, relatively stable, and not prone to significant changes after short-term interventions. Parenting styles within the family environment variable resemble trait-like variables and are also not easily subject to major changes in the short term. Divergent thinking, on the other hand, is closer to a state-like variable, and encouraging divergent and unconventional thinking is more likely to yield more divergent thinking outcomes. The creative preferences or attitudes is far from reaching the level of the Big Five personality traits in stability and is even lower than creative habits, leaning toward a state-like variable. For young children and adolescents whose personalities are still developing, there is greater plasticity, and creativity cultivation can influence their creative tendencies.

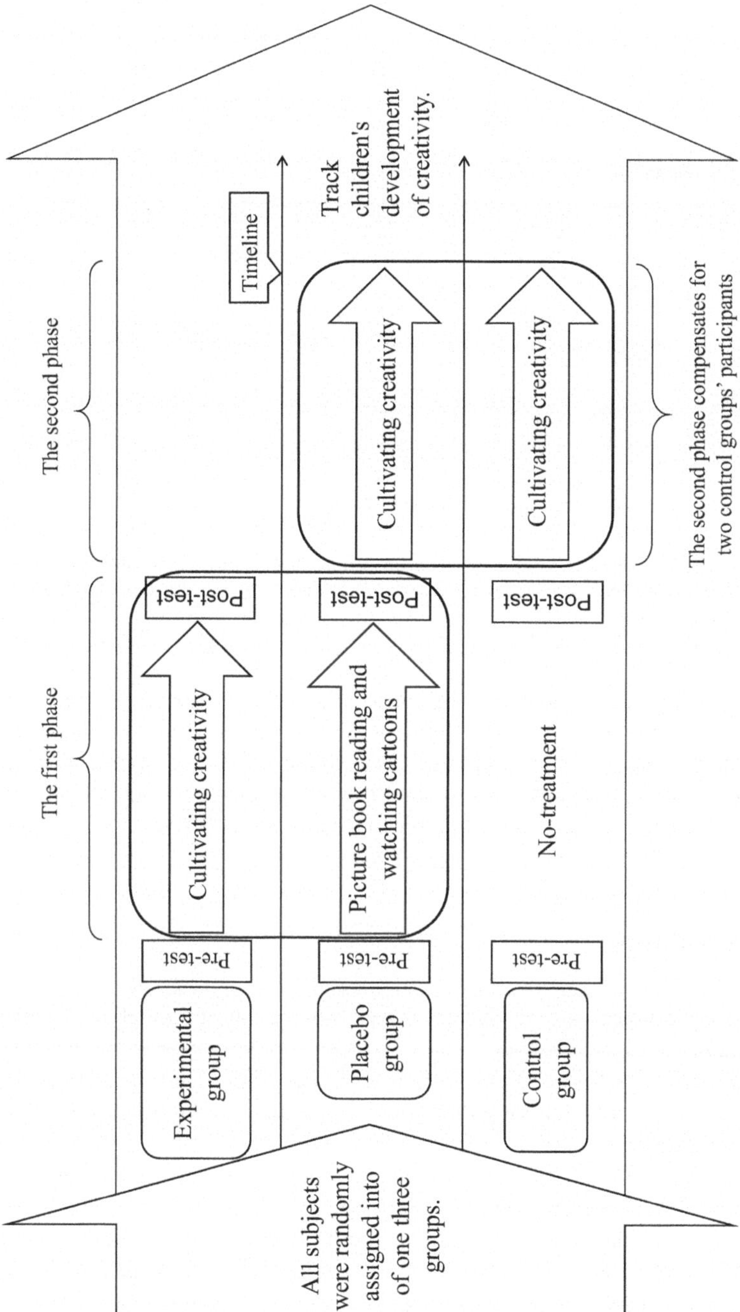

Fig. 21.5 Overall design of the educational experiment and flowchart

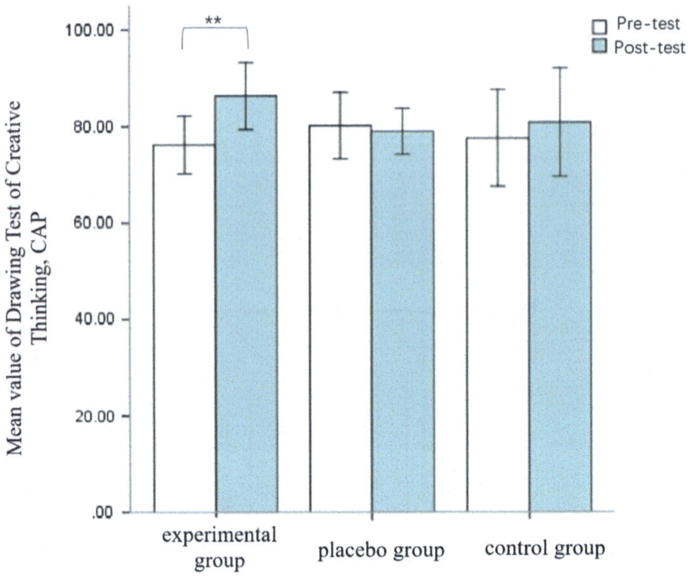

Fig. 21.6 In the 2018–2019 Fostering Creativity project at the Institute of Psychology, CAS, the scores of the experimental group, placebo group, and control group were measured before and after the Williams Drawing Test

For example, it can encourage children to enjoy exploration and prefer using unconventional methods to solve problems in the short term. However, the issue of creative tendency remains to be validated, as both educational experiment projects did not have a sufficient sample size to meet the requirements for statistical power. For the latest research findings, please refer to Zhang et al. (2025).

21.4 Innovations

- Design a creativity-fostering system based on the latest model of creativity development.
- Utilize the creativity-fostering curriculum system as a link for nurturing innovative talent in science and technology through collaboration among families, schools, and communities.
- Propose an institutional framework that encourages curriculum innovation and mutual growth between teaching and learning.
- Attempt to construct a value system for nurturing creativity, facilitating the sustainable development of the creativity-fostering curriculum system, laying a solid foundation for better collaboration among schools, families, and communities, and providing support for inclusive education.

Process / Year	Participation Sample size (whether grouped by age)	Grouping details	Fee structure	Intervention course (duration of intervention) (class size)	Measurement content and valid data volume		results √ indicates significant; × indicates not significant.
2018 - 2019 (Experiment location: Institute of Psychology, Chinese Academy of Sciences)	67 participants (Valid data varies across different tests) Grouped by age	Experimental group (34) Placebo group (15) Control group (18)	Charge a certain fee (No one takes leave casually; except for younger children who may skip tasks or not know how to do them, everyone takes the process seriously. The collected fees are used for materials and teaching assistant volunteers during the educational experiment.)	Conducting a 12-day creativity cultivation ~~program~~ (12 sessions of creativity cultivation courses, with a brief overview of each session and photos from the classes sent to parents for ~~reference)~~ Small class teaching	cognition	*Williams Drawing Test (67)*	√
						Edited Divergent-thinking Test (66)	√
						Edited Intelligence Test (60)	×
					Non-cognition	*Williams Prefer Measurement (33)*	√
					Family environment	*Robinson Parenting Styles and Dimensions Questionnaire (42~45)*	×
						Williams Scale for Rating Divergent Thinking and Feeling Behaviors (44)	×
2023 (Experiment location: China Science and Technology Museum)	39 participants (Valid data from 25 participants) No age grouping was applied; the participants included not only fifth and sixth graders but also third and fourth graders.	Experimental group (17) (Valid data 12人) Control group (22) (Valid data 13人)	Free of charge (There were frequent instances of participants taking leave, indicating a lack of commitment. Some students did not take the questionnaire seriously, resulting in careless responses.)	Conducting a 6-day creativity ~~cultivation program~~ (6 sessions of creativity cultivation courses, with photos from each class shared in the parent group) ~~Large class teaching~~	cognition	*Williams Drawing Test (25)*	√
						Edited Intelligence Test (25)	×
					Non-cognition	*Williams Prefer Measurement (25)*	×
						Big Five Personality Traits - Openness Subtest" (25)	×
						Science Motivation Questionnaire (25)	×
						Growth mindset and fixed mindset (23)	×
						Grit Scale (25)	×
					Family environment	*Williams Scale for Rating Divergent Thinking and Feeling Behaviors (17)*	×
						Parental Autonomy Support Scale (17)	×
						Creative Family Environment Scale (17)	×

Fig. 21.7 Comparison of two creativity training intervention experiments

21.5 Promotion

21.5.1 Current Cooperation Outcomes

This study has utilized a portion of the creativity-fostering curriculum system to conduct a two-week educational experiment in collaboration with the Exhibition Education Center of the China Science and Technology Museum and the Educational Center, Working Committee of Caring for Younger Generation, achieving certain results. In the future, we will continue to enhance cooperation, leveraging the strengths of all three parties to promote the in-depth advancement of creative teacher training and the promotion of the creativity curriculum system.

21.5.2 Current Creativity Cultivation Outcomes

Over the past decade of exploration, we have developed a curriculum system based on the Creativity Cultivation 2.0 version. The themes cover a wide range, including Stepping into the Door of Creativity, Exploring the Physical World, Unraveling the Mysteries of the Brain and Psychology, The Perspective and Vision of Astronomy, An Introduction to First-Order Natural History, The Secrets of Biomedicine, Playing with Chinese Characters, Creative Problem-Solving, and The Game of the Brave. Current curriculum system catalogue for creativity cultivation is as follows:

Stepping into the Door of Creativity

- *Introduction: Stepping into the Door of Creativity—Envisioning the Future through Creativity.*

Exploring the Physical World

- *The Art of Invisibility—Can We Truly Trust Our Eyes?*
- *Unraveling the Essence of Light*
- *The Mystery of Electromagnetic Forces: The Long-Distance Association of the Guiding Fish (a Magnetic Compass)*
- *Breakthrough in the Concept of Electricity—From Origin to Creation*
- *The Joy of Symmetry and Balance.*

Unraveling the Mysteries of the Brain and Psychology

- *Bengzi Painting–Exploring Three-Dimensional Space*
- *The Little Universe of Life—Uncovering the Secrets of the Brain*
- *Reconstructing the World*
- *The Evolution of the Brain—Where Are We Heading?*

The Perspective and Vision of Earth Science and Astronomy

- *Exploring the Earth—Seeing the World from a Different Perspective*

- *The Mysteries of Earthquakes—The Dual Nature of Earthquakes*
- *Exploring the Solar System—Landing on Mars*
- *Exploring the Starry Sky.*

An Introduction to First-Order Natural History

- *The Natural History of Flowers*
- *Exploring Plants—Seeds and Their Evolution*
- *Symbiotic Relationships—Seeking Connections between Humans and Nature*
- *The Wonders of Minerals*
- *Creating the Book of Life: Understanding the Journey of Evolution.*

The Secrets of Biomedicine

- *A Bird's-Eye View of Modern Genetics—Decoding the Mysteries of Genetics*
- *Exploring Teeth—Uncovering the Stories Behind Teeth*
- *Ancient Life—Discovering Dinosaurs*
- *Establishing Causality—From Ancient Medicine to the Establishment of the Germ Theory.*

Playing with Chinese Characters

- *Exploring Chinese Characters.*

Creative Problem-Solving

- *Transferring Ping Pang Ball Task*
- Exploring Architecture—From Construction to Deconstruction.

The Game of the Brave

- The Battle of Hot and Cold—Exploring Liquid Nitrogen in the Game of the Brave
- Exploring Gases—The Fusion of Water and Fire—Waterspouts, Dry Ice Cannons, and Fire Clouds.

21.6 Summary

This study explored a collaborative model involving schools, families, and communities to foster children's creativity. It aimed to develop a curriculum system dedicated to nurturing creativity and to investigate a corresponding model for teacher training.

21.6.1 Core Values

This study proposed that activating the creative dynamics was fundamental to cultivating creativity. Although many creativity training programs focused on cognitive strategies such as divergent thinking and problem-solving, they often failed to emphasize the role of intrinsic motivation in supporting children's creative drive. Without this internal motivation, the effectiveness of such methods was significantly diminished.

21.6.2 Significance of Applying Educational Theory and Innovating in Practice

Previous theories on fostering creativity were relatively diffuse, which made it difficult for curriculum designers to identify clear focal points. This study proposed a more targeted approach, by arguing that the core of creativity education lay in sparking students' intrinsic motivation to create. Moreover, the importance of designing learning experiences that were engaging and delightfully unexpected, while remaining logically coherent was exactly the experiences where educators should focus their efforts. Instead of relying on quick-fix solutions, the study advocated for a gradual developmental process: beginning with experience and imitation, followed by improvement, and ultimately leading to innovation. This step-by-step progression helped reduce the pressure to innovate, allowing teachers to slow down, remain curious, and pursue knowledge for its own sake. Such a grounded approach avoided empty slogans and supported the cultivation of genuinely innovative educators through empowered collaboration among schools, families, and communities.

21.6.3 Practical Applications and Future Research Recommendations

Well-developed creativity curricula can be expanded to a wide range of educational settings, such as science centers, museums, community venues, and homes, and tailored to meet the developmental needs of children at different age groups. The study recommended allocating funding to publish these curricula, which had been refined over the past decade, in order to provide ready-to-use models for innovative educators. As AI continued to show transformative potential, future research was encouraged to explore how it could be effectively integrated into creativity education, guiding both students and teachers in using AI as a tool for creative expression. The goal was to ensure that AI did not become a crutch that limited independent thinking, but rather an accessible aid—like trekking poles—that supports and enhances the creative journey. A crutch or walking stick replaces your legs, causing them to weaken over time; trekking poles, on the other hand, simply assist your walking, used only when needed, and help improve hiking efficiency.

Acknowledgements I am deeply grateful to all those who supported me throughout this journey. My sincere thanks go to Prof. Jiannong Shi and Mr. Ning Chen for their essential contributions to the development of the creativity cultivation curriculum. Without their strong support, this curriculum would not have come to life. I also appreciate the support provided by Ms. Lina Huang, Miss Yufei Zong, and Mr. Wei Kang during the process. Special thanks to Prof. Xingli Zhang for allowing me to participate in the competition from the very beginning. I would also like to thank the other members of our research team at the time: Miss Yanli Zhang, Miss Naili Bao, Miss Yaxin Diao, Miss Shuchen Li, Miss Xiaoqing Ding, Mr. Xu Chen, Mr. Chuan Wang, Ms. Wanjing Hong, and Prof. Zhen Zhang. Each of them contributed valuable ideas to the creativity development project in their own way. Finally, I want to express my gratitude to Xin Qi, Xiaona Ye, and Bowen Yang from the Exhibition and Education Center, China Science and Technology Museum, as well as Xueni Su and Jun Feng from the Working Committee of Caring for Younger Generation. Although they were not directly involved in the 6th Global Competition on Design for Future Education, they provided tremendous support for the creativity development project through collaboration among families, schools, and communities.

References

De Bono, E. (1970). *Lateral thinking: A textbook of creativity*. Penguin Books.

Horng, J.-S., Hong, J.-C., ChanLin, L.-J., Chang, S.-H., & Chu, H.-C. (2005). Creative teachers and creative teaching strategies. *International Journal of Consumer Studies, 29*(4), 352–358. https://doi.org/10.1111/j.1470-6431.2005.00445.x

Luthans, F., Avolio, B. J., Avey, J. B., & Norman, S. M. (2007). Positive psychological capital: Measurement and relationship with performance and satisfaction. *Personnel Psychology, 60*(3), 541–572. https://doi.org/10.1111/j.1744-6570.2007.00083.x

Ma, H.-H. (2006). A synthetic analysis of the effectiveness of single components and packages in creativity training programs. *Creativity Research Journal, 18*(4), 435–446. https://doi.org/10.1207/s15326934crj1804_3

Scott, G., Leritz, L. E., & Mumford, M. D. (2004). The effectiveness of creativity training: A quantitative review. *Creativity Research Journal, 16*(4), 361–388. https://doi.org/10.1207/s15326934crj1604_1

Shi, J., Qu, X., & Liu, T. (2007). Creativity and its cultivation. In A. G. Tan (Ed.), *Creativity: A handbook for teachers* (pp. 65–75). World Scientific. https://doi.org/10.1142/9789812770868_0004

Shi, J. (1995). System model of creativity. *Advances in Psychological Science, 13*(3), 1–5.

Wen, Z. (2017). Causal inference and analysis in empirical studies. *Psychology Science, 40*(1), 200–208. https://doi.org/10.16719/j.cnki.1671-6981.20170130

Zhang, Y., Chen, N., Chen, L. A., & Shi, J. (2021). Wisdom minds with creative wings: Igniting creative dynamics focusing on its interest cultivation. *Advances in Psychological Science, 29*(4), 707–722. https://doi.org/10.3724/SP.J.1042.2021.00707

Zhang, Y., Pan, Z., Huang, L., Zhang, X., Chen, L. A., & Shi, J. (2025). Teaching with surprise to enhance children's inclination for creative thinking and creative tendencies. *Educational Psychology, 1*(1), 1–23. https://doi.org/10.1080/01443410.2025.2464851

Chapter 22
The Sharing of Educational Resources Between Urban and Rural Schools Enabled by Real-Scene Classroom

Fei Hu, Yuhua Zhang, Long Li, Xiaoling Tao, Rui Xiong, and Xinyi Huang

Abstract Changleping Primary and Secondary School, spanning over 70 years, has faced significant developmental challenges, such as a severe outflow of excellent teachers and students. In 2019, Yichang Tianwen Education Group took full management of Changleping Primary and Middle School. It implemented a new strategy known as the "six inputs" (manpower, culture, training, curriculum, funds, brand) to revitalize rural education. This initiative led to a remarkable increase in the school's ordinary high school graduation rates, from 42 to 82%. The school has won the first prize for two consecutive years in Wufeng County comprehensive evaluation, leading to a stable return of students. However, the main factors restricting the development of Changleping Primary and Secondary Schools are outdated educational resources and an aging teaching staff. Tianwen School uses real-scene classrooms to enhance the efficiency of technology and equipment, effectively addressing the shortage of rural educational resources and teachers. Also, it complements the rural educational resources that urban schools lack, with the school and trusteeship school developing jointly. Online and offline interactions fully mobilize student enthusiasm, and the abundant resources make abstract cultural knowledge and scientific principles lively and engaging.

Keywords Real-scene classroom · Rural education · Educational technology · Teacher collaboration · Student engagement

F. Hu (✉) · Y. Zhang · L. Li · X. Tao · R. Xiong · X. Huang
Xiling Tian Wen Primary School, Yichang, Hubei, China
e-mail: 313316859@qq.com

Y. Zhang
e-mail: 28279235@qq.com

X. Tao
e-mail: 339451776@qq.com

R. Xiong
e-mail: xr0914@qq.com

X. Huang
e-mail: 417945216@qq.com

D. Liu et al. (eds.), *Empowering Educational Innovation Through Design: Volume One*, Smart Computing and Intelligence, https://doi.org/10.1007/978-981-95-4621-3_22

22.1 Background

Changleping Primary and Middle School is a rural nine-year educational institution located in Wufeng County, Yichang City, Hubei Province, for the first time in 70 years of operation, the school encountered teaching bottleneck—the severe outflow of standing students and staffs. In April 2019, Yichang Tianwen Education Group signed a cooperation agreement with the Wufeng County Education Bureau to fully manage Changleping Primary and Middle School, aiming to resolve the developmental difficulties the school has been facing in recent years. They focused on reforming classroom teaching methods through various approaches, mainly engaging on exploring the "six inputs" (manpower, culture, training, curriculum, funds, brand) to revitalize rural education, comprehensively developing a new path and improving the quality of rural education and teaching. Within two years of educational reform, the school graduation rate increased from 42 to 82%. Due to breaking this barrier, the school won the first prize in the comprehensive evaluation of Wufeng County for two consecutive years, and the number of students has steadily increased from 441 to 600.

22.2 Design Process

The main factors constraining the development of Changleping Primary and Secondary School are backwardness of educational resources, the age of teachers and slow updating of knowledge and technology. Tianwen Education Group decided to use real-scene classrooms, using real-world examples in the classroom to solve real-world problems (Cox, 2020) through "Little Ant" platform. Thus, the program needs two so-called "inner teacher" and an "outer teacher;" two "inner teachers" indicating the instructor in the classroom at both Gaoxin Tianwen School and Changleping Primary and Secondary School while the "outer teacher" is the teacher who connects with the class at the outdoor venue through live streaming. By effectively conducting this, rural children can benefit from the advanced and patterned urban teaching resources. Equivalently, the urban schools are interconnected and complemented by the rural teaching background, guiding urban school students to explore and learn the natural environment and labor practices in rural areas.

22.2.1 Science and Technology Empower Rural Schools

In the science class "Where Does Electric Energy Come From," teacher Rui Xiong from Gaoxin Tianwen Primary School serves as the "inner" instructor. He connects the classroom online to Changleping Primary School through the "Little Ant" platform, while the science teacher at Changleping Primary school leads the students

Fig. 22.1 Yichang real-scene classroom "Where Does Electric Energy Come From"

explore and operate synchronously. With real-time guidance from teachers in both locations, students from the two schools are capable of actively participating in in-class science experiments. This engagement mobilized their imagination to think about the principles of power generation and different methods of generating electricity. Through immersive learning in real-scene classrooms, students experienced vigorous science and technology development in China vividly (see Fig. 22.1).

This lesson begins with a topic depicting electricity consumption in daily life, helping students understand the pivotal role of electricity. By cleverly setting up a scenario without electricity, the lesson stimulates students' curiosity. The teacher combines humorous teaching methods with science knowledge, enabling students to gain a fundamental understanding of which electrical energy is generated by other power sources, such as chemical, nuclear, and solar energy.

To take the next step in conducting the real-scene classroom, teacher Rui Xiong exhibited an experiment, generating electricity through magnets, and the relationship between electricity and magnetism is hence explored. This allows students to experience the basic process of generating electricity from magnetism and construct a practical knowledge system for the conversion of various energy sources into electrical energy. Moreover, this approach helps students further appreciate the significant contributions of science and technology to humanity.

After the phase introducing theoretical knowledge in the course, the "outer" teacher Jianyu Wu, who was currently in Gezhouba Hydropower Plant, leading students visit the real power station through the real-scene classroom platform. Weikun Gu, the Director of Gezhouba Hydropower Plant, was invited to present the students with the construction scale of the hydropower station, its average annual power generation, the division of labor within each plant, and the relationship

Fig. 22.2 Director of Gezhouba Hydropower Plant Weikun Gu, is sharing

between Gezhouba Hydropower Plant and the Three Gorges Hydropower Station. He also introduced the process of generating electric energy using the hydro generator, detailed the working principle of the generator, and explained how electricity is transmitted to residents through reconversion and pressure reduction (see Fig. 22.2). After this introduction, students could genuinely realize the convenience that science and technology bring to their lives. During their visit to the Gezhouba Hydropower Plant, students explored the principles behind the generation of electricity with hydro-generator sets. Ultimately, Director Gu answered relevant questions and explained how the electricity generated is converted back into household electricity in real time.

22.2.2 Urban Sharing of Rural Resources

In the labor-oriented real-scene classroom "Spring Tea Picking Season," Xiaoling Tao, a teacher from Gaoxin Tianwen Primary School, served as the "inner" teacher and connected the classroom online to Changleping Primary School through the "Little Ant" platform. Fei Liu, a research teacher at Wufeng Changleping Primary and Secondary School, and Haiyan Zhang, a teacher from Wufeng Vocational Education Center both served as "outer" teachers. On the class, teacher Tao led the students in watching the video "Chinese Famous Tea" to learn about tea and tea culture briefly (see Fig. 22.3). Utilizing the data collected before the class by themselves, students from both places participated in a sharing session about famous Wufeng

Fig. 22.3 Yichang real-scene classroom "Chinese Famous Tea"

tea, experiencing the culture of Wufeng Tea. With the help of photography, they understood the different tea picking standards, the tools used for tea picking, and the techniques involved. Teacher Tao also introduced "lifting the hand," a common tea plucking technique. This technique not only ensures the quality of the tea, but also protects sustainability. In the real-scene class, teacher Tao then connected with teacher Liu, who introduced the methods for preserving fresh tea leaves, which encouraged the students to learn the classification of tea picking in the tea garden. Teacher Haiyan Zhang focused on tea ceremony. She explained the methods of brewing and tasting tea, as well as the tea ceremony culture behind it.

22.3 Implementation Outcomes

22.3.1 Student Learning Effect

In the process of real-world teaching, teachers utilize immersive teaching scenes, online videos, and on-site Q&A sessions, providing students with a comprehensive six-sense experience. This approach allows students to understand the teaching materials from a diversified perspective, which could improve their motivation for learning. It enables students to receive education in a state of subtle and silent immersion. The science class "Where Does Electric Energy Come From" leverages the advantages of real scene learning to allow students to virtually walk into the power station. Through these real-scenes, students can experience the entire process of hydroelectricity, understanding the intricacies of power generation. The online teaching collaboration with the trusteeship school ensures that students who stay at home have access to the same educational resources as those in urban schools. This approach addresses the shortage of high-quality teacher resources in rural schools, particularly in comprehensive subjects. By utilizing the geographical resources of

Wufeng comprehensively, the "Spring Tea Picking Seaon" real-scene classroom allows students to learn tea picking and tea making methods in real life. This simulation experience helps students develop a more detailed understanding of the tea picking process, refining their delicate picking techniques, mastering labor skills, and stimulating their interest in labor.

22.3.2 Teacher Capacity Improvement

For possessing the real-scene classroom skill, on the one hand, the instructors skillfully use the "Little Ant" platform, whiteboard interactive functions, network communication software, and other tools through individual training and teaching research sessions. On the other hand, teachers in the real-scene classroom will switch between "inner" and "outer" status at any time, teachers in Yichang and Wufeng need to comprehensively discover local teaching resources, carefully interpret teaching materials, and combine them with teaching aims to design efficient and engaging real-scene classrooms.

22.3.3 School Environment Construction

To keep pace with the times and help rapidly improve rural education modernize, the "Little Ant" multi-point interactive live broadcast system enables teachers and students to connect to online and offline classrooms anytime and anywhere. This system introduces advanced teaching concepts, optimizing resource allocation, and utilizing technology, strategies, and tools to better support student independent development. Real-scene classrooms are just a small part of the trusteeship school initiative. Through trusteeship, Changleping's comprehensive school operations, cultural construction, and reputation have significantly improved. The school's education has made substantial progress, taking the lead in rural education revitalization. Since its trusteeship began two years ago, the school has received several accolades, including "Advanced School of Municipal Education Modernization," "Top 100 School Cultural Construction in Hubei Province in 2020," "Cultural Construction Innovation Award," "Hubei Cultural Construction Annual Achievement Award," etc. Dr. Zheng Chen from the Institute of Educational Sciences at Wuhan University praised, "Tianwen entrusted Le Ping School is a model for education poverty alleviation and rural revitalization."

22.4 Innovations

In the "six inputs," the curriculum is the key. To address the main factors restricting the development of Changleping primary and secondary schools such as outdated educational resources and aging staff—Tianwen implemented strategies involving real-scene classroom empowerment, resource sharing, and the inclusion of renowned teachers. These approaches aim to foster technical equipment usage efficiency and narrow rural education resources and teacher shortage. The real-scene classroom is a collaborative effort between "inner" and "outer" teachers. The combination of online and offline interaction fully engages students (Yen et al., 2018), stimulating their curiosity and desire to explore. Abundant real-world teaching resources make abstract cultural knowledge and scientific principles vivid and engaging. Also, the initiative allows rural and urban schools to complement each other's educational resources, fostering joint development between the school and the trusteeship school.

22.5 Conclusion

Under the "Internet+" conditions, the real-scene classroom integrates textbook knowledge with the real-life scenarios, supported by rigorous structures and field engineers. Factories and enterprises interact with students in real-time, reforming the instructing mode, extending the teaching environment, stimulating students' interest in learning, and reflecting the latest advancements in the field. This approach broadens students' horizons and represents a significant development trend for future instructing models. Rural revitalization must begin with the revitalization of rural education. Improving the quality of rural education and transforming schools into cultural highlands, talent bases, and innovation fronts is a crucial duty and mission in achieving the second centenary goal and promoting endowment education. Leveraging modern urban education is both imperative and promising to trust and support rural education. The real-scene classroom is based on the theory of situational teaching, which refers to the new "multi-lecturer" teaching method. This method transmits real scenes from factories and enterprises, where the real teachers are located, to the school teaching classroom through signal acquisition equipment. It also feeds the real-time situation of the classroom back to the real teachers and engineers, so as to facilitate the interaction between students, teachers, real teachers and engineers. Pioneered by Beijing teacher Yinzhen Ren in 2016, this new teaching model leverages the context of the 5G era. It acts on students only after a high degree of reduction and purification of society and life. Examples of real-scene teaching include real-world traversal, literary real-scene copying traversal, literary real-scene copying, classroom competitions, role-based experience, poetry recitation, and picture book teaching, etc. These methods can be extended to various subjects and activity courses.

This project addresses the dual challenges of scarce educational resources and an aging teacher structure in rural schools, alongside the lack of practical learning resources in urban schools. It explores the use of Immersive Real-Scene Classrooms to achieve complementary resource sharing between urban and rural areas, enhancing rural education quality and promoting balanced educational development.

22.5.1 Research Objective

This project addressed the dual challenges of limited educational resources and an aging teaching workforce in rural schools, as well as the lack of practical learning resources in urban schools. It explored the use of Immersive Real-Scene Classrooms to facilitate the sharing of complementary resources between urban and rural areas, thereby enhancing the quality of rural education and promoting more balanced educational development.

22.5.2 Core Values

22.5.2.1 Technology-Enabled Resource Circulation

By using the "Little Ant" platform, the Immersive Classroom breaks geographical barriers. It enables two-way sharing of urban industrial resources (e.g., Gezhouba Hydropower Station) and rural natural labor resources (e.g., Wufeng Tea Plantation). This made a complementary model where urban students studied theoretical principles in the classroom, while rural students engaged in hands-on practice in real-world settings.

22.5.2.2 Dual-Teacher Collaborative Teaching Innovation

An indoor teacher led theoretical instruction, while an outdoor teacher (including industry experts) conducted practical demonstrations. The integrated approach of "online explanation + offline experience + real-time interaction" transformed abstract knowledge into tangible scenarios, stimulated students' curiosity and enhanced learning outcomes.

22.5.2.3 Sustainable Mutual Benefits

Rural students gained access to high-quality educational resources, resulting in a significant increase in their high school enrollment rate from 42 to 82%. Urban students acquired practical labor skills and local cultural knowledge through rural field experiences. This model effectively optimized the "cognitive structures and practical abilities" of students in urban and rural areas.

22.5.3 Theoretical and Practical Significance

22.5.3.1 Theoretical Innovation

The case expanded the "Internet + Education" framework, providing a new model for both situated learning and blended learning. It validated the feasibility of sharing educational resources in complementary way.

22.5.3.2 Practical Innovation

The case pioneered a rural education revitalization model that combined "hosted support + technology empowerment." Within two years, student enrollment increased from 441 to 600. Experts recognized the effect as "a replicable pathway for urban–rural educational collaboration," and it became a practical benchmark for reducing educational inequality and supporting rural revitalization.

22.5.4 Recommendations

22.5.4.1 Practice Promotion

This model expanded the subject coverage for Immersive Classrooms to include areas such as history and the arts. It also established a cross-reginal "Resource Sharing Alliances," which lowered technical barriers and extended the model into more under-resourced regions.

22.5.4.2 Future Research

Future studies should examine the long-term effect of Immersive Classrooms on students' core competencies. By exploring the AI integration with real-scene teaching, the model refined the "dual-teacher" training mechanism to enhance the efficiency of collaboration between urban and rural schools.

References

Cox, J. (2020) *Bring the Real World into the Classroom as a Teaching Strategy*. Future Educators. https://www.futureeducators.org/bring-the-real-world-into-the-classroom-as-a-teaching-strategy/

Yen, S.-C., Lo, Y., Lee, A., & Enriquez, J. (2018). Learning online, offline, and in-between: Comparing student academic outcomes and course satisfaction in face-to-face, online, and blended teaching modalities. *Education and Information Technologies, 23*, 1–13. https://doi.org/10.1007/s10639-018-9707-5

Chapter 23
Exploration and Practice of a Growth Education Evaluation Visualization Model

Yingying Xu and Ruiyuan Zhou

Abstract In response to the trend of integrating information technology with education, this study addresses issues such as the difficulty in capturing comprehensive process data for student assessment, the challenges in involving multiple evaluation stakeholders, and the unclear evaluation orientation. Our school has developed a Growth Education Evaluation Visualization System to shift student development from a knowledge-based focus to one centered on competencies. In this process, we actively explore the application of metaverse technology as an enabler. Built on technologies like VR, blockchain, AI, and computer vision, we enhance data collection through multiple modalities, strengthen the tracking and analysis of growth records, and reshape the spatial and temporal boundaries of growth evaluation. This further promotes the exploration of new evaluation models for growth education. The case design framework is guided by Dewey's concept of growth, utilizing holographic data theory and AI-driven software tools to create a visual, task-based, and game-like evaluation approach focused on children's competencies. A comprehensive visual system of elementary school students' overall development is constructed, centered on the values of "Joy in Wisdom, Health, Elegance, Labor, and Creativity." By leveraging information technology and big data, we aim to use technology to drive student evaluation and enhance student involvement in the evaluation process through data visualization. This approach transforms evaluation into a tool that inspires intrinsic motivation for learning and explores new paths for building a visual system of comprehensive student literacy evaluation.

Keywords Growth education evaluation · Visualization technology · Blockchain integration · Harmonious growth · Student literacy development

---

Y. Xu (✉) · R. Zhou
Shenzhen Bao'an Xiwan School, Shenzhen, Guangdong, China
e-mail: 407129816@qq.com

R. Zhou
e-mail: 121893688@qq.com

D. Liu et al. (eds.), *Empowering Educational Innovation Through Design: Volume One*, Smart Computing and Intelligence, https://doi.org/10.1007/978-981-95-4621-3_23

23.1 Background

On April 13, 2018, the Ministry of Education issued the *Education Informatization 2.0 Action Plan*, which proposed a transformation in views on "educational resources, technological literacy, educational technology, development dynamics, educational governance, and types of thinking."[1] It is clear that "Internet + " education and smart campuses are trends and inevitabilities for the future development of education. Visualization is one example. This concept was first introduced in the 1980s, and it has gone through three development stages: "data visualization—information visualization—knowledge visualization." It has increasingly attracted the attention of scholars in the field of education, particularly in areas such as visualized teaching design. However, its application in educational evaluation has been somewhat limited. Education evaluation plays a significant guiding role. A scientific educational evaluation system should not only focus on students' mastery of knowledge and skills, but also on the formation of their emotions, attitudes, and values. This approach transitions the focus from learning outcomes to the learning process. Traditional evaluation methods, driven by selection and developmental needs, are increasingly seen as limited. They often take a purely utilitarian approach, overlooking students as the central subjects of education and neglecting the cultivation of their intrinsic, independent personalities. By incorporating visualization technology, we can turn data into a more transparent, process-oriented, and proactive form of evaluation. This approach offers scientific evidence to inform teaching and provides clearer growth trajectories for students, addressing the shortcomings of traditional evaluation systems.

23.2 Design Concept

The exploration and practice of the Growth Education Evaluation Model centers on "harmonious growth" and is guided by Dewey's philosophy of growth (Hook, 1959). It relies on learning science theory, focusing on literacy development, and is based on the overall framework of smart campuses to plan the school's information technology infrastructure. By applying technologies like AI, VR, and blockchain, a Growth Education Visualization Evaluation System is created, forming an application architecture for a technology-powered growth education evaluation model (see Fig. 23.1).

The Growth Education Evaluation Model integrates information technology with education by using "visualization" to create clear educational and teaching contexts. It visualizes each student's growth process through the metaphor of a "sapling," incorporating elements like game progression, fun, and VR. The system is based on an energy accumulation model, supported by a title system and points exchange, and encourages collaboration between home and school in tracking students' growth

[1] http://www.moe.gov.cn/srcsite/A16/s3342/201804/t20180425_334188.html.

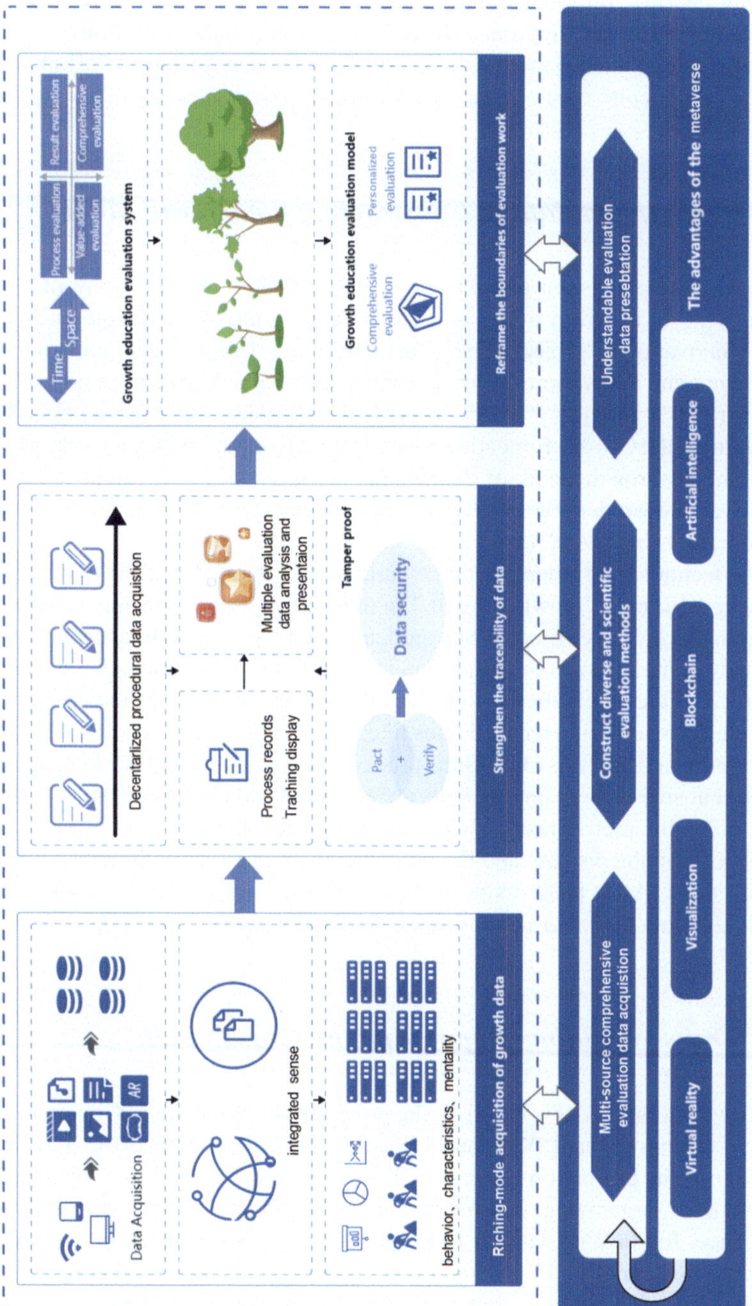

Fig. 23.1 Diagram of the technology-enabled student growth education evaluation model

milestones. The aim is to transform the educational process and outcomes into visible "tree-planting energy," turning external influences into an intrinsic motivation for students' development. This approach shifts from a simple, one-dimensional evaluation to a more comprehensive, multi-dimensional assessment, fostering the balanced development of students' core competencies through a rich curriculum.

23.2.1 "Idea" Implantation, Constructing the Growth Model

The growth evaluation system is built around the educational philosophy of "harmonious growth." It is driven by three key innovation engines: "technology-driven educational innovation," "project-driven curriculum innovation," and "governance-driven management innovation." These engines are aligned with the core values of harmony ("He") and joy ("Yue"), rooted in children's life experiences. The system introduces the Growth Education Visualization Evaluation Framework, which focuses on shared growth goals of "harmonious coexistence and joyful independence" and evaluates students across five key dimensions: "Joy in Elegance, Wisdom, Health, Labor, and Creativity" (see Fig. 23.2).

Using VR technology, students become gardeners of their own growth, with each student having their own "growth garden." In this garden, their growth is visualized as a sapling that grows over time. When students enter the school, they each plant a "tree." As they complete various evaluation tasks both inside and outside of school, intelligent records are automatically kept. The Growth Education Evaluation System serves as a platform, utilizing technologies like visualization, AI, and voice recognition. Data mining and analysis algorithms process evaluation data, generating energy balls of different sizes. These energy balls are collected and used as nutrition to help the sapling grow. Interactive tasks allow students to track their growth journey. The system reflects both the process and the outcome: the more energy the tree collects, the faster it grows. After progressing through five tree shapes, students will have successfully grown a tree after collecting "1000 energy points" (see Fig. 23.3).

23.2.2 "Traits" Building, Supporting the Growth Path

The growth evaluation model is based on the principles of "Wisdom + Governance," "Wisdom + Curriculum," and "Wisdom + Space," which together form the "Growth" engine and refine the evaluation indicators. The school employs "visible" evaluation tools and methods, organizing the growth evaluation system into five key modules: Joy in Elegance, Joy in Wisdom, Joy in Health, Joy in Labor, and Joy in Creativity. These modules encompass 15 evaluation indicators, including academic performance, extracurricular reading, punctuality, class participation, homework time,

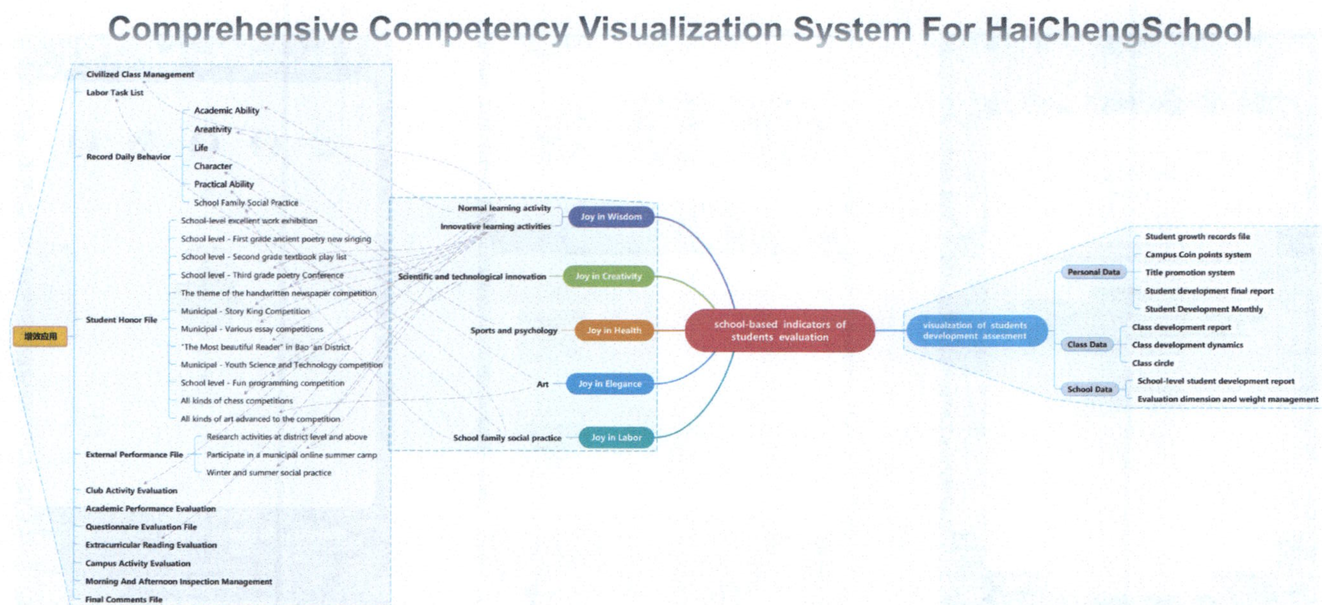

Fig. 23.2 Schematic diagram of the growth education visualization evaluation system

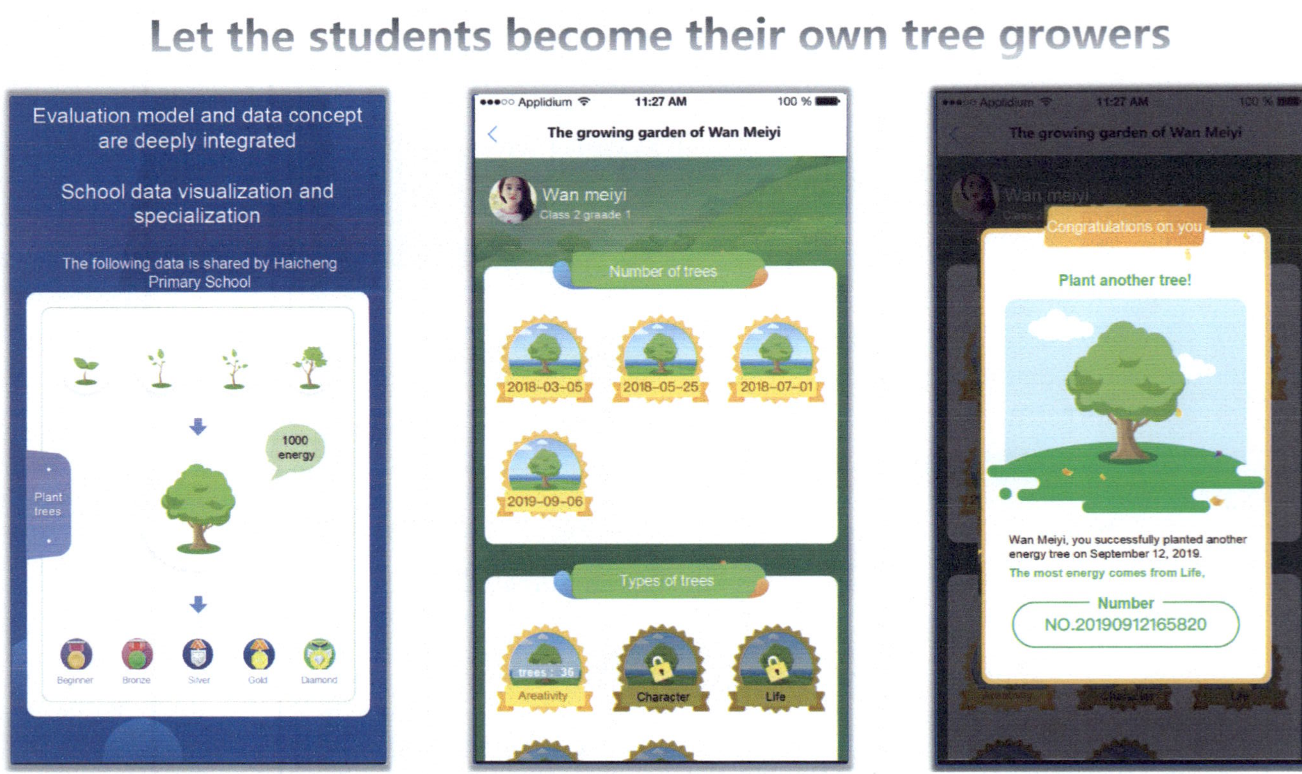

Fig. 23.3 Virtual student growth garden

exercise time, screen time, campus activities, club involvement, extracurricular practice, innovative behavior, classroom conduct, daily behavior, and Baby Star, among others.

23.2.2.1 Wisdom + Governance: Blockchain Technology is Applied to Enhance Growth Governance and Capture Growth Data in Real Time

DingTalk is used as the main platform for the school system, bringing together campus data, identity authentication, and access to software and hardware.

Based on the school's business scenario, growth evaluation data drives the development of routine systems, such as class performance evaluation, daily behavior tracking, labor task assessment, off-campus performance evaluation, teaching quality feedback, and student honor records. This helps reduce management workload. At the same time, a preset weight allocation system automatically maps the cleaned and standardized data to the relevant growth education evaluation indicators.

In the "Routine Growth Class" competition, the class quantitative evaluation system is used to assess students on aspects such as safety between classes, appearance, classroom cleanliness, physical exercise standards, and class culture. The duty monitors records data on a tablet, which is compiled in real time, allowing the class teacher to easily track progress. Titles are automatically awarded, and the moral education department can access the records at any time. This enables student self-management and significantly reduces the workload of the student council. When a class wins a title, it automatically captures the student's "Joy in Wisdom—Routine Academic Activity" growth data, along with 20 energy points.

23.2.2.2 Wisdom + Curriculum: AI Technology is Applied to Develop the "Harmonious Growth" Curriculum System and Capture Growth Data Through Multiple Modalities

In line with the growth characteristics of the classroom, the school has developed a digital-integrated teaching system based on diversified and personalized learning and human–computer interaction. This system reshapes the primary school teaching model by merging teaching, evaluation, and management into a unified framework. During the implementation of the national curriculum, it incorporates routine evaluations of learning habits, skill development assessments in subject expansion courses, stage-based literacy evaluations in literacy training courses, and diverse, comprehensive assessments in project-based learning courses. The system creates various educational application scenarios for students, providing scaffolding for different learning types and transforming technology tools into cognitive tools for students. This approach ultimately establishes a student growth education evaluation system rooted in the deep integration of information technology and education (see Fig. 23.4).

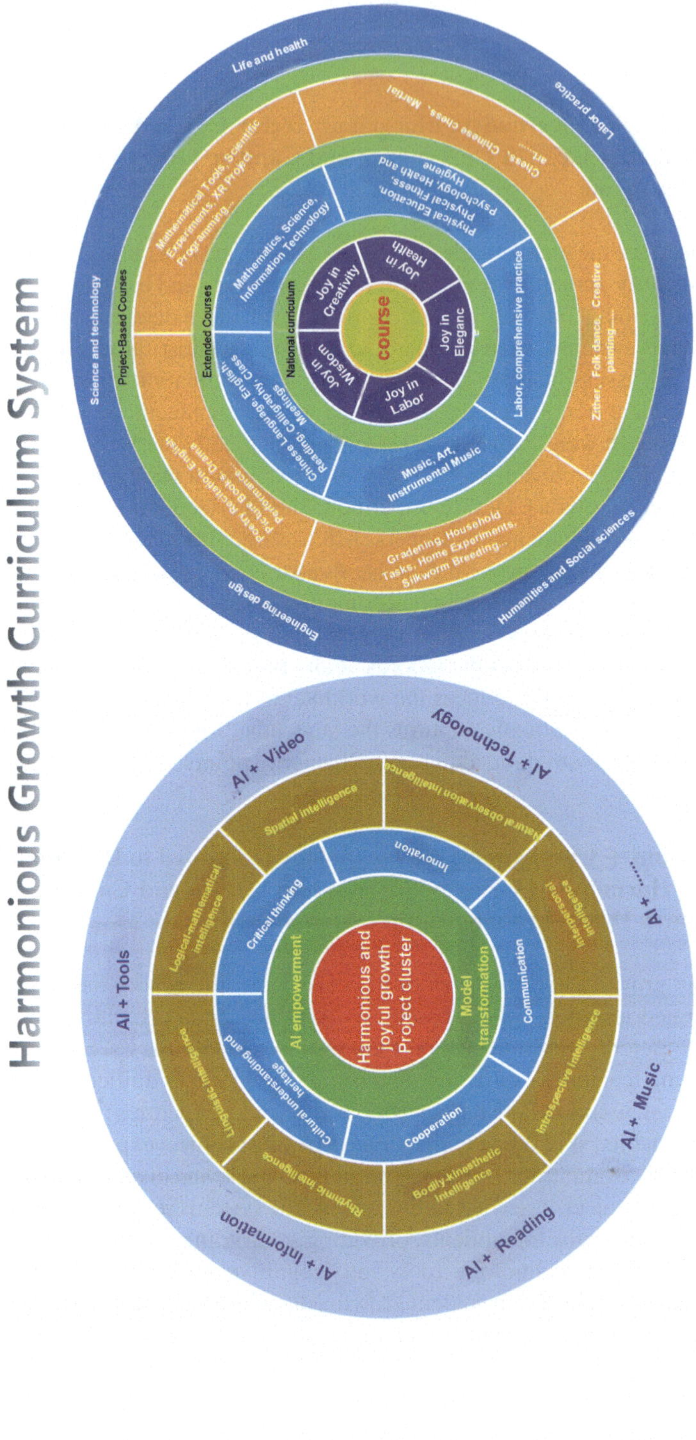

Fig. 23.4 Harmonious growth curriculum system

In the Joy in Labor curriculum's Eco-Park project, diverse evaluation activities are designed, such as planning the park's areas, selecting plants to grow, beautifying the park environment, and managing its order. Along with the project implementation rhythm, activities like "I'm a Plant Expert" and "I'm the Little Chef at Home" allow students to participate in tasks related to nutrition and cooking, as well as social service tasks. This promotes the involvement of family and society in the Joy in Labor dimension evaluation, realizing school-family-community collaborative education.

During implementation, teachers can use speech recognition, semantic analysis, and other technologies to verbally praise students in class, or directly issue "Joy in Labor" energy cards for process evaluation (see Fig. 23.5). When students complete activity tasks, they automatically gain 5 energy points for Joy in Labor. This approach makes evaluation time- and space-independent, allowing for multi-dimensional, multi-stakeholder collaborative evaluation.

23.2.2.3 Wisdom + Space: AI and VR Technologies Are Applied to Reshape the Spatial and Temporal Boundaries of Growth Environments and Capture Growth Data

To support diverse learning experiences, AI sports halls, painting studios, labor technology labs, and VR smart experience rooms have been built. Technologies such as VR, AR, and 3D enhance data collection, integrating smart access control, machine vision, smart bracelets, facial and voice recognition, and other advanced tools to track growth indicators. These technologies capture various types of learner data during evaluation activities, helping assess their overall abilities. This creates an immersive learning environment and meets diverse evaluation needs, ensuring both data accuracy and completeness.

23.2.3 "Incentive" Accompaniment, Triggering Self-Driven Growth

The incentive-driven approach incorporates game elements into the "Growth Tree" model, including energy accumulation, points exchange, and title advancement. It replaces raw data with "energy" to make growth status visible, substitutes traditional records with "games" for task tracking, uses "advancement" to illustrate progress, and presents evaluation results through charts instead of text for clearer visualization.

The student growth achievement title incentive system allows for advancement through five titles: Baby Beginner, Tenacious Bronze, Proud Silver, Clever Gold, and Flawless Diamond. Based on the energy system data collected at the student's energy center, the system automatically advances titles, providing motivation based on the student's overall performance.

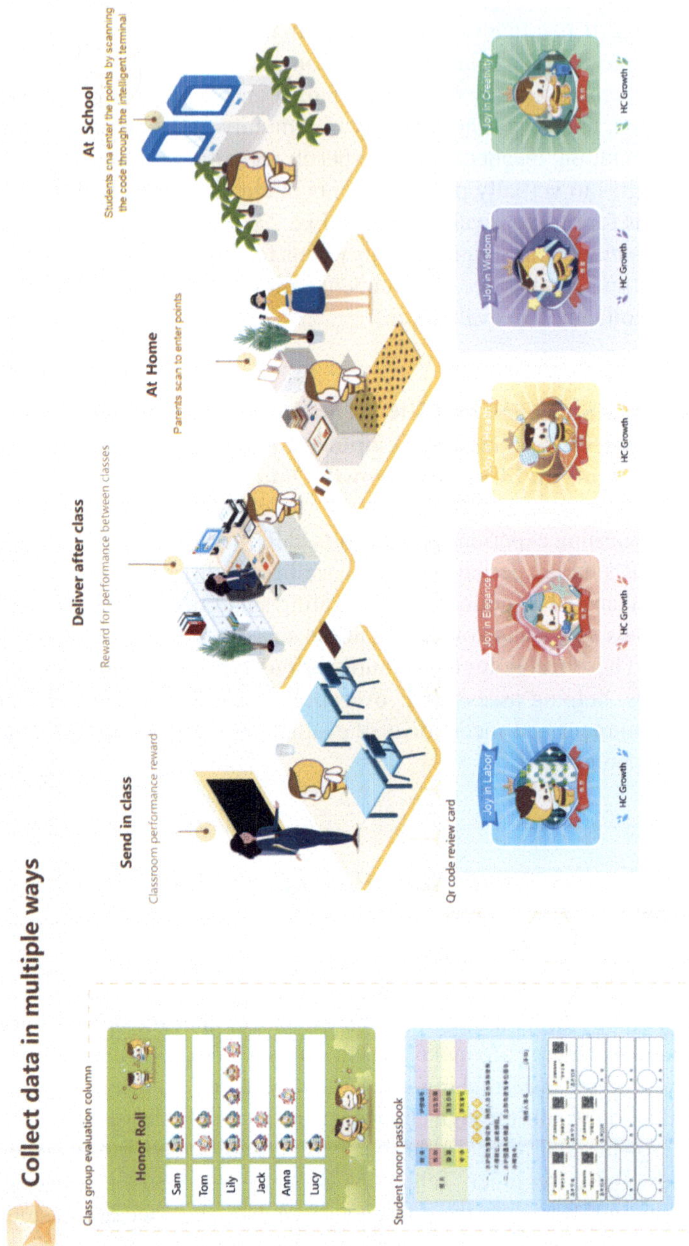

Fig. 23.5 Multi-modal data capture for growth—energy cards

At the same time, a point-based gift exchange incentive system is set up. According to the backend rules, the system calculates and displays the students' available exchangeable items, and the points required for the exchange. Items available for exchange include homework exemption vouchers, snack vouchers, game vouchers, gift vouchers, breakfast vouchers, photo vouchers, and exemption from punishment vouchers. This system enables student self-management, realizing growth incentives and promoting self-driven motivation for students.

From online title advancement to offline gift voucher exchange, this system not only allows students to see, play, and understand the process, but also gradually becomes an intrinsic motivation for triggering self-growth, thereby achieving an innovation in evaluation methods from the perspective of children.

23.3 Implementation Outcomes

The visualized growth education evaluation system incorporates child-friendly features, ensuring that evaluation is something students enjoyan explicit platform; creating dynamic interactive tasks, making evaluation something children usea dialogue platform; building visible data representations, making evaluation something children seea diagnostic platform; and focusing on comprehensive literacy performance, making evaluation something children growa growth platform. This approach shifts the focus from outcomes to processes, from individual data to interconnected data, and from viewing evaluation as an external force to seeing it as an intrinsic motivator. Evaluation is no longer just a tool but becomes a force for both teachers and students to self-correct and self-motivate.

23.3.1 Dynamic Generation: Data Supporting "Literacy" Development

By using visualization technology, evaluation helps children "grow." A child's literacy development is a continuous process, and evaluation should reflect this in real time. The student growth garden uses a quantifiable, cumulative energy value system to present growth data, turning both educational processes and outcomes into visible "tree planting energy." The five different growth stages of trees serve as visual representations, making the tasks of planting saplings an engaging way to show students' growth status and results.

This allows for the dynamic presentation of data, making student data "alive," providing a "growth trajectory," and offering "growth evidence." The visualization charts show how the five growth energy sourcesYueya (elegance), Yuezhi (intelligence), Yuejian (health), Yuelao (labor), and Yuechuang (creativity)are distributed, highlighting the student's growth state. The system also allows students to track their progress by reviewing all their energy acquisition over time using growth evaluation metrics.

23.3.2 Holistic Portrait: Data Laying the Foundation for "Lifelong" Development

Holistic evidence does not mean recording every behavior related to literacy performance but rather reflecting the overall information (Gallegos Nava, 2001). For data collection, we focus on both process-oriented and outcome-oriented data such as physical activity, sleep patterns, reading data, classroom performance, academic evaluations, course experiences, and physical and mental health. This data is then used to create a personalized portrait of the student's growth.

The student's personal growth literacy report automatically summarizes their total energy for the current semester, showing their total energy and the energy gained in each evaluation dimension. The system generates comparative charts for class average energy and semester energy (see Fig. 23.6).

Blockchain technology is employed to securely store growth dynamics. Blockchain ensures the security, authenticity, and traceability of growth records. All data undergoes a verification process, ensuring that the data is reliable and traceable (Lewis, 2018). Using visual technology, data relationships are analyzed and presented, allowing students' literacy development to be based on "evidence" and "trackable data," creating a personalized growth literacy report. This portrait promotes self-reflection, encourages the shift from "being evaluated" to "evaluating myself," and stimulates intrinsic motivation for self-growth, establishing the foundation for lifelong development.

23.3.3 Value-Oriented: Empowering "Holistic" Development with Data

Through virtual technology, the Harmonious Growth Island is constructed. The West Bay Virtual Growth Island map allows the school to monitor the growth states of all classes, track the growth dynamics of students, and see their rankings in terms of growth energy acquisition. With this data-driven approach, school leaders can accurately identify growth trends and problems, formulate targeted improvement measures, optimize resource allocation, and enhance educational quality.

Additionally, the West Bay Student Growth Data Hub provides a comprehensive record of students' developmental literacy, aiding in self-reflection and continuous growth. Through visual data displays and personalized growth route planning, students can clearly understand their growth potential in various areas and set higher growth goals. By breaking through traditional time and space limitations,

Fig23232323232323.....23.6 Zhang's personal growth literacy report

the system makes the evaluation process more dynamic, comprehensive, and visual. It ensures evaluation is both educational and motivating, guiding students toward holistic development (see Fig. 23.7).

23.3.4 Innovative Outcomes: Enhancing "Professional" Development

This project has been included in the China Internet White Paper, established as an educational planning project by the Ministry of Education, and recognized as a Guangdong Province Information Technology Integration Innovation Promotion Project and an Excellent Case. It has also been named a provincial pilot school for educational evaluation reform, a city-level model school for evaluation reform, and a key base for future education development, teacher professional development, and school-based research. The project has won four first-place awards in Guangdong and Shenzhen's evaluation case reviews. It has led to the publication of several works, including Children's Growth Made Visible and Visible Education: Brilliant Growth, and nearly 20 papers in journals such as School Administration and China Education News. It has been reported by media outlets like People's Education Press more than 100 times. Students have won 75 national-level awards, 55 provincial-level awards, and 102 city-level awards.

23.4 Innovations

Using information visualization tools, the evaluation system has been deeply integrated with the school's educational philosophy and culture. This approach promotes the development of student literacy, teacher professionalism, and school culture.

23.4.1 Building New Content: Comprehensive Profiles for Lifelong Development

The five-dimensional evaluation framework is complete and relevant, focusing on human-centered values. It emphasizes life growth and the cultivation of practical and innovative skills. The comprehensive profile of the child shifts the focus from utilitarian evaluation to a more process-oriented approach, highlighting the subjective aspects of education. By using "data" as a foundation, this system provides a clear path for children's growth and promotes the development of student literacy.

Fig2323232323.....23.7 West bay student growth data hub

23.4.2 Strengthening New Applications: Dynamic Results for "Whole Person" Development

The evaluation has shifted from "I am evaluated" to "I want to be evaluated," inspiring students' intrinsic motivation for self-growth and setting a foundation for lifelong development. Additionally, students' growth process files support future development planning, overcoming the previous limitations of passive evaluations. This new system uses evaluation results to inform the exploration of educational principles, leading to improved strategies and mechanisms that foster the comprehensive development of students—moral, intellectual, physical, aesthetic, and labor skills.

23.4.3 Creating New Methods: Technology Empowerment for Evidence-Based Teaching

Technologies like virtual reality, blockchain, artificial intelligence, and computer vision enable multi-modal data collection and enhance the tracking and analysis of growth records. These technologies reshape the boundaries of growth evaluation and use visualization to guide the evaluation process. Visualization allows students to see their progress and encourages peer interaction, promoting engagement with evaluation content and experiences. This data-driven approach supports evidence-based changes in teaching and learning methods.

23.5 Promotion

This application has been successfully implemented across multiple schools in the educational group, solving issues like multiple parents binding on a single student or one parent binding on multiple students. Parents find the system user-friendly, helping them monitor and analyze their children's progress, which motivates student growth. The school has been recognized as a provincial pilot for educational evaluation reform, a municipal demonstration school for deepening evaluation reform, and a research base for education in both the city and district. The system has been adopted by schools in Sichuan, Guangxi, Dongguan, and Luohu District, and has hosted over 30 school visits. Additionally, the team has been invited to deliver nearly 20 keynote speeches at major educational forums. Collaborations have also been established with schools in Dalian, Guangxi, Dongguan, Bao'an, Luohu, Futian, and Guangming, and the development of school-specific applications has been successfully carried out.

23.6 Summary

This case constructed a visualization system for growth-oriented educational assessment to address several issues in traditional educational evaluation: the difficulty in capturing comprehensive process data, the challenge of engaging multiple stakeholders in the evaluation process, and the lack of clear evaluative guidance. It aimed to shift the focus of student development from a knowledge-centric orientation to a competency-centric orientation. Guided by Dewey's philosophy of growth and drawing on holographic data theory and artificial intelligence (AI) technologies, the case developed a visualized assessment framework centered on five dimensions: "Joy of Wisdom," "Joy of Health," "Joy of Art," "Joy of Labor," and "Joy of Creativity." Using intelligent tools, the student's growth trajectory was presented visually as the development stages of a "sapling," and game elements were integrated to activate intrinsic motivation, thereby rendering the evaluation process dynamic, visual, and multifaceted. This approach suggested that education should focus on the child's entire lifelong growth, transforming assessment from a single knowledge test into a dynamic monitoring of holistic competencies. Through the use of advanced technologies, tools such as virtual reality (VR) and blockchain were applied in an innovative way to educational assessment, overcoming the spatial limitations of traditional evaluation. This enabled real-time visualization and data-based analysis of the assessment process, offering new ideas and practical approaches for reforming educational evaluation.

Future research aimed to further deepen the integration of technology and educational assessment by examining more diverse data collection methods to more comprehensively reflect students' competency development, including sentiment analysis and behavior prediction. Simultaneously, cross-regional and cross-cultural applied studies were encouraged to validate the model's adaptability and scalability in different educational environments. In addition, we recommended conducting longitudinal research to assess the long-term impact of this evaluation system on students' development, thereby providing empirical support for the ongoing refinement of educational assessment theory.

References

Gallegos Nava, R. (2001). *Holistic education: Pedagogy of universal love.* Foundation for Educational Renewal.

S Hook 1959 John Dewey-philosopher of growth The Journal of Philosophy 56 26 1010 1018 https://doi.org/10.2307/2021958

Lewis, A. (2018). *The basics of bitcoins and blockchains: An introduction to cryptocurrencies and the technology that powers them.* Mango Publishing.

Part V
Exploring Educational Policy Designs

Chapter 24
Enhancing Learning Style Transformation Through the Design and Application of Subject-Based Classrooms

Lili Cheng, Qing Zhang, Xinhua Pan, and Yongqing Tu Shen

Abstract This case is based on the practice of subject classrooms in Zhejiang Province, which conducts theoretical and practical innovation research on teaching space in the information age. Based on the requirements of curriculum reform for cultivating students' core competencies, this case proposes the concept and connotation of subject classrooms and constructs a "six-in-one" design paradigm from the dual perspectives of construction and application, describing the application methods of subject classrooms to promote the reconstruction of teaching ecology. The case design embodies the principles of "problem orientation, innovative thinking, social responsibility, and sustainable development," providing a comprehensive solution for implementing new curriculum concepts and plans, transforming classroom teaching and student learning methods, and cultivating students' core competitiveness.

Keywords Six-in-one · Subject-based classroom · Situational learning · Learning environment · Educational reform

L. Cheng (✉)
Department of Education of Zhejiang Province, Hangzhou, China
e-mail: 30280448@qq.com

Q. Zhang
Zhejiang Education Technology Center, Hangzhou, China
e-mail: 597136341@qq.com

X. Pan
Jiaxing Senior High School, Jiaxing, China
e-mail: 331624945@qq.com

Y. T. Shen
Zhejiang University Affiliated Middle School, Hangzhou, China
e-mail: xhstyq@163.com

D. Liu et al. (eds.), *Empowering Educational Innovation Through Design: Volume One*, Smart Computing and Intelligence, https://doi.org/10.1007/978-981-95-4621-3_24

24.1 Background

Since 2012, Zhejiang Province has initiated a series of educational reforms, including changes to the general high school curriculum, the pilot reform of the college entrance examination system, and the curriculum reform for compulsory education.[1] The introduction of the new curriculum and the reformed college entrance examination system has posed practical challenges to schools in terms of curriculum implementation and teaching methods, primarily including:

- The urgent need to innovate the teaching management mechanism for subject-based education. Specifically, the new elective course system for general high schools requires a shift from administrative classes to teaching-based classes.
- The urgent need to break away from the current classroom layout, which is functionally limited and environmentally monotonous, to create a flexible, personalized, and innovative subject-based educational space (Rands & Gansemer-Topf, 2017).
- The urgent need to form a new paradigm for subject-based education. This includes constructing a multi-level, diversified curriculum system; developing and supporting rich, accessible learning resources; innovating teaching organization forms; and deeply integrating technology with teaching.

The most influential educational experience is the environment in which children are placed. (Nair, Zimmer Doctori, & Elmore, 2020) With the deepening of educational reforms, advancements in cognitive science and neuroscience, and technological progress, the design and innovation of teaching spaces have become a focal point of educational transformation aimed at competence development. This case focuses on the implementation of basic education curriculum reform.

By designing and constructing subject-based classrooms that are rich in contextualization, resource availability, and learning path support, it aims to address the challenges of innovating teaching organization management mechanisms, implementing diversified curricula, and shifting learning methods in the context of curriculum reform. It provides a suitable and effective new type of subject teaching space for the new curriculum reform and promotes the transformation of teaching and learning methods (Yu & Wang, 2020).

Adhere to the principle of sustainable development; emphasize the future-oriented approach, and strengthen the deep integration of technology and teaching.

[1] http://jyt.zj.gov.cn/art/2012/2/16/art_1229266635_2378990.html.

24.2 Design Process

This case follows the fundamental principles and general approach of design, including "problem orientation, innovative thinking, social responsibility, and sustainable development." The design concept emphasizes the following key points (see Fig. 24.1):

- **Analyzing and Clarifying Problems**: Define and clarify the understanding and positioning of subject classrooms.
- **Systematic Analysis of Design Tasks and Elements**: Construct the basic paradigm for subject classroom design.
- **Focusing on the Innovation of Teaching and Learning Methods**: Clearly define the application approaches for subject classrooms.
- **Future-Oriented Approach**: Strengthen the deep integration of technology with education and teaching methods.

24.2.1 Basic Principles and Features of Subject-Based Classrooms

As a new type of subject-based teaching space in the context of the new curriculum reform, subject classrooms are designed to serve both teaching and learning for teachers and students. Guided by the concepts, ideas, and methods of their respective subjects, these classrooms aim to create an immersive learning atmosphere, connect subject knowledge with real-life applications, enrich students' learning experiences, and promote active learning driven by intrinsic motivation. Functional areas typically include teacher offices, teaching areas, experimental (experiential) zones, and communication/exhibition spaces, which support the coexistence of various learning methods. Subject classrooms combine teaching and office spaces, where subject teachers act as the primary administrators while also utilizing the classrooms as their workspaces. Subject classrooms generally have the following eight characteristics:

24.2.1.1 Professional Subject Contexts

The design of subject-based classrooms emphasizes the use of images, instruments, books, multimedia, and various technical and artistic methods to create rich subject elements. This approach constructs authentic subject contexts where students can practice, explore, and think like subject experts. For example, by using important laws, characters, and stories from the subject to decorate classroom walls, ceilings, and other environments, students can enter a cognitive state of subject understanding as soon as they step into the classroom.

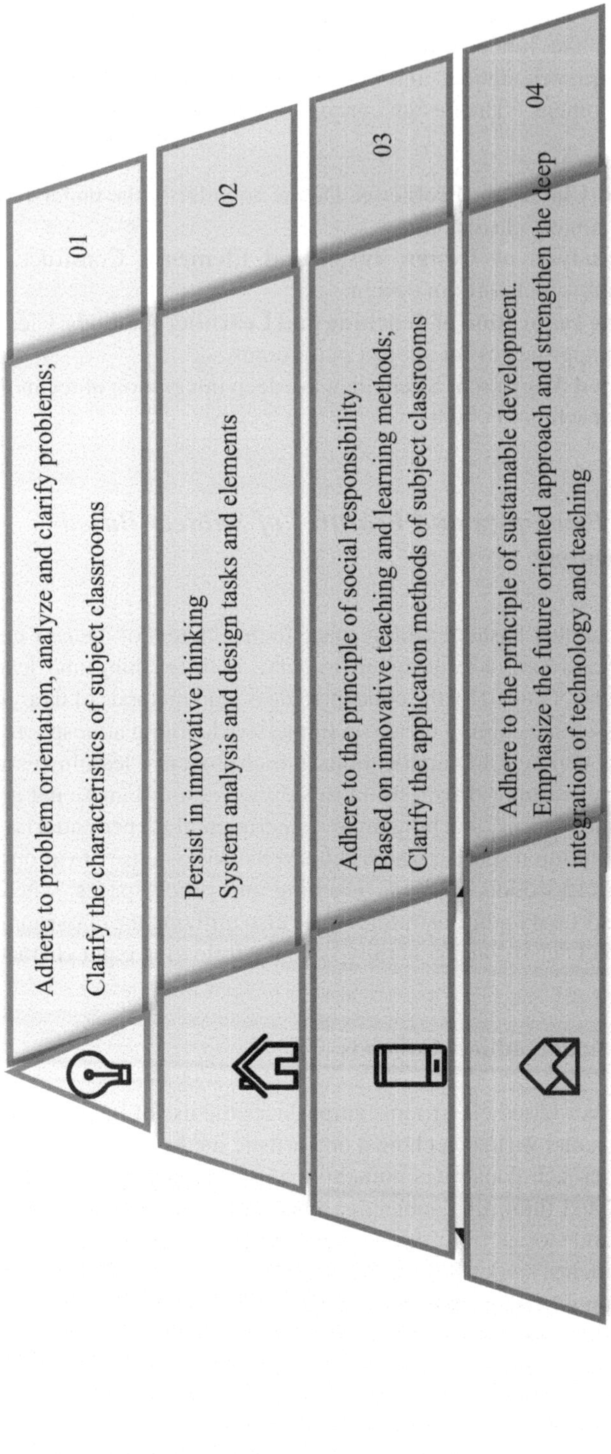

Fig. 24.1 Design concept of the case

24.2.1.2 Multidimensional Cognitive Design

Through the integration of environmental design, resources, functions, and technical equipment, subject classrooms can transform individual cognition into collective cognition, single-sensory cognition into multi-sensory cognition, and single-layered cognition into multi-layered cognition. They make abstract concepts concrete, simplify complexity, enrich knowledge representation, and reproduce the process of knowledge formation. For example, in a math classroom, a graphic calculator helps students visually understand the combination of numbers and shapes; in a physics classroom, a force sensor can measure centripetal force in circular motion, which was previously unmeasurable; and in a biology classroom, virtual reality technology can make abstract concepts like organ dissection and molecular structure visually tangible.

24.2.1.3 Teacher-Student Led Design and Construction

As an essential element of subject teaching, the subject-based classroom should reflect the unique values and ideas of the school's subject teaching and teaching research. Its design must transcend traditional "school-business" partnerships and the "turnkey" model of infrastructure construction. Instead, it should adopt a personalized design and management model where the subject teaching and research team leads the design, with the active participation of teachers and students in the design, construction, application, and daily management of the space. This ensures that every arrangement, facility, and detail in the classroom embodies the meaning and culture of teaching and learning.

24.2.1.4 Deep Integration of Information Technology

Information technology provides opportunities for creating immersive subject contexts and facilitates the transformation of teachers, and innovation of teaching content and methods. It also enhances both teachers' and students' information literacy. However, the use of technology should be guided by principles of accessibility and appropriateness for subject teaching.

24.2.1.5 Dynamic Construction Process

The construction of subject classrooms is not static, a one-time process. They should be adaptable to changes in teaching content, with timely adjustments in environmental setup and technical equipment. The classroom should be dynamically improved based on teachers' research, subject development needs, and teaching outcomes. This promotes the development of the school's subject system and boosts students' intrinsic motivation for learning.

24.2.1.6 Normalization of Teaching Application

As a key site for curriculum implementation, subject teaching, and the cultivation of subject-specific core competencies, the subject-based classroom should be regularly utilized in daily teaching. It should not merely served as a "showpiece" for external display or be reserved exclusively for special lessons. Instead, it should be an integral part of the daily subject teaching process.

24.2.1.7 Non-standardized Technical Equipment

Subject classrooms are designed by teachers and subject research groups based on their specific understanding, leading to diverse and distinctive layouts. While standardization is not feasible, certain common elements should be maintained, such as library corners, experimental areas, charts, multimedia, digital resources, and safety standards.

24.2.1.8 Orientation Toward Subject Competency Development

The fundamental value of subject-based classrooms is to promote the development of students' subject-specific competencies. Beyond merely teaching subject knowledge and skills, these classrooms aim to foster the understanding of the nature, characteristics, and fundamental laws of each subject. This serves as the foundational logic for constructing subject classrooms. For example, a math classroom should focus on cultivating core competencies such as mathematical abstraction, logical reasoning, mathematical modeling, computation, data analysis, and intuitive imagination.

24.2.2 The "Six-In-One" Basic Paradigm for Subject-Based Classroom Design

The design of a teaching environment is not just about setting up a "physical teaching context;" it also entails creating a "human teaching context," which includes the self-transformation of teachers into a "dialogue space" or "relationship space" (Zhong Qiquan, 2019). Subject-based classrooms can be transformed from traditional ordinary classrooms or constructed anew. Regardless of the method, the design must emphasize breaking away from the traditional, singular physical environment of classrooms. Instead, it should incorporate various elements of teaching activities, following the principles of "systematicity" and "integration." The basic paradigm must adhere to the "six-in-one" framework.

The design of subject-based classrooms must deeply incorporate the foundational concepts of teaching and learning reform. It should follow modern educational philosophies, teaching principles, and talent development rules, with a focus on universal access and the development of every individual. It should emphasize the individual construction of learning activities and collaborative networked learning, and support teaching methods and organizational forms centered on learning.

The subject-based classroom should integrate various educational technologies, including instruments, self-made teaching aids, and digital tools, to promote more flexible, diverse, and personalized learning approaches. It must also consider aspects such as lighting, temperature, humidity, and sound, to create a healthy and positive learning environment for students.

In designing the subject-based classroom, attention must be given to creating rich, positive, and creative learning environments through situational arrangements, functional zone divisions, and technical equipment. For example, key elements from the subject—such as famous figures, important events, iconic laws, and formulas—can be incorporated into the classroom's environment to stimulate students' interest in the subject.

The subject-based classroom should follow the principles of "convenience, accessibility, and richness" when planning and applying resources. These resources include regular curriculum materials, textbooks, specimens, student works, digital learning tools, and subject-related materials. In particular, digital learning resources and tools are crucial for significantly improving the learning process and providing strong support for students' learning.

Data-driven teaching is a key feature of the future development of subject classrooms and an essential aspect of their design. Data collection should encompass students' learning characteristics, learning processes, academic performance, as well as teachers' instructional behaviors and teaching resources. The collection and analysis of this data will support accurate student profiling, multidimensional assessment, personalized support, and targeted teaching research.

"Planned construction, regulated usage" should be a central aspect of subject classroom design. To truly become a learning center that promotes teacher-student interactions, dialogues, and reflections, subject classrooms must incorporate a unique cultural approach to classroom design, management, and application. This culture should orient around the subject, teachers, and students. Establishing robust systems for classroom design and application is essential to ensuring the sustainability and vitality of the subject classroom.

24.2.3 Application Methods for Promoting the Reconstruction of the Teaching Ecosystem

24.2.3.1 Teacher-Student-Centered Construction Model

An example from the High School Attached to Zhejiang University demonstrates a model where the subject teaching group creates the classroom construction plan and submits it for school approval. This is followed by the curriculum teaching department's verification and improvement of the plan. The logistics and information services department then budgets and submits the application for construction funds. After the classroom is built, it is inspected and delivered to the teaching group for use. This process emphasizes teacher and student involvement at every stage of construction, ensuring that the learning space directly aligns with the needs and objectives of teaching and learning.

24.2.3.2 Learning Methods that Reflect Real Learning Occurrences

At Jiaxing Senior High School, during a lesson on "Establishing a Correct View on Consumption," led by political science teacher Pan Xinhua, students engaged in a micro-project-based learning activity centered around the theme "What can we do for green consumption?" In this context, students used the national "Plastic Restriction Order" policy as a basis for research. The activity included a survey on the use of plastic shopping bags and a project focused on designing and producing multifunctional eco-friendly shopping bags, reflecting a real-world application of knowledge through STEM learning.

24.2.3.3 Multi-subject and Diversified Course Development Model

At Tiantai High School, the school has collaborated with Zhejiang Yinlun Machinery Co., Ltd. and other enterprises to develop vocational skill-based elective courses, such as physical chemistry control, brewing technology, automobile energy, and electrical engineering. This school-enterprise collaboration allows for more practical, industry-specific learning experiences. Similarly, Jiaxing Senior High School's political teaching group has developed multiple school-based elective courses centered around the subject classroom, further integrating diverse content.

24.2.3.4 Management Approaches that Respect Student Learning Needs

At Quzhou No. 1 High School, students compete for the right to use the biology subject classroom through a system of competitive application. Students submit

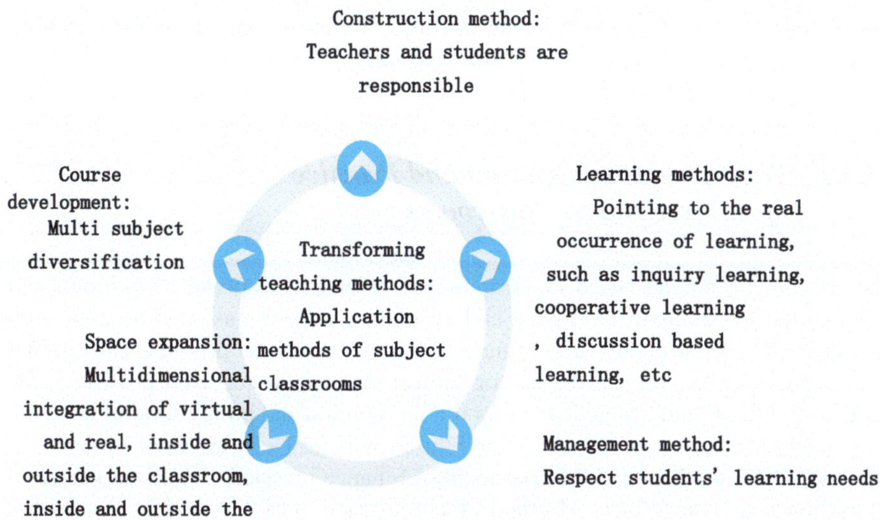

Fig. 24.2 Application methods for promoting the reconstruction of the teaching ecosystem in subject classrooms

proposals and defend their ideas in order to gain access to the space for their academic work. At Wenzhou Art School, the general technical subject classroom is managed in a "supermarket" style. It includes four major sections: courses, tools, materials, and instructors. Students can choose learning content, tools, materials, and even instructors based on their specific needs, offering a highly flexible and personalized learning experience.

24.2.3.5 Multidimensional Integrated Learning Space Expansion

Subject classrooms utilize multimedia teaching, online teaching platforms, and other technological tools to merge physical and virtual spaces, supporting a hybrid model of online and offline learning. The design also focuses on integrating formal learning spaces on campus with informal learning spaces, thereby expanding the learning environment beyond the classroom and making it more flexible, interactive, and adaptable to various teaching and learning needs (see Fig. 24.2).

24.3 Application Outcomes

As an innovation in teaching spaces under the new curriculum framework, this initiative has been deeply practiced in the educational technology and equipment work of Zhejiang's primary and secondary schools in recent years. It has made positive contributions to transforming learning methods and promoting classroom teaching reform. The following are the significant outcomes:

24.3.1 Widespread Recognition and Practice of Subject-Based Classrooms

The concept of subject-based classrooms has gained widespread recognition and practice across Zhejiang's primary and secondary schools, as well as education administrative departments, becoming a conscious action to promote educational reform. In 2014, the pilot program for subject classrooms received positive feedback from schools and teachers, earning full recognition and policy support from the Provincial Department of Education and the Provincial Department of Finance. In 2016 and 2017, the Provincial Department of Finance invested 10 million yuan and 18 million yuan respectively to support the construction of 914 subject classrooms across the province. According to the educational technology equipment statistics for Zhejiang's primary and secondary schools, as of now, there are a total of 16,165 subject classrooms in the province, covering 54% of schools, with this number increasing annually.

24.3.2 Enhancement of Students' Learning Motivation and Potential Development

Subject-based classrooms create rich, engaging, and inviting learning environments through immersive subject culture and contextual designs. These environments stimulate students' motivation to explore subject knowledge, skills, and values. For example, in the history subject classroom at the high school attached to Zhejiang University, an electronic display board at the entrance shows "On this day in history," where students can see significant events that occurred on this day in various years. If they wish to learn more, they can touch the event on the screen for detailed interpretations, related images, and historical reflections. Inside the classroom, walls are filled with historical materials, including "Historical Major Events," "Student History Exhibits," and a collection of over 400 history books and various teaching tools, all of which greatly enhance students' interest and motivation to learn history (see Fig. 24.3)..

Fig. 24.3 History subject classroom at the high school attached to Zhejiang University

24.3.3 Shifting from Passive Learning to Subject-Based Practice

In traditional classrooms, subject learning mainly occurs through "static" activities such as listening, reading, thinking, memorizing, and practicing. The design and application of subject-based classrooms, with the rich contextualization and resource offerings, place subject knowledge in real-world situations, enabling students to apply interdisciplinary thinking and methods to analyze and solve problems. This approach links students' prior experiences and real-life situations with social development, thus promoting authentic learning. For example, Jiaxing Senior High School uses its political science subject classroom, known as "The Home of Thinkers," for annual mock Chinese People's Political Consultative Conference (CPPCC) sessions. During these sessions, teachers and students engage in activities such as data retrieval, group learning, proposal discussions, and sharing, giving students first-hand experience of the proposal formation process and rules of procedure in the CPPCC. Additionally, project-based and inquiry-based learning have become common methods in this classroom where students have used equipment like 3D printers to design and create over 300 national utility patents. Yiting High School of Yiwu has developed a distinctive "technology-based education" model by building subject-based classrooms focused on fields such as electrical appliance research, information technology, and other subjects.

24.3.4 Flexible Implementation of Multiple Learning Methods and Experiences

The ability to meet students' diverse learning needs is crucial for driving the transformation of learning methods. Subject-based classrooms, through multifunctional space designs and a rich supply of learning resources and tools, allow both teachers and students to engage in flexible learning approaches such as lectures, independent inquiry, group collaboration, and blended teaching. Interactive, heuristic, exploratory, and experiential learning, as well as boundary-free learning that connects the classroom with online and offline resources, are common in subject classrooms. For example, the physics subject classroom at the high school attached to Zhejiang University is equipped with professional knowledge charts, subject-specific books, and experimental apparatus for group experiments. The classroom also includes a digital learning area with online physics labs, virtual physics museums, and micro-course platforms, where students can independently read, search, conduct virtual experiments, and engage in online learning. Additionally, QR codes related to the curriculum are printed on "knowledge curtains," encouraging students to scan them for personalized learning.

24.3.5 Promoting Schools' Unique Development

The construction and application of subject classrooms provide a favorable environment for the implementation of the new curriculum's emphasis on cultivating subject literacy. They also serve as powerful engines for schools to enhance subject teaching quality and their unique educational characteristics.

24.4 Main Innovations and Reflections

This innovative study provides a comprehensive solution for implementing new curriculum concepts and plans, promoting changes in classroom teaching methods and student learning approaches, and fostering the development of students' core competencies.

24.4.1 Main Innovations

- Proposing the principles and characteristics of subject-based classrooms, driving the transformation of traditional classrooms.

- Constructing the basic paradigm for the design of subject-based classrooms, providing effective guidance for local implementation.
- Initiating practical actions for the construction of subject classrooms in primary and secondary schools across Zhejiang Province, forming a set of validated application methods that provide a powerful engine for curriculum reform and changes in education models.
- Exploring feasible paths and effective methods for promoting the deep integration of information technology with education and teaching.
- Providing a Zhejiang model for national basic education curriculum and teaching reform.

This study is free from objective limitations in educational application and innovation, allowing for replication and promotion of results across various schools and regions.

24.4.2 Areas for Further Research and Reflection

Despite achieving positive outcomes, further research and reflection are necessary in some areas for the exploration and practice of subject-based classrooms.

- The support of educational theories from disciplines such as pedagogy, psychology, neuroscience, and design are needed.
- The design, construction, and application of subject classrooms require teachers to have a profound understanding of subject knowledge and teaching methods, posing higher challenges to teachers' capabilities.
- The effective application of subject-based classrooms requires changes in teaching activities, educational equipment, teaching methods, and assessment methods. Innovations in policies and systems, particularly in educational equipment standards and government procurement systems, need to be reformed first.

24.5 Summary

The purpose of this case was to create a meaningful learning environment for subject teaching, particularly for cultivating students' subject literacy, by reconstructing learning space, and promoting the transformation of subject learning methods. At the theoretical level, the case proposed a new understanding of the concept and characteristics of subject classrooms. It developed a "six-in-one" construction model with an application strategy to support the reconstruction of the teaching ecology. At the same time, the case was piloted and promoted in primary and secondary schools in Zhejiang Province, where it cultivated a series of typical cases and achieved positive outcomes in teaching reform. It became an important component of policy related to curriculum reform and basic education technology equipment in Zhejiang Province.

Next, it was planned to engage with emerging research in areas such as neuroscience and artificial intelligence, and to carry out further studies and practical work on the development of subject classrooms, especially the integration of physical and virtual learning spaces.

References

Lifeng, Y. & Huijun, W. (2020). Continuously deepening the construction and application of subject classrooms and innovative laboratories. In H. Wang & Y. Zhu (Eds.), *Zhejiang Education Yearbook* (p. 206). Zhejiang Education Press.

Nair, P., Zimmer Doctori, R., & Elmore, R. F. (2020). Learning by design: Live, play, engage, create. Education Design Architects.

Qiquan, Z. (2019). *Cumiculum logic*. East China Normal University Press.

Rands, M. L., & Gansemer-Topf, A. M. (2017). The room itself is active: How classroom design impacts student engagement. *Journal of Learning Spaces, 6*(1), 26–33.

The manufacturer's authorised representative in the EU is Springer Nature Customer Service Centre GmbH, Europaplatz 3, 69115 Heidelberg, Germany. If you have any concerns regarding our products, please contact ProductSafety@springernature.com

Printed and bound by CPI Group (UK) Ltd, Croydon, CR0 4YY
21/09/2026
02219988-0010